Praise for *Spite: and the Upside of Your Dark Side*

'[A] thorough and entertaining book, which poses a provocative thesis . . . McCarthy-Jones is a funny, playful writer, especially for a psychologist . . . an illuminating examination of an under-discussed topic.'

New York Times

'An interesting and at times provocative exploration of an emotion that has to this point been underexplored and, if McCarthy-Jones is right, significantly underappreciated.'

Independent

'An informative, evidence-based page-turner. A rare pleasure.'
Richard Stephens, author of *Black Sheep*

'*Spite* is an eye-opening examination of humanity's nastier impulses – from Achilles to Trump. An erudite and eloquent guide, McCarthy-Jones deftly examines cutting-edge psychological research and evolutionary theory, with some truly startling insights for our personal relationships, business and politics. You will never look at your human nature in quite the same way again.'
David Robson, author of *The Intelligence Trap*

'*Spite* is a fascinating insight into how we all behave in a world of big egos and thin skins.'
Michael Cockerell, award-winning political documentary-maker

'With rigorous science, penetrating analyses, colourful and enjoyable prose, and an astonishing breadth of knowledge – Simon McCarthy-Jones has delivered a book that will undeniably be appreciated by many.'
Frank Larøi, Professor of Psychology at the University of Bergen

Also by Simon McCarthy-Jones

Spite: and the Upside of Your Dark Side

*Can't You Hear Them?: The Science
and Significance of Hearing Voices*

*Hearing Voices: The Histories, Causes and
Meanings of Auditory Verbal Hallucination*

FREE
THINKING

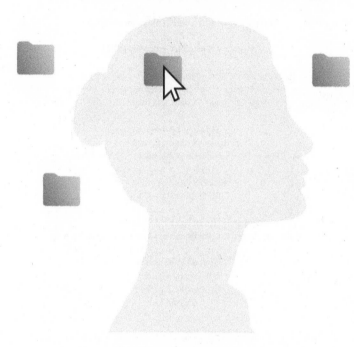

SIMON McCARTHY-JONES

ONEWORLD

A Oneworld Book

First published by Oneworld Publications in 2023

ISBN 978-0-86154-457-8
eISBN 978-0-86154-458-5

Typeset by Hewer Text UK Ltd, Edinburgh
Printed and bound in Great Britain by Clays Ltd, Elcograf S.p.A.

Oneworld Publications
10 Bloomsbury Street
London WC1B 3SR
England

Stay up to date with the latest books,
special offers, and exclusive content from
Oneworld with our newsletter

Sign up on our website
oneworld-publications.com

MIX
Paper from
responsible sources
FSC® C018072

For Daisy and Ken, neither gone nor forgotten,
and Rose, E and Z, who continue the battle.

The most dangerous man to any government is the man who is able to think things out for himself, without regard to the prevailing superstitions and taboos. Almost inevitably he comes to the conclusion that the government he lives under is dishonest, insane, and intolerable and so, if he is romantic, he tries to change it. And even if he is not romantic personally he is very apt to spread discontent among those who are.

H. L. Mencken[1]

What would the individualists and freethinkers of the eighteenth and nineteenth centuries say could they but see what idols a man must now worship, to what jackboots he must now pay homage, if he is to escape being hunted and stoned?

Bertrand de Jouvenel[2]

[E]very thought which does not serve and does not conform to the ultimate purpose of a machine whose only purpose is the generation and accumulation of power is a dangerous nuisance.

Hannah Arendt[3]

To become a vital part of the living Constitution, a value must have more than a strong historical and analytical foundation. The value must also succeed at the level of rhetoric; it must have its great quote.

Vincent Blasi[4]

CONTENTS

ACKNOWLEDGEMENTS

Above all, I must thank my wife Rose and my family. Nothing of value gets done alone in this world.

This book would not have been possible without the legal scholars who have blazed a trail for the right to freedom of thought, including Susie Alegre, Marc Blitz, Jan Christoph Bublitz, John and Leslie Francis, Gabriel Mendlow, Patrick O'Callaghan, Ahmed Shaheed, Bethany Shiner and Lucas Swaine.

I am also grateful to all the people with whom I have communicated about this topic over the years, including Olga Cronin, Sinéad O'Sullivan, Brendan Kelly and Joel Walmsley.

My agent, Bill Hamilton, and the supportive, patient and thorough team at Oneworld Publications, including Sam Carter, Rida Vaquas and Hannah Haseloff, also have my immense gratitude, as does my copy-editor, Kathleen McCully.

Naturally, none of the problems in these pages are the fault of anyone named above. Furthermore, my acknowledgement of these names in no way means they endorse the ideas in these pages. Enough caveats, let's proceed.

INTRODUCTION

Controlling bodies gives power but controlling minds grants dominion. For this reason, there has always been and always will be a battle for our minds. It is a fight we cannot afford to lose. Evolution has stripped us of traditional mammalian defences and left us with instincts that take us only so far. Our programming is incomplete. In its place, we have been gifted a new tool for life: thought. To endure, discover, govern ourselves and flourish we must cogitate and deliberate, reflect and reason, imagine and remember, ponder, muse, mull things over, figure life out. We think and thereby invent our lives.[1] To let others think for us is to let others live for us.

As a psychologist, I knew this in theory. But it would take a combination of personal and political events to awaken me to the battle for our minds that rages around us. Before 2016, people often told me malevolent forces controlled their minds. All were amidst mental health struggles, often traceable to a treacherous world.[2] Then things changed. People began telling me that sinister forces were controlling *other people's minds*. And these accounts weren't coming from people seeking support for their mental health. Instead, they came from Hillary Clinton supporters and anti-Brexit campaigners.

The following year, my mind came under someone else's control. A short article I wrote went viral.[3] Or, at least, it went viral by academic standards, meaning it was read by more than two people unrelated to me. At the time, I still had a Facebook account. My

1

mind kept urging me to 'check Facebook, check Facebook'. It wanted to know if I'd been rewarded with a red notification button showing someone had commented on or shared my article. I could cope with the predictably disastrous results of perching my laptop on a treadmill to monitor Facebook whilst jogging. But Facebook made things personal when it began stealing my mind away from family activities. I finally started to pay attention to the forces buffeting our minds.

The mainstream media were already telling a story about the contemporary battle for our minds. Their account ran roughly as follows. The twenty-first century had seen a new combatant in the battle for the mind: social media companies. These companies had not just seized our minds, they had polluted them. Echo chambers were engineered, hate amplified and elections swung. Social media, it was said, had brainwashed Britain out of Europe and America into Trump.[4] But if the Western mind was in trouble, things were worse in the East. The Chinese Communist Party's (CCP's) use of surveillance technologies and artificial intelligence promised to access, control and punish the thoughts of a billion people.[5] The media were clear; something had to be done. But what?

To solve a problem of this magnitude, a god was needed. Luckily, the United Nations (UN) knew where to find one. The UN journeyed to the land where the secular gods reside – international human rights law. The deity they summoned was the right to freedom of thought, sired in 1948 by the Universal Declaration of Human Rights.

As with any god worth its salt, the right to freedom of thought was omnipotent. Human rights law gave it the status of an absolute right.[6] In contrast, most other rights, such as privacy, free speech and assembly, are qualified rights. Governments can permissibly restrict them for national security or public safety reasons. But absolute rights, like the right to freedom of thought, can never be violated, under any circumstances, for any reason.

There was, however, a significant problem. The only thing absolute about the right to freedom of thought was its neglect. The right to free *speech* had been endlessly debated, discussed and developed in the courts. Free *thought* – not so much. The right to freedom of thought turned out to be the Napoleon III of rights, a man of whom Bismarck said, 'At a distance it is something, but close to it is nothing at all.'

The law had put the cart before the horse by placing speech before thought. 'Tell me, Mr Anderson', runs a line from the 1999 film *The Matrix*, 'what good is a phone call if you are unable to speak?' Likewise, what good is free speech if you are unable to think? Speech in the absence of free thought is platitude, received wisdom, the party line. A government that completely controlled citizens' thoughts could safely grant them complete freedom of speech, as no one would have anything destabilising to say. Without thought, speech is mindless.

It was beyond bizarre that the international community had not spelt out what the right to freedom of thought meant in practice. In part, this was a failure of imagination. Law review article after law review article, court after court pondered free speech when free thought was clearly the issue. Articles that claimed to plunge into the ocean of free thought often ended up promptly swimming for the firm land of free speech. There was some pioneering scholarly work on the right to freedom of thought[7] and its potential relevance to technology, neuroscience[8] and the digital world,[9] but this was rare. It was less that the law had put the cart of speech before the horse of thought and more like it had left this cart alone in a vaguely horse-smelling field.

By deciding that the right to freedom of thought should be inviolable, the UN had given it a titanium skeleton. But the law had never put flesh on the bones of this beast. What form could or should this right have in practice? The right's lifeless bones lay on a slab, under a half-century of cobwebs, awaiting its legal Geppetto, Prometheus . . . or Frankenstein.

On 19 October 2021 the UN finally dusted off this 'forgotten freedom'[10] and began to build it a body. Ahmed Shaheed, a UN Special Rapporteur, presented a report on freedom of thought to the 76th Session of the UN General Assembly.[11] This was the first attempt, at the level of the UN, to spell out what the right to freedom of thought should involve. The process of deciding what this fundamental human right should look like is ongoing. It is happening now. When consensus is reached, judges will wield a new weapon. They will free us from the machinations of social media corporations and the tyranny of totalitarianism. Our minds will be free again.

This is a compelling story, perhaps too compelling. But is it cast correctly? Take the villains of the piece. Are the traditional media right in portraying social media and autocracies as the primary threats to free thought? Or, by shining a light on these figures, are they leaving other threats to gimble and gyre in the darkness? Then there is the hero of this story. Will a freshly forged right to freedom of thought sweep us all to safety? Or, like all good heroes, will it have a dark side? And will all our thoughts find shelter under this hero's cape? In deciding what the right to freedom of thought covers, the UN is also deciding what this right doesn't cover. Important facets of thought could be left flapping unprotected in the wind.

We could easily shrug our shoulders and leave the legal profession to protect whatever conception of thought they settle upon, however limited, strange or contorted it may be. Yet if the law promises to protect our thought, we should hold it to its word. The job of law, said one of its great thinkers, is to 'divine the form of what lies confused and unexpressed' and 'bring to light the substance of what is half surmised'.[12] To do this in relation to freedom of thought, the law cannot but benefit from hearing perspectives on thought from psychology, neuroscience, philosophy and anthropology.

All of us must have a chance to contribute our voice to the development of the right to freedom of thought. Due to the political consequences of the nature and scope of the right that emerges, we must have a public conversation about this process. This book aims to stimulate such a conversation. In a democratic society, such discussions are essential. Democracy is where the people decide on the laws that bind them.[13] Democracy is self-government. If we are to be bound to fulfil the responsibilities entailed by the right to freedom of thought, as well as to enjoy its fruits, we must all have a say in its design for it to have any legitimacy. If politics is too important to leave to politicians, then human rights are too important to leave to lawmakers.[14] Yet, as we will see later, having human fingerprints all over human rights is not necessarily a good thing. Human rights are better born through virgin birth.

In these pages, I will set out, as a mere psychologist, how I think this right should look. My central belief is that we should demand protection for our thought in all its rich forms. We must reject impoverished conceptions of thought that limit it to the happenings inside our heads. We must recognise that thought happens between people as well as within them. Yet, if we are really serious about protecting thought, as I think we should be, we will need to change not only our laws, but also our culture and societal structures. To think freely requires a new enlightenment that goes beyond a focus on individuals. It requires a deep enlightenment.

Why do we need a right to freedom of thought?

Despite the need for widescale input into developing the right to freedom of thought, this issue is not garnering much attention. When the UN's Special Rapporteur issued his report, it barely raised a ripple in public awareness. That day, many things trended on social media: 'NBA is back', 'Angelina Jolie', 'Xbox Mini Fridge', but not 'freedom of thought'.

Yet this report should have created shockwaves. We should have come out in a cold sweat at the realisation that freedom of thought, lauded by luminaries from Voltaire, Thomas Jefferson and Hannah Arendt to Noam Chomsky, George Orwell and the US Supreme Court, was not *already* defined in law in crystal clear terms.[15] But maybe we are rushing ahead of ourselves. After all, why exactly should we care?

Where rights come from is a thorny issue. This question is ultimately about who has the authority to pronounce on rights. There are three common answers. The first is that they come from a divine being, which implies that theologians have the authority to determine our rights. The second answer is that rights exist naturally and that we can use philosophical reasoning to uncover them. This effectively gives philosophers the authority to pronounce on our rights. Such thinking led to the idea that as rational, sociable creatures with instincts for self-preservation, we have natural rights to life, liberty and property. This idea was seen as both consistent with reason and God.[16]

Unfortunately, it has been convincingly argued that the idea of natural or divinely inspired rights is a 'tiresome illusion' (Walter Lippmann), 'at one with belief in witches and unicorns' (Alasdair MacIntyre) and 'nonsense upon stilts' (Jeremy Bentham).[17] As the philosopher Jeremy Waldron has said about natural rights, 'no one now uses the phrase except in a disparaging sense.'[18]

To reject the idea of divine or natural rights is not to say that we shouldn't have rights. Rather, a third option is to recognise that we make a political choice to create rights. We don't just discover them in nature. A realist definition of a right comes from the American journalist Walter Lippmann. He defined a right as 'a promise that a certain kind of behavior will be backed by the organized force of the state.'[19] This definition helps us see how important it is, in a democracy, for the people themselves to have a say in decision making surrounding rights.

As part of our democratic discourse, we need to understand why we should have a right to freedom of thought. We should not accept something simply because a judge or court has previously ruled that way. These are the arguments from authority against which free-thinking has always rebelled. I don't want precedents; I want arguments. Simply shouting 'thought police!' is not an argument against interfering with people's thoughts. If we shouldn't punish people's thoughts, why is that precisely? Are such arguments valid today? Were they ever? Yet, as we will see, tradition can have its silent reasons, which the claws of inquiry may need to respect.

If we are not clear on the fundamental importance of the right to freedom of thought, we may only embrace it when it supports our partisan political goals. The right to freedom of speech has already suffered this fate, being used by activists on both ends of the politi-cal spectrum to achieve their aims and then tossed aside when proving inconvenient. To be committed to free thought means being committed to others' free thought, not just our own. We are only likely to take this stance if we fully appreciate the importance of the right.

Offering reasons for why freedom of thought matters will be important in this process. But reasons alone may be insufficient. As the philosopher Richard Rorty claimed, reasoning may do little to convince people of others' rights. For Rorty, stirring emotions was the way to go. As he put it, 'the emergence of the human rights culture seems to owe nothing to increased moral knowledge, and everything to hearing sad and sentimental stories'.[20] In getting people to buy into the right to freedom of thought, stories of mental manipulation may be just as effective as any reason we can give, or even more so.

Knowing why we should have a right to freedom of thought means knowing what this right is trying to achieve. Only then can we design a right that achieves these aims. In my view, we need a right to freedom of thought to enable, protect and support our personal autonomy and political self-government. Freedom of thought is necessary for us to have an autonomous mind that is sovereign over itself. Such a mind can decide for itself (in consultation with others) between wrong and right, truth and falsity, beauty and ugliness, friend and enemy, rather than being tricked, manipulated or forced into positions that others would have it hold. Without mental autonomy, we would simply follow what authorities told us was the case. We would become what the poet John Milton called heretics in the truth.[21]

By releasing us from the yoke of fallible authority, freedom of thought gives individuals and society the best chance of finding truth. Many widely held beliefs are unlikely to be the gospel truth. Time, US Supreme Court Justice Oliver Wendell Holmes Jr observed, has upset many fighting faiths.[22] This necessitates that we possess a device for winning new truth.[23] Freedom of thought is such a device.

Yet while we need freedom of thought to help us win new truths, paradoxically, without the ability to restrict thought we are unlikely to succeed. As scholars have pointed out in relation to free speech, when we want to reach truth we don't just make speech as free as possible. Instead, we place rigid restrictions and limitations on speech, such as stringent criteria for publishing in academic journals.[24] The search for truth justifies both the right to freedom of thought *and* its limitation.

The English writer G. K. Chesterton once argued that free thought was the best of all safeguards against freedom. He claimed that if you teach someone to worry about whether they want to be free, they will not free themselves.[25] Yet by helping us discover truths, the right to freedom of thought gives us the greatest possibility of living a free life. But what is a free life? As Alexander Meiklejohn argued, 'To be free does not mean to be well governed.

It does not mean to be justly governed. It means to be self-governed.[26] Self-government is only possible if, when voting (or 'exiting'),[27] we can make informed decisions based on the actual state of the world.

By helping us win new truths, freedom of thought lights the path to meaningful self-government. This is why the manipulation of our minds is wrong – it represents 'the destruction of self-government'.[28] Self-government is not possible without strong protections for free thought.[29] Without it, we are condemned to be 'governed by apathy, impulse, or precautionary conformism'.[30] As US Supreme Court Justice Robert Jackson put it back in 1950, 'our Constitution relies on our electorate's complete ideological freedom to nourish independent and responsible intelligence and preserve our democracy from that submissiveness, timidity and herd-mindedness of the masses which would foster a tyranny of mediocrity'.[31]

Not only can freedom of thought support self-government, it may also help preserve it. The Ancient Athenians argued that freedom of thought mattered because it ensured their independence by keeping them strong. The playwright Aeschylus emphasised that the Athenian navy defeated the Persians at the Battle of Salamis because they were 'men elevated and inspired by the freedom to speak their minds and govern themselves.'[32] Similarly, as the Athenian statesman Pericles put it in his famous funeral oration, 'the great impediment to action is, in our opinion, not discussion, but the want of knowledge that is gained by discussion preparatory to action.'[33] Free thought helped the Athenians survive in a hostile world, at least for a while.

Freedom of thought also supports our human dignity.[34] The French thinker Blaise Pascal even claimed that 'All the dignity of man consists in thought'.[35] This relation becomes clearer if we understand dignity as 'the presumption that one is a person whose actions, thoughts and concerns are worthy of intrinsic respect, because they have been *chosen, organised and guided* in a way which makes sense from a distinctively individual point of view'.[36]

Not only that, but by enabling mental autonomy, freedom of thought is also arguably what allows us to be us at all. Our ability to think freely is so essential to our identity that violating it has been argued to deprive us 'of personhood altogether'.[37] Consider the French philosopher René Descartes' (1596–1650) famous saying, 'I think, therefore I am'. If you are not thinking, if the thoughts you have are put there by someone else, then do you still exist?

The importance of mental autonomy remains in spite of the autonomous individual being a cultural construction,[38] an 'invention'.[39] This fiction has been so naturalised that autonomy is regarded as a basic psychological need.[40] Psychologists argue we are creatures who are 'born to choose'[41] and claim that we feel this way because it has adaptive consequences. These include a longer life[42] and greater resilience to adversity.[43] By promoting autonomy, freedom of thought helps us survive.

Indeed, freedom of thought is central to our survival. The philosopher and 'novelist' Ayn Rand stressed the need for us to think to survive. As she put it, 'Man's life, as required by his nature, is not the life of a mindless brute, of a looting thug or a mooching mystic, but the life of a thinking being – not life by means of force or fraud, but life by means of achievement'.[44] Following Rand, the economist Murray Rothbard also argued that freedom of thought is our very instrument of survival. As he put it, humans, 'not having innate, instinctive, automatically acquired knowledge' of our 'proper ends, or of the means by which they can be achieved, must learn them'.[45] To do this, we must think.

But thinking does more than just enable us to survive. It helps us truly live. Thinking, as Alexander Meiklejohn puts it, 'fits us for the making of ourselves'.[46] Meiklejohn distinguishes between the body, with its desires and search for comfort, and the spirit, with its ideals and quest for excellence. He argues that the fulfilment of the spirit should matter more than the fulfilment of the body. As thinkers, we hold ideals and strive for them. If freedom of thought both keeps our body alive and lets our spirit strive, it should be sacrosanct.

What is thought and what makes it free?

Once we are happy that we need a right to freedom of thought, we next need to agree on what counts as thought. International law does not define 'thought'.[47] What should count as thought is entirely up for grabs. In my first chapter, I will argue that traditional conceptions of thought, which view it as *that which happens in our head*, are far too narrow.

I believe that some of our speech, such as public reasoning together, is a form of thought. Not only that, but I see this as the most powerful form of thought. Thinking in one's head is thought-lite. Thinking together is full-fat thought. Thought is primarily a social process, not a private one. Speech and thought are not distinct; they merge and meld. By creating a clear legal distinction between speech and thought, the law has made a conceptual mess, albeit one that keeps courtrooms clean. We need to focus on protecting and promoting public thinking. We need to make thinking aloud allowed again.

If we are serious about protecting thought, we must safeguard all its protean forms. For this reason, I want to see the right to freedom of thought acknowledge a range of external activities, such as writing in diaries and some internet searches, as being thought. Such an approach would have significant implications. If your diary is protected only by your right to privacy, this right can be permissibly violated. But if you can get legislators or judges to agree that your diary counts as 'thought', you have effectively made it invulnerable. In that case, no one, whether a cop, a court, or a king, gets to open it without your permission. Ever.

Yet courts are currently not addressing basic questions about our right to freedom of thought, even when this right is clearly relevant. Let me give you an initial example. In 2020 a case came before a judge in London. Mr Harry Miller was contesting the actions of the police. The police had investigated him for a series of alleged transphobic tweets and, as part of this, a police officer had visited Mr Miller's place of work. Mr Miller claimed the officer said, 'I need to check your thinking' (the police officer denies saying this) and warned him that if his behaviour escalated it could move from a non-crime hate incident to a crime and that the police would then need to deal with it. Ultimately, the judge ruled that 'the police's treatment of [Mr Miller] . . . disproportionately interfered with his right of freedom of expression.'[48]

What intrigued me about this case was that despite Mr Miller raising both 'thoughtcrime' and the 'Thought Police', the court's judgement did not mention the right to freedom of thought. There were plenty of opportunities to do so. For example, the judge observed that some rights are 'absolute and can never be interfered with by the state'. The judge then gave freedom from torture, inhuman or degrading treatment or punishment as an example of an absolute right. But why did the judge not mention the right to freedom of thought on his list of absolute rights, given the claimant argued the case was about thinking? The judge went on to note that other rights 'such as the right to . . . freedom of assembly and association are qualified and require a balance to be struck between the rights of the individual and those of the wider community'. But the right to freedom of thought should not be balanced against anything. Wasn't this right at least worth mentioning, even if only to argue why it wasn't relevant? It was as if the right did not even exist.

Moving on, we also need to be careful not to solely use a Western conception of thought when we develop the right. Typifying a Western approach, Aristotle argued that to make people behave better, they had to change themselves into more virtuous creatures.

Yet an Eastern perspective would stress the need also to change the person's environment to make them behave better. A Western approach to protecting free thought may emphasise teaching individuals to become critical thinkers. Yet an Eastern approach would stress creating surroundings that support free thought. Both approaches are necessary.

So, whilst the question 'what is thought?' might seem like a mind-numbing academic discussion (and trust me, it can be), it is not. It is a contentious scientific, political and legal discussion that will establish what is sacred, untouchable and inviolable. It really matters.

Once we have reached agreement about should count as thought, we need to decide on what makes thought free. A crucial question will be what we wish to define as acceptable and unacceptable ways to influence other people's thinking. The answers to such questions will profoundly impact how government and business are permitted to interact with us, as well as how we interact with each other.

We will then run into the question of whether the right to freedom of thought should be 'absolute'. It seems odd to deny that there are any circumstances under which our right to free thought might need balancing against other considerations. All this gives us plenty of questions to ponder. As the answers given will affect us all, we should all have a voice in these deliberations. Worryingly, the societal effects of the answers we reach are extremely hard to foresee.

Who threatens freedom of thought?

When we have established what our right to freedom of thought should look like, we need to work out who threatens this right. Have the media, as discussed earlier, correctly identified the critical threats to free thought? Autocratic governments clearly pose a threat to free thought. But we are programmed to spot such obvious villains. It is easy to point at other countries, such as China, as violators of free thought. Such countries fit neatly into the totalitarian-shaped hole that George Orwell cut for us in *1984*. But we must see what other shaped holes have been cut in the cloth of liberty. We must see how our liberal democratic societies threaten our ability to think freely. We need new nightmares.[49]

All of China's methods of 'mind control' have analogies in liberal democracies. European rulers have a long history of controlling public opinion by monopolising information and suppressing criticism.[50] Yet the work done by the government in China is today performed by corporate power and citizens themselves in liberal democracies. In China, the government pays citizens to create social media posts to distract the rest of the population from controversial issues.[51] This aims to stop the Chinese people from having discussions that could lead to large-scale collective social action. The extent of this operation is phenomenal. Estimates suggest that the CCP pays two million citizens to create half a billion social media posts. As a result, one in every 178 social media posts on commercial websites is a governmental Trojan horse.[52] The US government doesn't need to pay for such posts. The corporate media already provide a distraction economy. Why would the US government want to pay Americans to create distracting social media posts when *Love Island* is on TV?

The Chinese state controls news sources, blocks information and provides patriotic education.[53] In the West, news sources are controlled by private corporations who block advertiser-unfriendly thinkers and cheerlead for capitalism. China has a state surveillance system, centrally operated, that inspires fear, threatens people's jobs, and prevents questions from being asked. The West has a social media surveillance system – a 'crowdsourced panopticon'[54] – operated by its citizens' ire, that inspires fear, threatens people's jobs and prevents questions from being asked. 'Thou hypocrite', says the Gospel of Matthew (7:5), 'first cast out the beam out of thine own eye; and then shalt thou see clearly to cast out the mote out of thy brother's eye'.

The idea that autocratic regimes are swaying the minds of millions of voters in liberal democracies is itself an information operation conducted by autocracies.[55] The very idea of mind control is an attempt at mind control. Autocracies have realised that making citizens of open societies believe lies is unnecessarily hard work. It is easier to make people doubt the truth than believe the lie. Confusion is easily sown. Autocracies' information operations undermine trust within liberal democratic societies, making it hard to think clearly. Western journalists who overhype Russian and Chinese influence campaigns are useful idiots for autocracy.

For example, after Democrats released details of Facebook adverts created by the Russian Internet Research Agency in the run-up to the 2016 US presidential election, the *New York Times* promptly put one advert on its front page. This advert portrayed Satan arm-wrestling Jesus, with the caption 'SATAN: IF I WIN CLINTON WINS! JESUS: NOT IF I CAN HELP IT!' In doing so, the *Times* gave the impression of a powerful Russian meddling campaign, making Americans wonder what information they could trust. In reality, the original ad was displayed for one day, seen by seventy-one people, and clicked on by fourteen Americans.[56] The *Times* did the Russians' work for them.

The media is correct to alert us that social media companies, such as Facebook, pose problems for free thought. Yet much of the backlash against social media reeks of a moral panic. Claims about social media's impact on our minds have run far ahead of the evidence. Once careful and time-consuming scientific research is performed and published, these exuberant claims are pulled back like a sprinter on a bungee-cord. For example, the evidence does not support the idea that fake news on social media is destroying democracy.[57] As a recent study concluded, 'the origins of public misinformedness and polarization are more likely to lie in the content of ordinary news or the avoidance of news altogether . . . [than] they are in overt fakery.'[58]

Look closer at the hand that points at social media. Traditional legacy media, television and newspapers, are incentivised to flag social media as the key threat to our minds. After all, that is *their* job. As MSNBC host Mika Brzezinski once said of Donald Trump, 'he's trying to make up his own facts . . . he could have undermined the messaging so much that he can actually control exactly what people think . . . that is our job'.[59] The legacy media are not embedded reporters in an information war: they are combatants.

We know the power of existing legacy media to influence people. Famously, in the 'Fox News effect', Fox News moving into a new town's cable service leads to more people voting Republican. Such effects can swing elections. The 2000 US presidential election came down to whether George W. Bush or Al Gore won the state of Florida. Bush won by 537 votes and assumed the presidency. The estimated number of Republican votes created by Fox News's move into Florida that year was 10,757.[60] When asked, people say that TV coverage and offline discussion with friends are more impactful on their voting decisions than anything they do online.[61] The legacy media hide their power and, in doing so, increase it. As the political scientist Samuel Huntington once put it, 'power revealed is power reduced; power concealed is power enhanced.'[62]

The owners and creators of social media are convenient villains. It is simple to locate societies' problems in the profitable activities of a much-maligned young billionaire. It is harder to concede that the roots of our problems grow from the stony rubble of our prideful, wrathful and slothful hearts. We embrace content that makes us feel both right and righteous. We lash out at others for thinking differently. We fail to check information's provenance and lap up what we are spoon-fed. Blaming others for pulling our strings abnegates our responsibility for our strings. If social media vanished tomorrow, the eternal problem of our fallen nature would remain. This is not to say we should pin-point all societies' problems inside us. As we will see when we look at social control in 1920s Imperial Japan, being encouraged to locate problems within ourselves, rather than our social system, is an effective method of social control itself. Yet we must be willing to look both within and without for the causes of our current malaise.

Focusing on social media and autocracies pulls our attention away from other significant threats to free thought. In liberal democracies, government remains an ever-present threat to the minds of citizens. A government's potential to act in its self-serving interest rather than the interest of the people, combined with its control of the education system and monopoly on legitimate violence, whip up winds that blow against a freethinking citizenry. As the stunningly frank 1975 *Crisis of democracy* report authored by the elite Trilateral Commission made clear, our ruling powers do not see a thinking and engaged citizenry as essential for democratic government, but rather as a threat to it.[63]

The Commission's report argued that the 1960s had led to an 'adversary culture'. In this culture, people no longer felt 'the same compulsion to obey those whom they had previously considered superior'.[64] This led to the challenging of established institutions and the promotion of the idea that justice meant equality. This was a problem, claimed the report, because the effective operation of democracy needed some people and groups to show a degree of 'apathy and non-involvement'. Predictably, the rich and powerful were not required to be apathetic. It was the new political activity of 'marginal social groups, as in the case of the blacks', which troubled the report's authors. The involvement of the masses in politics, thought the authors, was not the effective operation of democracy but rather an 'excess of democracy'.[65] In their view, this posed as serious a threat to society as fascism, monarchy and communism.

So, who did the Commission think should govern? Their report looked back fondly to after the Second World War when the US President ruled, working with 'the Executive Office, the federal bureaucracy, Congress and the more important business, banks, law firms, foundations, and the media'. The questioning of 'hierarchy, coercion, discipline, secrecy and deception', which the report defended as 'inescapable attributes to the process of government', posed a threat to this ruling coalition. A freethinking citizenry meant a freefalling state.

Back in 1720, when a right to free speech was first proposed, the discussion focused on governmental threats to free speech.[66] Today, corporations pose a more significant threat to free thought than the government. Legal scholar Jeffrey Rosen's observation that Google executives 'exercise far more power over speech than does the Supreme Court' also holds true for thought.[67] Yet, strictly speaking, it is wrong to say that corporations are a greater threat to free thought than government. This is because these two entities are not independent. This was explicitly identified way back in 1901 in the mainstream *Bankers' Magazine*. The author noted that:

> The businessman . . . seeks to shape politics and government in a way conducive to his own prosperity. As the business of the country has learned the secret of combination, it is gradually subverting the power of the politician and rendering him subservient to its purposes. That government is not entirely controlled by these interests is due to the fact that business organization has not reached full perfection.[68]

Since 1901, business organization has increased and government has increasingly become a tool of big business, following the bidding of its corporate paymasters. Politics is 'the shadow cast by business over society'.[69] We are encouraged to stay in Plato's cave, take shadows for reality and not turn around to see who is casting them. In the twentieth century, capital had to ensure that democratic self-government was not allowed to threaten its interests. This required a vast propaganda effort.

Obviously, corporations seek to influence the public mind to sell products to customers. Capitalism is about producing desires in our minds. As one American advertisement executive put it, 'what makes this country great is the creation of wants and desires, the creation of dissatisfaction with the old and outmoded'.[70] But corporations are also continually selling capitalism itself to citizens.[71] Capitalism is about capital creating more capital. In capitalism, a human is a dollar's way of making another dollar.

Unsurprisingly, this is not an idea everyone will spontaneously embrace.[72] As a result, the 1920s and 30s saw the use of techniques initially developed to encourage American citizens to support US involvement in the First World War, to destroy the US labour movement.[73] The 1930s and 40s then saw businesses, through organisations such as the National Association of Manufacturers, undertake propaganda campaigns to link democracy with free enterprise. Such propaganda portrayed big government as evil precisely because government had the potential to respond to the will of the people and threaten corporate interests.

Although the US government's ability to influence citizens' thoughts is restricted by the First Amendment, corporations are largely free to influence people's thoughts as they see fit. Yet free thought cannot exist when corporations have open season on citizens. The idea that our thought is free is a bad joke when, as historian Fara Dabhoiwala puts it, 'under American law, any private entity can censor, fire, or punish people as much as it likes, for exactly the same forms of expression that any "government" may not inhibit.'[74] The most impactful way we can increase free thought today is to change the balance of power between labour and capital, employees and employers, government and business. There is an urgent need for citizens, through their governments, to demand that corporations respect freedom of thought. Government, so long a sword to free thought, must now be its shield.

Yet to solely focus on governments and corporations is to overlook another key player in the battle for thought. Governments and corporations often act as handmaidens to this third force, shutting down discussion, debate and dissent at its bidding. This force had historically been contained. Its range of influence had once been limited to those in earshot. In 1440, Gutenberg's printing press increased the reach of this force. In 1695, deregulation of printing presses in England empowered this force further.[75] But with the creation of the internet, this force could pulsate around the globe at the speed of light. This third threat to free thought, dear reader, is you and me.

There was once a dream that was the internet. This new frontier, wider than the sky, would let us all raise our voices. We would be able to circumvent the gatekeepers of traditional media, gain access to previously hidden and suppressed information, expose power and create a more just world. We would both hear and be heard. And what did we, status-seeking primates that we are, do with our voices? The exact same thing that chimps do with their teeth. We ripped the living flesh off each other in a war of all against all. The internet became a Hobbesian hell.

We are a menace to freethinking in both ourselves and others, suppressing, strangling and silencing each other's thoughts. Deeper and deeper we sink into a censorious, virtue-signalling swamp of sanctimonious pus that has made public thinking near-impossible. We cry out to governments and corporations to punish others for their thoughts. In doing so, we have sacrificed the lion of thought on the altar of feeling.

As we silence others in a quest for vain-glory, pride shuts down our ability to think. We believe our opinions rain down from pure white clouds of rationality. Yet, in a phenomenon known as the Dunning–Kruger effect, the less competent we are, the more we think we know. For example, when surveyed, a third of US adults thought they knew more about the causes of autism than doctors and scientists. The participants who objectively knew least about autism, as assessed by a quiz, were most likely to think they knew more than experts.[76] They weren't just ignorant about autism, they were ignorant of their own ignorance. The only thing Socrates is said to have known is how little he knew. If we too can replace pride with humility, and try out our ideas with others, we may come closer to truth.[77]

This may be a vain hope. The classic arguments for free thought and discussion proposed in 1859 by the philosopher John Stuart Mill are near incomprehensible to the modern mind.[78] For Mill, even a wrong-headed idea was valuable to the truth-seeker and should be heard. Errors could either contain a snippet of truth or help the truth stand out clearer in contrast. But we have lost all tolerance for statements that err. We condemn where we should converse, excoriate where we should explore and inflame where we should inquire.

In doing so, we have forgotten the lessons of history. In the early 1950s, during the Red Scare in America, Senator Joseph McCarthy scoured the country for communists. He wanted to root out communist thought, expose it and punish the thinker. Just as the Russians would go after what they called 'inako-myslyashchie' ('those who think differently'), so did the Americans.[79] As a result of McCarthy's actions, livelihoods and lives were destroyed. Around 600 teachers lost their jobs due to their political beliefs.[80] The 'black silence of fear' that McCarthyism created also led many others, including scholars, to suppress their opinions.[81] After just a few years though, McCarthy was being roundly condemned by the good people of America. But we are all McCarthys now.

To think in public today, one must be a fool, a madman or a saint.[82] This is partly because we have lost the willingness to differentiate between thoughts and their thinker. There are no longer bad ideas, only bad people. As a result, when we crucify a thought, we make sure to hammer our nails all the way into the thinker. In the 1500s, the Catholic Church introduced the idea that one of its lawyers should act as the devil's advocate and argue against the canonisation of an individual. Yet no one mistook that person for the devil (with the possible exception of when Christopher Hitchens performed the role, at the Catholic Church's request, when Mother Teresa was being considered for sainthood).[83] Today though, anyone who thinks an idea, or presents arguments for a position, is deemed to endorse it. This is the complete opposite of the approach that led to the flourishing of our society. We achieved progress, in the words of the philosopher Karl Popper, by 'letting our hypotheses die in our stead'.[84] Yet we have reverted to playing the man and not the ball.

We have lost the willingness to differentiate between a thinker and their thoughts because there are powerful incentives to act this way. In an increasingly atomised, competitive world, we can bene-fit from maligning, cancelling and destroying others.[85] In particu-lar, social media incentivises people to engage in moral grand-standing. This involves talking about moral issues to try to boost our social standing in other people's eyes. Such moral grandstand-ing is driven by desires for prestige and domination. It leads people to 'ramp up' their ideological positions, taking increasingly extreme stances, to try to appear more 'virtuous, respectable, or worthy of admiration than others in their group'.[86] This incentivises social media users to push others down, by attacking, shaming, insulting and demeaning them, in order to raise themselves up.[87] It is not about ideas winning and losing, it is about people winning and losing. It is a marketplace of people, not ideas. And, unfortunately, as the American thinker Thomas Sowell has emphasised, you get what you incentivise.[88] The fact that most social media users have never met those whom they attack makes it easier for the bile to flow. We were never meant to live this way. Ironically, the individualist ideology of liberalism, which gave birth to the right to freedom of thought, is contributing to the development of the fraught and atomised conditions that are attempting to strangle the very right it birthed.[89]

As a result, there is no longer a space for the free play of ideas, except in private circles of trusted friends. This 'crisis of thinking' is a disaster for society.[90] Feelings have safe spaces, but thought has no home. Naked, exposed and vulnerable, thought, if not housed soon, will perish. That said, we should be careful about viewing safe spaces for feelings as antithetical to truth-seeking thought. The philosopher Derek Anderson argues that safe spaces actually 'promote the unimpeded pursuit of free thought and discourse'.[91] Anderson claims that the ability of members of marginalised groups to think freely and pursue truth is enhanced in such safe spaces. The resulting inclusion of what would otherwise be suppressed or excluded thoughts benefits everyone in the pursuit of truth.

In short, the threats to free thought are myriad. The right to freedom of thought must protect our minds from enemies both old and new, foreign and domestic, silicon and carbon.

Is the right to freedom of thought the answer?

The right to freedom of thought may be part of the solution to the threats our thought faces, but it is not a panacea. Clearly, there are benefits to building this right. The UN Special Rapporteur's report on freedom of thought is a positive step forward and to be welcomed. Yet we must not let our senses be overwhelmed by opportunity. We must scan the sky for the inevitable problems flying towards us low on the horizon. Every action has unintended consequences. It would be ironic if, failing to think ahead, we endangered free thought by trying to protect it.

Leaving an international body to decide on the meaning of our right to freedom of thought may seem unproblematic. But, on reflection, how does it make you feel to know that a global body of experts (including me)[92] have their sticky legislative fingers not only in your mind, but also in the minds of future generations? The power of the right being constructed is fearful. If you can convince a judge that your actions are really speech, you will get away with most things. If you can convince a judge that your speech is really thought, you will get away with anything.

What mortals could be trusted to make a god? Anyone familiar with the horror genre knows that our attempts to create new life-forms typically result in creatures that kill their makers. When constructing a right to freedom of thought, we must consider whether our creation could harm what we meant to save. Failing to do this could lead to a Frankenstein-right rising from the slab.

We may avoid such unintended consequences if we ponder whether a good reason may exist for why this right is not already well defined. Consider this parable from G. K. Chesterton. Two people walk down a road until they are blocked by a fence. The first person, not seeing a reason for the fence to be there, proposes simply removing it. However, his wiser friend says, 'If you don't see the use of it, I certainly won't let you clear it away. Go away and think. Then, when you can come back and tell me that you do see the use of it, I may allow you to destroy it.'[93] Applied to our current situation, we may ask why there isn't already a well-marked fence around our minds.

One answer is political. Ruling powers in both autocracies and democracies are interested in keeping thought within contained bounds. Why would these powers create problems for themselves by promoting a right to free thought? Furthermore, violations of free thought are more explicit, indeed brutally explicit, in autocracies. It is hence hard for the UN to promote a right that will inevitably cause tension between its members.[94] It is already hard enough to keep the world on the same page, without unnecessarily emphasising areas of clear difference. Pushing a clearly defined international right to freedom of thought could come at the cost of damage to international relations or worse. It may be that the only way to establish a universal right to freedom of thought is to make it so vague that it has no identifiable content.[95]

Another reason why the right has not previously been developed is because of a failure of imagination. Our minds have not been fenced off because everyone thought their boundaries were already impenetrable. The US Supreme Court put forward this idea in 1942, claiming that: 'freedom to think is absolute of its own nature, the most tyrannical government is powerless to control the inward workings of the mind'.[96] If this were true, developing a right to freedom of thought would seem pointless. What use is a right that no one would ever need?[97] Creating a right to freedom of thought in this situation would be like putting a fence round a nuclear bunker.

But what if the US Supreme Court had things entirely backwards? What if the elites' ability to control people's minds, rather than their inability, is what prevented the right to free thought from being developed? What if elites felt a robust and meaningful right to freedom of thought was too dangerous and destabilising to be made accessible to all? True freedom of thought would allow citizens to pull on strings that could unravel powerful interests and even the fabric of society itself. Not every thought that is dangerous advances truth, but every thought that advances truth is dangerous. Could it be that even in liberal democracies, perhaps especially in liberal democracies, thought must be kept in clearly fenced lanes?[98]

Another reason the right may have lain fallow is that it was foreseen that free thought could be a threat to free speech. Suppose those in power decide that certain ways of speaking are 'manipulative' of other people's minds. The right to freedom of thought could cause such communications to be made illegal. There will be clearcut cases where this makes sense, such as existing bans on subliminal advertising. But there will also be a large grey area. Reasonable people may disagree about what speech should be deemed manipulative of other people's thought. The friends of free thought may emerge as the enemies of free speech. If anyone offers you new freedoms, hold on tight to those you already have.

The potential for the right to freedom of thought to prohibit some speech means we will run into a variant of the twentieth-century debate that economists had about coercion. Economists argued whether the free market was 'free' or coercive. Neo-liberals argued it was free. Marxists argued it was coercive. The neo-liberals won, and we now generally accept capitalist exchange as non-coercive. The free market exists as neo-liberals wanted. The same debate needs to occur concerning freedom of thought. Are we, for example, happy to accept traditional marketing and advertising practices as being non-coercive and harmless to our freedom of thought? Powerful financial interests want the answer to this question to be 'yes'. But are you happy with this? Perhaps you want a right to freedom of thought that will rip up much of existing capitalist practice, even at the potential price of economic growth. The point is that an absolute right to freedom of thought in the hands of lawfare activists could radically change society. This could give us a society that is freethinking yet neither secure nor economically prosperous. If you protect anything unreservedly, much else will die in the shadows.

Not only do we need to consider the dangers of a militant right to freedom of thought, but we also need to think about whose interest this right will serve. Walter Lippmann argued that the 'system of rights and duties at any time is, at bottom, a slightly antiquated formulation of the balance of power among the active interests in the community'.[99] Even if you are not a Marxist, you will probably concede the truth of Karl Marx's assertion that 'The ruling ideas of each age have ever been the ideas of its ruling class.'[100] The right to freedom of thought risks being drafted in a form that will serve the interests of an elite.

Even if this doesn't happen, the right to freedom of thought will remain tightly tied up with politics. The right will sit on a shelf until a global power can use it to combat another global power. Human rights were of relatively little interest to the global community until the US decided it could undermine the USSR using the concept in the 1970s.[101] The right to freedom of thought is only likely to have any impact on the global stage if it becomes important to the foreign policy of a superpower. Unless we are careful, this right will be developed or used to serve superpowers and elites rather than us. Even if we are careful, this will probably still be the case. So it goes.

In short, if the first serious international attempt to define the right to freedom of thought does not terrify you, you're not paying attention. The right to freedom of thought will be developed, for good or for ill and most likely both. What is decided will have seismic effects on society for generations to come. As such, we need to get this right *right*.

Why is the process of developing the right to freedom of thought happening now? I have two answers: one technological, the other political. The dawn of the twenty-first century has seen a quantum leap in technology. Driven mainly by private capital's quest for profit, we are all lab rats for Manhattan Projects of the mind. Neuroscience has increased its ability to identify, manipulate and enhance the neural basis of thought. Behavioural science is armed with the potent combination of big data, machine-learning and artificial intelligence. All this is drawing back a veil that has hidden human thought for millennia. As always, new understandings foreshadow new forms of control. Freethinking was never easy, but now it threatens to be impossible. As legal scholars have pointed out, we must think about these technological developments through the lens of the right to freedom of thought.[102]

In addition to the rise of the machines, there are also political reasons why the right to freedom of thought is being considered today. As the legal scholar Julie Cohen has emphasised, 'Human rights discourses are always polemical and ultimately unintelligible if one does not understand the political stakes in historical context.'[103] From the left's perspective, today's political discourse is reminiscent of 1948, when the UN created the Universal Declaration of Human Rights. Now, as then, a fear of fascism hangs in the air. In 2016, many on the left felt that the people were voting wrongly. The result of the 2016 Brexit referendum, in which the UK voted to leave the EU, followed by Donald Trump's victory in the 2016 US presidential election, went against the liberal consensus. Many worried that this was a prelude to, at best, a new authoritarianism or, at worst, a fascist revival. Given the horrors of the mid-twentieth century, we should be alert to this threat. Late Generation X-ers and early millennials were raised breathing the fumes of fascism. Popular culture, such as films like *The Matrix* (1999), *Fight Club* (1999) and *Gladiator* (2000), gave them the figure of a new man who would redeem a decadent world.[104] Today, the anonymous right-wing author and podcaster Bronze Age Pervert continues the script.[105]

Yet such fears skewed how the left saw the rise of Trump and Brexit. The left did not see Trump's victory as resulting from voters' valid social and economic concerns or feelings that they were not being represented by elected officials. Instead, many on the left claimed Trump voters had been 'electronically brainwashed'.[106] Trump and Brexit voters were alleged to have been the passive victims of mind control campaigns perpetrated by Big Tech, Russia and a 'global plutocracy'.[107] The world was full of Manchurian voters. Some on the left went as far as to claim that millions of white Republican Americans needed 'deprogramming'[108] and their children sent to re-education camps.[109] The atmosphere echoed 1950s China, where the Communist Party used a programme of thought reform to try to re-educate millions of citizens into 'new people'.

Based on such concerns, one journalist claimed that we now lived in a 'managed' democracy.[110] There are valid concerns about how the powerful control and shape information for their own agenda. But, given the people had voted against the consensus of elite expert and media opinion, it looked less like we had moved into a managed democracy and more like we had always lived in one.[111] What the left seemed to object to was that they were no longer the ones doing the managing. We should be concerned about how others are shaping our thinking, whichever side of the political spectrum is ascendant.

Free thought has also become a political issue due to the recognition that it poses a national security issue for liberal democracies. Vladislav Surkov, personal adviser to Russian President Vladimir Putin, recently responded to accusations of Russian interference in other countries' elections by stating that 'it's much more serious than that. Russia is meddling in their brains, changing their consciousness, and they have no idea what to do about it.'[112] Such claims represent an attempt at 'perception management', aiming to convince Western citizens that their information ecology has been polluted and is untrustworthy. In the 2018 US midterm elections, the Russian Internet Research Agency claimed it was operating thousands of fake accounts to try to sway the election. As alluded to earlier, in reality it wasn't. It was simply trying to 'poison the well' of public discourse so people couldn't trust each other.[113]

Incidentally, the way Russia sowed confusion in the US follows a trail blazed by the US itself. After the heavy-handed approach of Senator McCarthy had been discredited in 1954, the US deployed the FBI to interfere, confuse and impair the activities of a range of

US political groups. The FBI's COINTELPRO programme tracked and interfered with thousands of American citizens to defuse threats to the political status quo. This included the FBI sending anonymous letters and spreading rumours to get politically active citizens, such as those in the Black Panther movement or the American Communist Party, to turn against one another. It was highly effective.[114] Such ideas have not gone away. Prominent American legal scholars have proposed that 'government should engage in cognitive infiltration of the groups that produce conspiracy theories'.[115] In practice, this would involve government agents or their allies, potentially anonymously, 'planting doubts' about the theories, their premises and the evidence for them.

China is also undertaking large-scale online social media influence operations. Unlike Russia, China primarily exercises soft power by using online adverts to showcase itself.[116] Between 2018 and 2020, Facebook adverts made by the Chinese state were seen 655 million times.[117] The more people see these ads, the more positively China is viewed.[118] China is also playing a long game to influence Western public and elite opinion by creating Confucius Institutes on Western university campuses, influencing Western classroom discussions of China and shaping academics' views on the subject.[119] China has less military hardware than the US, so it has significantly invested in psychological warfare. As one of its planning documents notes, 'even the last refuge of the human race – the inner world of the heart – cannot avoid the attacks of psychological warfare'.[120] Although often exaggerated, such actions by autocratic regimes have highlighted the need for Western governments to defend and support the freethinking of their electorates.

Will this right really make any difference?

You may wonder whether a robust right to freedom of thought would add anything to the already well-developed right to free speech. The US Constitution protects free speech so vigorously that a comprehensive right to freedom of thought won't make legal any speech that is currently banned. The First Amendment states that 'Congress shall make no law ... abridging the freedom of speech'. The US Supreme Court interprets this to mean that the government cannot restrict speech simply because it is 'upsetting' or 'hurtful', even if it gives 'offense' or inflicts 'great pain'.[121] The Court has largely rejected the idea that free speech should be balanced against other concerns.[122] For the Court, an essential function of free speech 'is to invite dispute', and it may best serve its purpose when it causes 'unrest' and 'stirs people to anger'.[123] US citizens are expected to respond to challenging speech with either more speech or by leaving, rather than violence. The US may be the only country that still puts thoughts before feelings.

Yet even in the US, you can't legally say anything you want. When the First Amendment states that Congress shall make *no* law abridging freedom of speech, its meaning seems straightforward. It doesn't say Congress shall make only the minimally necessary number of laws restricting speech. Nor does it say there are clearly some situations where freedom of speech needs to be curtailed by law. It says *no* law. Yet despite this, the US government can legally restrict a number of forms of speech. These include true threats (using speech to threaten others, intending to make them fear bodily harm or death),[124] incitement (speech that aims to incite imminent lawless action and is likely to produce such action),[125] defamation (making false statements that harm the reputation of another), some types of obscene speech and threatening the President. If societal attitudes shift, there is no reason this list couldn't grow. As has been noted, 'we kid ourselves if we believe our legal freedoms will survive if our free speech culture is undermined'.[126] The right to freedom of thought may help prevent new types of speech from being added to this list of exceptions. Furthermore, it could lead to new protections in the US if we deem writing in diaries and certain internet searches as 'thought'. Such activities would then gain absolute protection and privacy.

In Europe, speech faces significantly more permissible restrictions than in the US. This is evident from Article 10 of the European Convention on Human Rights. After granting a right to freedom of expression, Article 10 stipulates that governments can restrict this right for many reasons. These include 'in the interests of national security, territorial integrity or public safety, for the prevention of disorder or crime, for the protection of health or morals, for the protection of the reputation or rights of others, for preventing the disclosure of information received in confidence, or for maintaining the authority and impartiality of the judiciary'. This is quite the caveat. The phrases 'prevention of disorder' and 'protection of . . . morals' are particularly suspect.

The European Court of Human Rights believes it has the authority to tell us what we can and cannot think. It stipulates that certain beliefs are 'not worthy of respect in a democratic society' and that courts decide what these beliefs are. These ideas seep into society, chilling thought. In 2020 an employment tribunal in the UK ruled that the belief that 'even if a trans woman has a Gender Recognition Certificate, she cannot honestly describe herself as a woman' was not worthy of respect in a democracy.[127] A subsequent Employment Appeal Tribunal would overturn this ruling, clarifying that the sort of beliefs meant to fall into this category were those 'akin to Nazism or totalitarianism'. The Appeal Tribunal did not feel that the claimant's gender-critical views were akin to such beliefs.[128] The point here is not about the best way to understand the complicated issue of gender. The point is that European courts reserve the right to decide what its citizens can and cannot think. All this would make most Americans queasy and cause some to reach for their Second Amendment rights. Americans drink their freedom neat, Europeans on the rocks.

The European Court of Human Rights has also upheld the idea that it 'may be considered necessary in certain democratic societies to sanction or even prevent all forms of expression which spread, incite, promote or justify hatred based on intolerance'.[129] The EU's proposed laws on hate speech will lead to the censorship of what it deems impermissible hate as well as perfectly legal speech. For example, consider the effect of making digital platforms legally responsible for what people post on their sites. If faced with a decision whether or not to delete a comment that could get them in trouble, companies will err on the side of caution and delete it.

Such self-censorship is already well underway in Europe. Jacob Mchangama, the director of a Copenhagen-based human rights think-tank, and his colleagues studied 63 million Facebook comments in Denmark. They found that the Facebook algorithm classified 1.4% of these comments as 'hateful attacks'. Yet less than 0.01% of these comments violated Danish law concerning crimes such as threats, hate and glorification of terrorism.[130] Another study Mchangama was involved in found that for every comment deleted from the Facebook pages of Danish news media (being deleted either by Facebook, the news media site, or the original author of the post) that would be deemed illegal under Danish law, thirty-six legal expressions of opinion were deleted.[131] Mchangama also points out that whereas European national courts typically take more than two years to rule on hate speech cases, tech platforms will be asked to make equivalent decisions regarding comments on their sites in hours or days.[132] Reframing some speech as thought, and therefore giving it absolute protection, would have widespread consequences in Europe.

What will this book try to do?

This book will probe the questions that we have begun to raise above. Chapter 1 will dive into what should count as thought. Chapter 2 will then look at which of our abilities support free thought. I will argue that we need to focus on attention, reflection, reason and courage, which I refer to as the ARRC of free thought. Chapter 3 will leave our inner world and examine the external threats free thought faces. Here I will discuss the technological threats to free thought, including 'brain-reading' and 'behaviour-reading'. Yet I will argue that these are not free thought's primary challenges. For me, the main threat to free thought comes not from silicon but from carbon. In addition to discussing the global threats to free thought emanating from autocracies, I will also examine what Western democracies add to this problem. Chapter 4 will then explore the existing state of the right to freedom of thought, including four elements of the right outlined by the UN Special Rapporteur's report. By the end of this chapter, we will be armed with a conception of what thought is, how thought can be free, the threats to free thought and how the law currently conceives of the right to free thought.

Moving into the book's second half, chapter 5 will begin sketching what a right to free thought could look like. I will use an analogy with sculpture. Just as a sculptor must obtain clay, the would-be thinker must get information with which they can work. Ruling powers have always regulated the information that goes into the minds of the population. Following Johannes Gutenberg's invention of the printing press in the fifteenth century, governments quickly licensed and registered presses, destroyed unauthorised works and imprisoned authors and printers. Our challenge today is less about obtaining information and more about getting trustworthy, reliable information upon which our thinking can act. Chapter 5 will explore the process of obtaining the clay of thought.

As sculptors need tools to shape their clay, thinkers need tools to shape their thoughts. Chapter 6 will examine what these tools are. Following a line used in two of Shakespeare's plays, it is often said that 'thought is free'.[133] Admittedly, one of the plays this line appears in is *The Tempest*, a key theme of which is the manipulation of minds by the nobleman and sorcerer, Prospero,[134] but let us turn the page. Thoughts may be free, but thinking is profoundly difficult. We must be able to control our attention. If we cannot focus our attention, and others steer and harvest it, our thoughts will lie in chains. We must also have the ability to reflect upon our thoughts. If we cannot, we will be carried along by the information we encounter rather than working it up into our own thoughtful content. To process information effectively, we must be able to reason. We must have the ability to think logically. Part of this involves knowing how our mind works and the rocks of logical fallacies onto which our minds would lure us. The right to freedom of thought must protect our abilities to attend, reason and reflect.

Having a good grip on the intellectual arts and being able to reason one's way to conclusions is necessary to become a freethinker, but it is not sufficient. The freethinker must also be courageous. Society will 'always try to tyrannize thought'.[135] This means the freethinker must be able to defy dogma, tackle taboos and neglect norms.[136] Whilst the truth may set you free spiritually, it often leads to confinement physically. Socrates was poisoned, Jesus crucified, Bruno burnt. If we separate our thinkers from our fighters, our fighting will be done by fools and our thinking by cowards.[137]

Finally, a sculptor needs a workshop to practise their craft. Working your potter's wheel in a jostling bar or on a rolling ship is unlikely to produce a quality product. Similarly, thinkers need an environment conducive to thought. Chapter 7 will explore what this environment could look like and how a right to freedom of thought may protect and promote it. Clearly, we need private spaces where we can test out ideas safely alone. But thinkers need more than ivory towers. They need other people to engage in debate and dialogue to feed, fire and correct their thinking. Thinking is both a private and a social process, as we think both alone and together. To think freely, we must be free to roam both into and out of company. We need both inner and outer workspaces for thought.

It would be ironic for a book on the right to freedom of thought not to think critically about the concept of a right to freedom of thought. Concerns about free thought have been raised ever since Plato in Ancient Greece. Back in 1713, it was suggested that promoting freethinking 'will produce endless Divisions of Opinion and by consequence Disorders in Society'.[138] For this reason, chapter 8 will explore arguments against the right to freedom of thought. Things will get lively.

We will finish by looking to the future of free thought in chapter 9. We face two possible futures, one in which we enthusiastically embrace thought and another in which we apathetically surrender it. Embracing free thought will be a two-step process. The first will be developing a clearly defined right to freedom of thought. Yet whatever we decide this right to mean, it will have a limited impact if it inhabits a culture that does not value free thought. Culture can kick away any crutch that legislation provides. For free thought to flourish, we need a culture that supports this process, which extends from the home to the workplace to government. Creating legal solutions to protect thought is much easier than developing a culture of openness and trust where people feel safe to think. We cannot create a culture of humility, patience and gratitude by simply putting new words on a page. We need virtue more than we need law. The call goes out for a new St Francis, not just another St Thomas.

To embrace free thought we need to (re-)create a culture that values and protects free thought. The Ancient Greeks had a social contract that could provide a safe space to speak truth. They called this the parrhesiastic contract. We have ripped up this contract, burnt it and unconcernedly watched the smoke drift far away. Part of this has resulted from our focus on rights at the expense of duties. There cannot be a meaningful right to freedom of thought without recognising a corresponding duty to respect other people's attempts to think, reason and seek the truth. A well-developed right to freedom of thought can aid cultural change by giving the culture something to aspire to and a red line it cannot cross, but it cannot guarantee free thought. Intuitively this all sounds desirable. But what would it look like to live in a society that gave complete protection and support to the freethinking of its citizens? Once we see what such a world could look like, we may reconsider whether we want free thought.

And yet cultures that do not support free thought are terrifying. If we abandon our nature as thinking creatures, someone else will think for us. And the people thinking for us may be monsters. As one German who lived through the Nazi era has observed:

> Most of us did not want to think about fundamental things and never had. There was no need to. Nazism gave us some dreadful, fundamental things to think about – we were decent people – and kept us so busy with continuous changes and 'crises' and so fascinated, yes, fascinated, by the machinations of the 'national enemies,' without and within, that we had no time to think about these dreadful things that were growing, little by little, all around us. Unconsciously, I suppose, we were grateful. Who wants to think?[139]

When we abdicate our responsibility to think, we grease the hinges of the doors to hell. Not thinking is not an option.

1

THOUGHT

As the hands of clocks across Ohio passed beyond midnight, figures scurried in the dark outside the Cleveland Museum of Art. It was Tuesday, 24 March 1970. Outside the museum, atop a marble pedestal, squatted a six-foot-tall, half-ton-heavy bronze statue. The figure was hunched forward, elbow on knee, chin on hand. Created by Auguste Rodin, it was known to the world as *The Thinker*. Cleveland's version was one of only ten full-size models ever made under Rodin's supervision. In the darkness, a pen was produced; and on the pedestal these words appeared: 'Off the Ruling Class.' Abandoning the pen, fingers began fumbling with a metal cylinder. They attached it to *The Thinker* before vanishing into the still and silent night.

Seven and a half minutes later, an orange-yellow explosion lit up the night sky, followed by a boom heard for miles around. *The Thinker* lay face down on the ground, his lower legs mangled. The Cleveland Police never identified the bombers, though they suspected the left-wing terror organisation, the Weather Underground. Returned to its pedestal, the dynamite-damaged thinker still ponders there today. It is a monument to the persistence of thought in the face of force.

There is much that Rodin's *Thinker* gets right about thought. Thinking can be a solitary process, requiring space, silence and security. The figure portrayed is the poet Dante sitting before the gates of Hell. Like Dante, much of our thinking is spurred by

39

encountering the discordant, the problematic and the painful. *The Thinker* also offers more challenging truths. The muscular metal statue looms intimidatingly over the mortal viewer. We may have cause to fear. Thinking can be a tool of domination. Reason is not always the opposite of violence; it can be its continuation by other means.

Nevertheless, there is also much that *The Thinker* gets wrong about thought. *The Thinker* is alone. Yet we often think with others. His furrowed brow emphasises that thought happens inside our head. But thinking isn't limited to the confines of our skulls. *The Thinker* sits, yet movement improves thought. The bowed head of *The Thinker*, looking inwards, suggests thought is about interpreting the world. But, as Marx recognised, thought is for changing the world. *The Thinker* is muscular. Yet reasoning, argued the German philosopher Nietzsche, is a weapon of the weak, a 'last-ditch weapon in the hands of those who have no other weapon', a hammer for those who cannot command.[1]

To create a meaningful right to freedom of thought, we must once again explode *The Thinker*. We need to detonate the erroneous conceptions of thought represented by Rodin's statue and in its place erect an accurate representation of the creature we are trying to save. Continuing with an oversimplified concept of thought, such as 'words in our head', is like relying on a definition of whales as big, blue and blubbery. If we don't know what we are trying to save, then our preservation efforts are doomed to failure. To save thought, we must first find it.

Thinking with each other

In the half-foot between our ears, miracles occur. We consciously experience words, images and feelings. Yet it is a profound mistake to believe that this is the only place where thinking occurs. The idea that we have an inner world seems natural and intuitive. As the philosopher Charles Taylor puts it, 'who among us can understand our thought being anywhere else but inside, "in the mind"?'[2] But this idea is not natural. As with all human concepts, someone had to invent it. It was in the fourth century that Christianity created the distinction that the law today embraces between our inner world (the *forum internum*) and the outer world (the *forum externum*). As the philosopher Alexandru Ovidiu Gacea argues, it was then that 'a new form of identity' arose. People became defined by being able to observe and intro-spect an inner world. Our true selves and God were to be found in this inner world.[3] The inner became the place to go. As St Augustine said, 'Do not go outward; return within yourself. In the inward man dwells truth'.[4]

The Ancient Greeks hadn't been averse to retreating within to find truth. With the exception of Plato, they loved a good cave.[5] But the Athenians believed life was to be lived in public, with truth accessible through the cut and thrust of public discussion. They sought harmony from conflict. For Plato, thinking was not 'some mysterious, peculiar, and immaterial process that gets its virtues from taking place inward'.[6] Instead, he conceived of thinking as happening in a material space inside us. For him, thought took place in the air inside us, which we then breathed out as speech. Thought was physical. It was not a mysterious inner substance (*res cogitans*), as Descartes would later claim.

When we stop viewing thought as occurring via a magical inner substance or through a network of neurons in our brain, it is easier to conceive of it happening outside our heads. We can now take a more communal view, in which thought can be both a private conversation with ourselves and a public conversation with others. Thought becomes a 'stream of meaning flowing among and through and between us'.[7] As I will argue later, we should understand free thought as happening between us, not within us. The freest thought is social, not individual.

To understand how thought occurs outside us, it's helpful to consider how thought gets inside us. Take the thinking we do in our heads, specifically that which uses words. Such 'thinking in words' is what psychologists call inner speech. These words can whizz through our heads at a rate of about 4,000 words per minute. They vastly outstrip our out-loud speech, which runs at a measly 140 words per minute.[8] But how did these words get inside our heads?

The philosopher Thomas Hobbes (1588–1679) argued that out-loud speech was created to transform our silent inner speech into utterances for all to hear. But three centuries later, the Russian psychologist Lev Vygotsky (1896–1934), working in a noisy one-room apartment with his wife, two young children and tuberculosis, turned Hobbes upside down. Vygotsky's developmental account of inner speech argued that we first learn to talk aloud with others and then learn to speak silently to ourselves. The psychologist Charles Fernyhough gives an example involving a child trying to do a jigsaw with its mother.[9] At first, there will be an out-loud dialogue between the child and the mother. The child asks where the piece should go, and the mother answers, perhaps suggesting turning it around and putting it in the corner. Both the mother and child contribute their perspective to the dialogue. They think together. When the mother is no longer present, the child will continue this out-loud dialogue without her. The child may say out loud: 'Where could this go . . . ah yes, over here'. This is called private speech. It is not meant to help others. Here we speak aloud but only for ourselves.

Next, the child breathes this private speech into themselves. Private speech goes underground and becomes our silent inner speech. As it does, it keeps its back-and-forth form, its multiple perspectives, consistent with the interactions from which it was born. It retains its conversational quality. Fernyhough calls this 'dialogic thinking' to capture how inner speech can first take one point of view, then another, creating a dialogue.[10]

Thinking begins as a public act, which we then partially privatise. When we think aloud together as adults, thought returns to its roots. Thought occurs outside our head both before and after it occurs inside. Yet we continue to view speech as the *expression* of thought, rather than as the process of thought itself. We need to change this view. Clearly, speech sometimes *is* the expression of thought. Indeed, we caution people to think before they speak. However, on other occasions, there may be no thought before our speech, with thinking taking place through speaking. We all recognise this, often referring to ourselves or others as 'thinking aloud'. Well-known quips, such as 'how can I tell what I think until I see what I say?', capture the idea that speaking is sometimes the only way to access our thoughts. The writer and entrepreneur David Perell usefully refers to speech as the 'first draft of thinking'.[11]

A stumbling block to protecting our thinking aloud together is prejudice against this type of thought. Whilst there are exceptions, such as our embrace of trial by jury, the idea of doing anything in a group, thinking included, is often disparaged today. Glancing through psychology textbooks, you could be forgiven for believing that only bad things happen when people come together. The idea of the madness of crowds has a long history. Most famously, the French psychologist Gustave Le Bon (1841–1931) claimed that whilst people were rational as individuals, people in groups acted more like animals. As he put it, when a civilised person joins a crowd, they descend 'several rungs in the ladder of civilisation' and become 'a barbarian . . . a creature acting by instinct'.[12]

This 'individual-good, group-bad' perspective, encouraged by the events of the French Revolution, was reinforced by the horrors of the Holocaust. After the Second World War, a slew of psychological studies claimed to show how individuals were corrupted by being in groups. In the 1950s, Solomon Asch claimed to have found that individuals often ignored the evidence of their own eyes to go along with the erroneous judgements of a group. Yet Asch appears to have overestimated how common such conformity is.[13] In the 1960s, Stanley Milgram claimed to show how people would set aside their moral scruples to obey an authority figure's instructions to harm others. However, over half a century later, psychologists still can't agree on what Milgram's results mean.[14] Most pertinently for our purposes, in the 1970s Irving Janis put forward the idea of groupthink. By echoing Orwell's language in *1984*, Janis intended this term to have negative connotations.[15] Central to groupthink was the idea that groups of people tended to make bad decisions because their 'strivings for unanimity override their motivation to realistically appraise alternative courses of action'. Groups were alleged to be more interested in harmony than truth. Initially applied by Janis to account for the Bay of Pigs fiasco, psychologists would later apply it to situations including Watergate and the *Challenger* disaster.

Henry Kissinger once defined an expert as someone who articulates the consensus of the powerful. In a society where elites fear collective mass action, those who argue that groups lead to problems are elevated to expert status. Marx long ago noted that the dominant ideas of society represent the ruling class's interests. These interests incentivised the view that accurate thought happens inside an individual's head. People in groups either failed to think, resorting to base instincts instead, or thought poorly. This idea resonated with the longstanding Christian conception of a pure, spiritual inner realm and a corrupted outer world.

Yet there can be wisdom in group deliberations.[16] We can see productive thinking together in Ancient Greece. In his walks across Athens, from the port of Piraeus to the Ancient Agora, barefooted Socrates would reason aloud with his fellow Athenians. Unfortunately, the argument that large-scale thinking together took place in the

Greek Assembly, where (some, male) citizens deliberated together, runs into problems. The Athenian historian Thucydides tells us that the strength of Athens lay in 'knowledge which is gained by discussion preparatory to action' and much is made of how the Greeks deliberated together in the Assembly.[17] Yet this 'deliberation' seems to have largely consisted of individual thought. People would listen to the minority of people who actually made speeches, before voting on policies, rather than having meaningful dialogues with each other and thinking together.[18]

Nevertheless, in Ancient Athens, attendees at the Assembly were at least exposed to other people debating and deliberating on issues of policy. This stands in contrast to the Assembly in another Ancient Greek city, Sparta. Those attending the Assembly in Sparta did not see or have the chance to take part in debate and deliberation. Instead, they were presented with laws which they acclaimed by shouting. If Athens was thinking together, Sparta was thinking alone.

The idea that thinking aloud together is important has echoed down the ages. In his famous 1784 essay 'What is Enlightenment?', the German philosopher Immanuel Kant worried about how well we would think 'if we did not think as it were in community with others to whom we communicate our thoughts, and who communicate theirs with us'. Kant claimed that any external power 'which wrenches away people's freedom publicly to communicate their thoughts also takes from them the freedom to think'. As a result, he concluded that 'the public use of one's reason must be free at all times, and this alone can bring enlightenment to mankind.'[19] In short, free thought is only possible if we are free to think together.

Former US Supreme Court Justice Louis Brandeis often invoked this idea of public reasoning. Brandeis would quote from the Book of Isaiah (1:18), 'come let us reason together.'[20] This quote captures the idea that we think aloud together, rather than merely speak together, and that we 'seek the truth collectively, not individually'.[21] As the US President who nominated Brandeis to the Supreme Court, Woodrow Wilson, once put it, one 'cannot be said to be participating in public opinion at all until he has laid his mind alongside the minds of his neighbors and discussed with them the incidents of the day and the tendencies of the time.'[22] Wilson, hardly a hero of free thought, was nonetheless right to identify how speaking allows us to lay our minds together.

A good example of this comes from the Farmers' Alliance, born in the United States in 1877.[23] Before the planting season began, farmers needed to acquire seeds and equipment. To pay for this, they would give creditors the right to part of their future harvest. Due to factors such as high interest rates, low crop prices, and droughts, many once-independent farmers ended up as indebted tenants to these creditors. The Farmers' Alliance was a way to fight back against what had become a 'modified form of slavery'. The co-operative alliances that farmers built with each other gave them a 'place to think in'. They could deliberate together, share experiences, and build a counterpower to the forces that were oppressing them. As a result, they were able to build a political movement that valued independence and community, and proclaimed that 'wealth belonged to him who creates it'. This was only possible because the farmers were able to think together.

Yet, just as there have always been those who have valued the thinking together associated with Athens, there have also been those who hailed the lone thinkers of Sparta. The classic example is the French philosopher Jean-Jacques Rousseau. Rousseau believed that citizens should make up their minds alone, without entering into dialogues with their fellows. As he put it, 'each citizen should think only his own thoughts'.[24] His reasoning here was that thinking together would create factions that would prevent a 'general will' from emerging. Ultimately though, the value of thinking together is an empirical question.

Today, an avalanche of evidence shows that thinking together, formally termed 'socially distributed cognition', can be beneficial. Crowd-sourced thinking has solved problems that have confounded individuals for centuries. For example, the Polymath Project has crowdsourced mathematicians' expertise and led to a range of breakthroughs.[25]

There are also examples of crowd-sourced thinking helping us better solve problems that we mere mortals have a chance of understanding. An excellent example of this comes from a study examining the 'Linda paradox'. In this, you first read the following information:

Linda is 31 years old, single, outspoken and very bright. She majored in philosophy. As a student, she was deeply concerned with issues of discrimination and social justice and also participated in anti-nuclear demonstrations.

You are then asked which is more likely, 1) that Linda is a bank teller; or 2) that Linda is a bank teller and is active in the feminist movement. The correct answer is option one. The second option is a subset of the first, meaning Linda cannot be more likely to be in it. To use an analogy, a missing sailor cannot be more likely to be found just off the coast of Hawaii than he can be to be found somewhere in the Pacific Ocean. Plenty of us get this question wrong. Psychologists found that when people did this problem alone, fifty-eight percent gave the wrong answer. But when people did it in pairs, only forty-eight percent got it wrong. And when people did it in trios, only twenty-six percent got it wrong.[26] What we see here, the performance of a group being better than the performance of its best member on their own, is called the assembly bonus effect.[27]

If free thought is reasoned thought, and we reason better together than alone, then maximal freedom of thought is to be found in groups, not individuals. This has devastating implications for the idea that the right to freedom of thought aims to protect something unobservable inside our heads. The idea that thought is something that happens in our heads starts to look like an ideological construct that serves some other interest than getting to the truth.

Given that thinking in groups can clearly go wrong, our challenge becomes identifying the factors that give groups the best chance of reaching truth. We don't think better in groups simply because we become more motivated to be better individual thinkers. It isn't that individuals try to show off to the group and think better as a result.[28] Instead, *the group itself thinks*. When people come together to try to solve a problem, everyone may have some truth and some error in their view. The group can refine this, keeping each person's truths and discarding their errors, to reach the correct answer. For a group to display collective intelligence, it really helps if its members understand what each other is thinking. The better individual members can recognise what other group members are thinking or feeling, the greater the group's collective intelligence.[29]

Group thinking is more likely to go wrong if, at the outset, everyone already agrees.[30] A lack of disagreement means there is no motivation for anyone to try to create counterarguments. People end up only proposing ideas in favour of the shared position. This makes the group even more committed to the position it already holds. Imagine that a group of people, all moderately in favour of legalising cannabis or euthanasia, come together to discuss the issue. By the end of their discussion, they will probably have heard new arguments for cannabis/euthanasia legalisation. But they will be unlikely to have heard new arguments against legalisation. Furthermore, to show they are a 'good' member of the group, people are likely to exaggerate the group's basic tendencies. As a result, the group will end up more in favour of cannabis/

euthanasia legalisation than when it started. This is the phenomenon of group polarisation.

To have the best chance of finding truth, groups need to be
made up of people with a diverse range of views. Everyone then
needs to be made to debate their ideas, as juries do.[31] For deliberative democracy[32] to exist, in which policy decisions are informed
by the results of citizens coming together to deliberate, rather than
being left to the whims of elites, a pervasive culture of free thought
is required. Such a culture is needed to ensure that when citizens
come together to deliberate, they bring a diverse range of view-
points and ideas, giving their deliberations the best chance to reach
truth. Individual free thought helps thinking together, creating a
mutually reinforcing feedback loop. The right to freedom of
thought must protect both private and public thought.

Recognising that some speech should be viewed as thought will
have significant implications. Courts currently deem what we say
to be protected by our right to freedom of speech which, as we have
seen, can be legally restricted. But if certain speech is thinking-in-
progress, it should get the absolute protection accorded to thought.
Clearly, not all speech should be deemed thought. If I seriously
threaten violence against you, we wouldn't want to say that I am
thinking aloud and could thereby claim legal protection for this
speech. We are not looking to justify barbarity.

The crucial question then becomes what speech do we want to
count as thought, and what criteria could we draw upon to make
this distinction? Let's call speech that is thinking-in-progress
'thoughtspeech', which we can contrast to 'mere-speech'. The
former would be protected by the absolute right to freedom of
thought, the latter by the limited right to freedom of speech.

For speech to be deemed thoughtspeech it should support the
same things that make thought valuable. As we saw in the
Introduction, these include reaching truth and promoting autonomy. For me, the most straightforward criterion is whether the

speech is performed with truth-seeking intent. One feature strongly suggestive of speech with truth-seeking intent is when it has a dialogic form. If speech is dialogic, it invites or already latently contains multiple perspectives. Such speech invites conversation. It encourages the person being spoken with to turn thoughts around and examine them from various sides. Dialogic speech does not simply assert but is formed to suggest openness to responses and questions. As the French philosopher Jacques Ellul has written, 'propaganda ceases where simple dialogue begins.'[33]

The concern that it would be difficult for courts to assess the truth-seeking intent of a speaker is simple to address. Courts already routinely evaluate the intent of defendants. They frequently assess whether someone intended to kill another person. This allows courts to distinguish whether the scrutinised act constitutes murder or manslaughter. Similarly, we could feasibly task a court with assessing truth-seeking intent to determine whether a speaker was engaged in speech or thoughtspeech.

It may not only be our fellow human beings with whom we think. If thinking relates to a stream of meaning, then there is no reason this need only flow between people. A relatively recent development is our thinking with computers. We need to reflect more on how we can benefit from thinking with computers and artificial intelligence, as well as manage the potential risks.[34]

Yet we must not forget that we can think with both carbon and silicon. We may think with trees or rivers or elephants. The environmental activist and writer Derrick Jensen makes this point well. Someone had told Jensen that he was just as responsible for deforestation as the CEO of a timber company. Jensen decided to ask a tree about this.

> And so I did, I asked a tree. You know, am I as culpable? And the tree said, Look, you are an animal, you consume things, get over it. And then I realized, Yes, I am actually culpable for

deforestation, but not because I use toilet paper. I'm culpable for deforestation because I consume the flesh of a tree, but I don't fulfill my end of the bargain by stopping [timber companies] ... So what I need to do is, I need to stop [timber companies].[35]

To take another example, Patrick Lionel Djargun Dodson, an Aboriginal Australian, has observed that whereas many Australians 'find it hard to think with the land[, w]e Aboriginal people find it hard to think without the land.'[36] This suggests that the taking of land from aboriginal peoples can be seen as a violation of their right to freedom of thought (which, to stress again, is an absolute right that there is no justification for violating). The stream of meaning can flow through everything in our biosphere. We think both on and with the planet we share. To truly think, we must be prepared, to adapt a phrase from John Locke (1632–1704), to jumble heaven and earth together.[37]

Thinking with technology

At the end of the second book in Douglas Adams's six-book trilogy, *The Hitchhiker's Guide to the Galaxy*, Arthur Dent and Ford Prefect face a sizeable problem. A series of unfortunate events, including but not limited to an argument with a robotic 'tea' dispenser, a poor choice of transport as the universe ends, and tragically counter-productive contempt for telephone sanitisers, have left them stranded on pre-historic Earth. For reasons we need not go into, Arthur and Ford possess the answer to life, the universe and everything. It is forty-two. But Arthur and Ford don't know the precise question to which this is the answer. This information is hidden in Arthur's mind, but he can't consciously access it. They try to get at Arthur's thoughts by getting him to pick out

Scrabble letters from a bag at random. The tiles they pick spell out the question: 'what do you get if you multiply six by nine?' Genius. The point for our purposes here, and there is one, is that Arthur and Ford are in some sense thinking with the Scrabble tiles. Similarly, when we play Scrabble, and move around the seven tiles on our rack, we too are thinking with tiles. Thought is happening outside our heads.

A second reason to pull apart the conception of thought portrayed by *The Thinker* is that he represents a dated style of think-ing. The only thing *The Thinker* holds in his hand is his head. He has no iPhone, no keyboard at his fingertips and not even a pen. Yet today, our minds are spread across the gadgets around us. Technology has extended our minds.

In 1998, philosophers Andy Clark and David Chalmers proposed the idea of the extended mind.[38] For them, human thought was not bound by skin and bone to the space between our ears. Instead, they argued our minds extend out into the world. For Clark and Chalmers, if 'part of the world functions as a process which, were it done in the head, we would have no hesitation in recognizing as part of the cognitive process, then that part of the world is . . . part of the cognitive process'. If something looks, feels and smells like thought, then it is thought, irrespective of where it is found.

Following this reasoning, using a pen and paper to perform calculations constitutes thinking. For someone with Alzheimer's, a notebook may function the same way that memory does. Indeed, in Orwell's *1984*, the novel's protagonist Winston Smith thinks by undertaking the forbidden act of writing in his diary. To write can be to think. As the cognitive scientist Peter Carruthers clarifies, 'one does not first entertain a private thought and then write it down; rather the thinking is the writing'.[39] In such ways, the site of thought extends into the world through our use of external objects.[40]

Clark and Chalmers proposed three criteria for what makes a technology part of our extended mind. First, the technology needs to be a constant in our life, which we would rarely act without consulting. Second, the information from the technology is available to us directly and easily. Finally, we automatically endorse the information we retrieve from the technology. Even before we think about what technologies fit these criteria, we can see that a trusted friend may fulfil all three. But let's get back to the tech.

In 1998, Clark and Chalmers argued that our internet use would probably fail to meet their three criteria for being part of our extended mind. Yet today it meets all three. Take Clark and Chalmers's first criterion, the constant presence of the technology in our life. Beyond the ubiquity of smartphones today, our minds are set up to look to Google for answers. Once we start using search engines to answer questions, we automatically turn to these engines in similar situations in the future; even to answer questions we are quite capable of answering ourselves.[41] When asked hard trivia questions, our minds automatically start thinking about internet search engines.[42] In terms of Clark and Chalmers's second and third criteria, information is accessible through a search engine or familiar websites without difficulty, and we tend to trust it (we will come back to the problems our trust in search engines raises later). Much of our internet use now constitutes thinking. The internet has digitally expanded our minds.[43]

In this light, googling can be remembering, which makes it literally thinking. Intriguingly, we don't experience Google-related remembering as involving *Google* remembering for us. Instead, we feel that *we* are doing the remembering. When people are allowed to use Google to answer trivia questions, they report feeling smarter, even though they clearly can't take credit for the remembering.[44] A strange form of symbiosis is underway. Our skin–skull boundaries blur as we seep into the machines around us (or they seep into us).

Many of us use apps to store, develop and retrieve our thoughts. If this software crashes, or if the company providing it collapses, it can erase part of our minds almost as effectively as a stroke. There is little practical difference between a faulty internet connection and a blockage in the arteries of our brain. If we want to protect our thinking, we must protect our use of technology. When, in 2010, the Finnish government made access to a broadband connection a legal right, it fulfilled part of its duty to support its citizens' right to freedom of thought.[45]

Not only are we increasingly outsourcing memory to external technologies, we may be doing this without realising it. Take a study done by the late American psychologist Daniel Wegner and colleagues.[46] This asked people to type forty facts into a computer, such as 'an ostrich's eye is bigger than its brain'. Half the participants were told the computer would save their work; half were told it would be erased. Those who believed the computer would save their work were much worse at later remembering the facts. Even when people were asked to try to remember what they typed, those who thought the computer was saving the information were still worse at remembering it later. Similarly, if we take digital photos whilst visiting a museum, we remember fewer exhibits than if we'd left our camera at home.[47] We are outsourcing our memory to machines, often without even knowing it.

More than a few score years ago, Abraham Lincoln declared that 'the dogmas of the quiet past are inadequate to the stormy present . . . we must think anew and act anew'.[48] Now is another of those times. The programming prodigy Aaron Swartz once observed that 'there's a battle going on right now, a battle to define everything that happens on the Internet in terms of traditional things that the law understands.'[49] We must start thinking of much of our internet activity outside traditional legal categories. This means recognising some internet activity as thinking and then protecting this, absolutely, with the right to freedom of thought.

Thinking with movement

The next reason to strap metaphorical dynamite to *The Thinker* is that it portrays thinking as a stationary activity. Thinking, as represented by *The Thinker*, is something we do on our backsides. Fittingly, there is a famous picture of the comedian Robin Williams reaching up to *The Thinker* holding out a toilet roll.[50] Rodin actually wanted us to see the whole body of his statue involved in thinking. As he put it, *The Thinker* 'thinks not only with his brain, with his knitted brow, his distended nostrils, and compressed lips, but with every muscle of his arms, back, and legs, with his clenched fist and gripping toes.'[51] Nevertheless, Rodin still has his statue sitting down. Yet thought is supported by our ability to move. As philosophers put it, thought is *embodied*. If we were only to be a head-in-a-vat, our ability to think would not be what it once was.

Many great philosophers, and Jean-Jacques Rousseau, have emphasised the link between walking and thinking. 'All truly great thoughts', claimed Nietzsche, 'are conceived while walking'.[52] The Danish philosopher Søren Kierkegaard reported that 'I have walked myself into my best thoughts'.[53] Philosophy happens on the move.[54] Pascal may have been right when he claimed that all our problems stem from our inability to sit quietly in a room alone. But such problems are less likely to be solved if we sit back down again. *Solvitur ambulando*, said St Augustine – it is solved by walking.

Psychology backs up these philosophers. We now have evidence of the benefits of walking and movement in general for our thinking.[55] Annie Murphy Paul offers an excellent treatment of this in her book *The Extended Mind*.[56] Paul introduces us to evidence that radiologists are better at analysing scans when they do this on a treadmill and that children learn better if pairing a story with movement. She also describes how our attention, problem-solving abilities and memory all improve after exercise. Walking outside emerges as a boost to creative thought. This is not simply due to having the wind in your hair or ever-changing surroundings; exercise itself is important. When researchers compared the creative thinking abilities of a group of volunteers who agreed to be pushed outside in wheelchairs to another group of volunteers who walked outside, they found more creative thinking in the walkers.[57] Our legs don't just carry our minds; they rouse them.

Not only does being able to move help our ability to think, but so does our use of gestures. Paul notes that we both speak and think less fluently if we can't gesture. Of course, we don't want to fall into an ableist narrative here. Obviously, it is still possible not only to think well but to think some of the most complex, original thoughts ever conceived of whilst immobilised (take Stephen Hawking, for example). Nevertheless, for a mobile individual, movement enhances thinking. Our right to freedom of thought needs to reflect this.

Thinking with mindware

The Thinker looks like he is engaged in slow, laborious thought. Yet this is only one of the systems of thought we have. Understanding these two different systems is crucial to protecting our minds from those who would manipulate them. The world has loaded our minds with software. This 'mindware' includes knowledge, rules, procedures and strategies that help us solve problems and make decisions.[58] Some of our mindware comes pre-loaded by evolution, and some we download from our culture.[59] Mindware has been the topic of many recent books, including Daniel Kahneman's *Thinking, Fast and Slow*, Steven Pinker's *Rationality* and Richard Nisbett's *Mindware*.[60] Whilst we don't need to recapitulate such work in detail here, we do need to understand the basics of our mindware to become aware of how others may hack it. We must then design the right to freedom of thought to defend us from such attacks.

Psychologists distinguish between one type of thinking that is fast and automatic and another that is slow and laboured. These two types of thought are called System 1 and System 2 respectively.[61] System 1 thought is quick, automatic, evolutionarily ancient and happens outside consciousness. It uses heuristics (rules of thumb) and doesn't show us its workings. Answers just pop into our heads. In contrast, System 2 thought is conscious, slow, deliberate, involves language, makes use of reason and is a relatively new invention unique to humans. When we think in a slow, deliberate, reasoned train of thought, this is System 2 in action. As I can never remember which system is which, I will call our fast, automatic System 1 thought 'rule-of-thumb thinking'.[62] The slow, conscious thought of System 2, I will call 'rule-of-reason thinking'.

Both systems have their pros and cons. Rule-of-thumb thinking gives us quick answers, but at the cost of greater error-proneness. For example, we often make judgements using something called an availability heuristic, in which we estimate the likelihood of an event based on how easily examples come to mind. Plane crashes frequently make headlines, but tuberculosis deaths don't. As a result, we overestimate how many people die in plane crashes and underestimate tuberculosis deaths.

Another example of rule-of-thumb thinking is the anchoring heuristic. In this, we use the first piece of information we get as the basis for our future decisions. This leads to some weird effects. In one study, psychologists Amos Tversky and Daniel Kahneman asked people to guess what percentage of countries in the United Nations were African. Half the people doing this task had just come in from watching a roulette wheel spinning, where the ball had landed on 10. The other half had just seen a roulette wheel where the ball had landed on 65. The roulette task had no logical bearing on the later task involving African countries. Yet people who had seen the roulette ball land on 10 estimated that twenty-five percent of UN countries were African, whereas those who had seen the ball land on 65 guessed forty-five percent. A completely irrelevant piece of information had 'anchored' people's decisions to high or low numbers.

All this might sound a bit detached from the real world, but once someone knows how your rule-of-thumb thinking works they can manipulate you. Take credit cards. What effect do you think seeing the meagre minimum repayment figure appearing on your statement has on your decision about how much to pay off your card? None? Wrong. Seeing this low number subtly influences you to anchor your choice of repayment amount at a lower figure than you would otherwise choose. More debt is left on your card, meaning more interest for the credit card company.[63]

Even though rule-of-thumb thinking can cause errors, this doesn't mean we shouldn't use it. The gut feelings it gives can be helpful. Indeed, it can make us millions. Consider a study that looked at the performance of financial traders in the City of London. These traders had no guaranteed salary. All their money came from their trades being successful. Researchers measured how much the traders listened to their bodies by seeing how well they could guess their heart rate. The study found that the better the traders were at this task, the more profits they made.[64] As the former Chairman of the US Federal Reserve Alan Greenspan put it, 'the gut-feel of the 55-year-old trader is more important than the mathematical elegance of the 25-year-old genius'.[65] Of course, this is not to say there is no role for rule-of-reason thought in trading. A later study found that traders used rule-of-reason thinking when things had become difficult and they were trying to find a way out of a hole.[66]

Indeed, rule-of-reason thinking can help correct rule-of-thumb thought. John Locke warned that ideas that pop into our head, if 'freed from all restraint of reason, and check of reflection', become 'heightened into a divine authority, in concurrence with our own temper and inclination.'[67] Being able to override rule-of-thumb thought and engage rule-of-reason thought is a crucial skill for a freethinker.

Yet rule-of-reason thinking has its own problems. It can fall victim to our own laziness.[68] As we will discuss in the next chapter, evolution wants us to expend as little effort as possible to make decisions. This encourages us to be 'cognitive misers' who minimise our use of rule-of-reason thinking. Someone who never engages their rule-of-reason thinking could be said to not have 'a mind of their own'.[69] Unless we try to engage our rule-of-reason thought, we are likely to slip back into automatic thought patterns. Another problem with rule-of-reason thought is that it can talk us into bad decisions made by rule-of-thumb thinking, rather than challenging them. This idea was captured in the definition of philosophy offered in Aldous Huxley's 1932 novel, *Brave New World*, as 'the finding of bad reason for what one believes by instinct'.

When we look at what happens in the real world, as the psychologists Hugo Mercier and Dan Sperber note, we make most decisions using rule-of-thumb thought. When we do use rule-of-reason thinking, we often make bad decisions.[70] The reason for this is intriguing, and I'll elaborate on it in the next chapter. But to give a preview here, Mercier and Sperber argue that rule-of-reason thinking did *not* evolve because it allowed us to reach the truth. Instead, it developed because it helped us persuade other people of our viewpoint. As a result, rule-of-reason thought 'pushes people not towards the best decisions but towards decisions that are easier to justify'.

To make this more concrete, imagine a bar that sells a $4 high-quality beer ('beer A') and a $2 low-quality beer ('beer B'). The bar owner wants his customers to buy more of beer A and comes up with a cunning plan. He introduces a new high-quality beer to his bar, which costs $6. When he does so, sales of beer A increase. Why is this? Well, when he only offered two types of beer, people only had one reason to buy beer A: it was better quality than beer B. Now he offers three beers, people have two reasons to buy beer A; it is better quality than beer B and cheaper than the new $6 beer. Beer A may be neither the most rational choice for customers, nor what fits their pre-stated preferences, but it is what they can best justify to others.[71] This phenomenon, in which the introduction of a third option pushes us towards one of the first two options in a not strictly rational way, is called the attraction effect. It shows how rule-of-reason thinking can be manipulated by those who know how it works.

In short, if someone knows how your mind works better than you, they can design products and services to manipulate your rule-of-thumb and rule-of-reason thinking. In liberal democracies, this is called advertising and is an accepted capitalist practice. In Russia, it is called reflexive control and is an accepted psychological warfare tactic.[72] The right to freedom of thought must take a stance on which of these perspectives makes the most sense.

Thought as mind wandering

When *The Thinker* sits down, he appears to be thinking about something. As the Philadelphia Museum of Art puts it, he is 'clearly focussed'. But often we are not focused. Our minds wander. As the philosopher John Locke put it, there is 'nothing more testy and ungovernable than our thoughts'.[73] Psychologists differentiate between the thoughts we have about immediate happenings in the world, and those that are about worlds and people not present. So, for example, when you are in the shower, your thoughts may not be driven by the immediate things around you, such as soap, towels, or water temperature, but rather by thinking about a meeting you have upcoming later that day. When you reach that meeting, rather than thinking about what your boss is saying, you may drift off and start thinking about the pleasant shower you'd rather be back in.

We spend between a quarter and a half of our time thinking about events in our inner world, rather than attending to the external world immediately around us. Psychologists call this 'mind wandering', 'stimulus-independent thought' or 'self-generated thought'. Much of this content relates to our past or future concerning current worries or goals. However, we are more likely to find our minds wandering into future events rather than past ones.[74]

Mind wandering can cause us problems, being more likely to make us feel bad than good.[75] A wandering mind takes in less information and remembers less. How much your mind wanders during tests and exams is related to your results, and not in a good way.[76] Mind wandering is also associated with depression, an inability to delay gratification, and being in car accidents.[77] Having your mind wander more is even associated with faster ageing.[78] This makes it essential to know when your mind is wandering. If your mind wanders and you are aware of this, you have 'tuned out', but if you are unaware, you've 'zoned out'. We perform worse on tasks when we zone out than when we tune out.[79] Having good mental control means being aware of where your head is at and stopping your mind wandering when you have a job to do.[80]

Yet good mental control also means freeing your mind to wander when appropriate. There are benefits of mind wandering. Minds wander because we have goals that aren't tightly linked to what is immediately happening. Someone whose thoughts spontaneously drift towards their future goals will be better able to work out how to achieve these goals than someone who hasn't thought about them in advance. Mind wandering can also help creative thinking and problem-solving. It can boost positive feelings of meaningfulness by placing past and future events into a meaningful narrative, which in turn helps feelings of well-being and health.[81] Guarding our capacity for mind wandering, and supporting our awareness and control of this ability, is something for which the right to freedom of thought should strive.

The Roman writer Horace once counselled, 'rule your mind or it will rule you.' Today, it appears better to advise, 'rule your mind or others will rule it for you'. We must understand the protean nature of thought, and how our minds work, to develop a meaningful right to freedom of thought. As we have seen, thought ain't just in our heads.[82] It happens out loud with others. It happens through external devices, from diaries to internet searching. Thought happens when we sit still, but is facilitated by movement. Thought can be both engaged in the world and separate from it. We may use slow, deliberate, rule-of-reason thought or fast, hidden, rule-of-thumb thought. Thinking involves the ability to let our minds wander as well as to pull them back again. The right to freedom of thought must protect and support all these facets of thought. I have started to show how, once others understand these properties of thought, they can use this knowledge to manipulate us. Recognising this will be essential when we consider what it means to protect thought.

We now face a choice. Should the right to freedom of thought protect all forms of thought or just some? Are we serious about protecting thought wherever we find it? Or are we merely wanting to offer protection for thinking that in no way disturbs the world, which we could call the Prufrock Option?[83] This is a hard call to make. A robust and comprehensive right to freedom of thought will change society. Yet no one knows what the ensuing society would look like. If we genuinely buy into the idea that free thought is as important as it is made out to be, we should plough ahead and offer absolute protection and support for thought in all its myriad forms. If the ensuing undamning of thought causes social turmoil, and we decide that security, comfort and noble lies are more important, at least we can be honest about this.

Now we have a better understanding of what thought is, let's turn next to the question of what makes thought free.

2

THE ARRC OF
FREE THOUGHT

It is a bright, cheerful Tuesday morning and no less beautiful for being purely hypothetical. You and a friend have decided to volunteer for a study of thinking. Arriving at a leafy university campus, you eventually find your way to the inevitably hideous concrete laboratory building. A scientist greets you and shows you into the lab. You are led to one end of a long table and sit with a computer monitor in front of you. At the other end of the table sits your friend.

'What's on your screen?' your friend calls out to you. 'Nothing yet', you reply. The scientist explains that she will put sensors on your friend's head and then ask him to talk about anything he wants to. The scientist tells you that your friend's words should appear in real time on your screen.

The experiment begins. Your friend starts talking. Sure enough, as he speaks, his words appear in text form, scrolling across the screen in front of you. You aren't impressed. Apps on your phone can transcribe speech this quickly. As your friend begins babbling about goats, your mind starts wandering. As it does, you notice a small box on the table, just off to your side. It has two buttons. One is blue, pressed down and lit up. The other is red and unlit. You wonder what would happen if you pressed the red button. Glancing around, you can't see the scientist anywhere. Your fingers start to tap on the desk. To hell with it, you think and press the red button.

At first, nothing seems to have happened. No lights flash. No sirens sound. No one falls through a trapdoor. Your eyes return to the screen in front of you. Text continues to scroll. But then you notice that something *has* changed. The text on the screen no longer matches what your friend is saying. You seem to have messed up the experiment. Feeling guilty, you reach to press the blue button to restore normality. But, before you can do so, you notice something that causes the hairs on your arms to stand up. Your friend's words are still appearing on the screen, but they are now appearing before your friend says them.

As you look closely at your screen, the words 'The great thing about goats' appear in black and white text. You raise your eyes and look at your friend just as he opens his mouth and says, 'The great thing about goats . . .' As you watch in mounting horror, the screen continues to anticipate what your friend will say. You do not like this. As you look at your friend, he now seems uncanny, less human, more machine-like. Reaching out, you press the blue button. The text on the screen jumps slightly and returns to a familiar, reassuring real-time read-out. Your friend is your friend again. The scientist comes back into the room. You pretend nothing has happened because, frankly, you want no part of this devilry.

That night, lying awake in bed, you struggle to work out what happened. A verse from the Book of Proverbs comes into your mind: 'Even before a word is on my tongue, behold, O Lord, you know it altogether.' There seems to be a new god in town. You realise that the feeling of choosing our thoughts, of being in control of what we think, must be an illusion. The feeling that we are free to decide what to think next is merely that, a feeling. It is rooted in ignorance. As a clock ticks monotonously in the distance, you also realise, by extension, that our feeling that we choose what actions to do next must also be illusory. Our feeling of freedom, our feeling that we make choices, is made possible by our ability to imagine alternative futures, coupled with our inability to know which of these will come to pass. Our ignorance of the future lets us feel free.

Whilst this scenario is imaginary, it is theoretically possible. The idea behind it comes from studies showing that our brain initiates what we will do next, long before we have the illusory experience of consciously making this choice ourselves.[1] In short, the next thought that will pop into your head is already determined.

The most famous phrase in Western philosophy, René Descartes' 'I think; therefore I am', is hopelessly wrong. 'I' do not think at all. 'I' am merely a silent witness to thinking. 'I' am a spectator, not an actor.[2] Whatever is creating the thoughts I experience lurks in the inaccessible basement of my brain. 'I' don't control it. Its control is so total that not only do we not sense it lurking there, but we actually feel free. The brain's best trick was making us think it didn't exist. We need to get away from the idea that we, perched behind our eyeballs, just above our nose, are a magical being that can cause any thought it wants to materialise. My brain thinks, I witness. *Cerebrum cogitat, ego testor.*

What is the point of conscious thought if it is the witnessing of thought rather than the creation of thought? The answer may be that conscious thought is an output of the brain that can also be used as an input. To think is to create something new in the world to which we can react. Conscious thought creates a new layer in the universe. This boosts communication, both within ourselves and with others.

Consciousness allows us to communicate with ourselves. One of the leading models of consciousness is the Global Workspace Theory. This proposes that our brain has distinct islands of processing that can't communicate with each other. To overcome this problem, our brain creates a workspace where parts of the brain can place information, which then becomes accessible to all parts of the brain. Consciousness is like our brain's inter-office memo service. The *lingua franca* through which these sub-systems can communicate is inner speech.[3] We are conscious of whatever inner speech our brain promotes to the workspace. By putting content into this new workspace-world, consciously experiencing it, then reacting to it, we gain the ability to control our behaviour 'from the outside'.[4] For example, by thinking to ourselves 'keep going' as we get a stitch on the treadmill, or imagining ourselves approaching the end of a marathon, we create a new stimulus to react to, one not previously present in the world. We can respond to this thought by putting in extra effort. This allows us to keep going longer than we could have without the ability to create a new world to react to.

Consciousness also allows us to communicate with others. The psychologist Roy Baumeister and colleagues have argued that we often define conscious thoughts as those we can report to others. The implication of this, notes Baumeister, is rarely heeded, namely that 'the purpose of conscious thought is precisely for enabling people to tell their thoughts to one another'.[5] Conscious thought may allow us to communicate better with others, with obvious evolutionary benefits.[6] All this fits nicely with the literal meaning of *conscientia*: 'shared knowledge'.

Interesting as this may or may not be, what does it have to do with freedom of thought? One implication is that the right to freedom of thought is not trying to help 'you' think freely. 'You' don't think at all. 'You' are just a dumb witness. 'You' are the burning sensation that occurs when your brain coughs up and swallows its own vomit. Given this, any attempt to increase freedom of thought is an attempt to get your brain in control of your mental life rather than other people. You do not control your attention. Either your brain points it somewhere congruent with a goal your brain has selected, or it is dragged elsewhere by others. The right to freedom of thought should strive to improve the ability of non-conscious processes in your brain to control your thinking. But what are the characteristics of a brain that thinks freely? What does it mean to be mentally autonomous?

Autonomous behaviour requires autonomous thought

Life is a constant struggle, with no opponent being more ubiquitous or relentless than the one inside us. We wrestle urges to eat
and to sleep. We strive to focus on the matter at hand. We battle a
body that wants to give up. All this happens because we strive for
long-term goals that transcend the immediate dictates of our
bodies. To succeed, we need the ability to self-regulate; to cogni-
tively control our emotions and actions.[7] For example, we need to
be able to override our desire to see what others are saying about us
on social media to focus on our life goals. Overall, particularly
when compared to non-human primates, we are remarkably good
at self-regulation.[8] This ability plays a crucial role in our development of autonomy.

'Autonomy' comes from the Greek *autós* ('self') and *nomos*
('law' or 'custom').[9] The Greeks used this term to describe city-
states that made their own laws. When we apply it to an individual,
autonomy means the ability to form our will independently and
control our behaviour through reasons and rational arguments.[10]
Autonomy does not require an absence of external influences but
rather the ability to accept or reject them.[11]

The mentally autonomous person does not have a marionette
mind. As the American psychologist John Dewey put it, 'freedom
of mind means mental power capable of independent exercise,
emancipated from the leading strings of others'.[12] We will examine
these 'strings of others' in the next chapter. Here, I want to consider
what kind of thinking someone sitting alone in a room would be
undertaking if we wanted to describe their thought as free. What
internal properties does our thinking need to be 'free'?

Dewey again gets to the heart of this when he argues that true
freedom 'rests in the trained power of thought' and particularly in
the ability to 'turn things over'. By turning things over in one's
mind, says Dewey, we escape enslavement to 'appetite, sense, and

circumstance'.[13] Here we see two critical elements of free thought: reason and reflection. I want to add two other ingredients: attention and courage. Taken together, I call Attention, Reflection, Reason and Courage the ARRC of free thought. Society does not naturally bend to the ARRC of free thought. It will only do so if we force it to by creating and enforcing a right to freedom of thought.

Attention

To think freely, your brain must be able to influence where your attentional spotlight points. Your attention must not be at the mercy of whatever glitters. Now, if you're not familiar with the basketball pass task, I'd recommend stopping reading right now and going to the YouTube clip in the footnote below.* I say this because once you see what I'm about to describe, you won't be able to unsee it.

One of the most famous experiments in psychology involves a person dressed as a gorilla. The video I am referring to is a variant of this where a person in a bear costume moonwalks through a crowd who are passing a basketball to each other. The bear walks into the scene from the right, pauses in the middle of the group, waves to the camera and then exits stage-left, still moonwalking. Despite the presence of a waving, moonwalking bear, roughly half of people don't notice the furry intruder when they first watch the video. This is because they have been asked to count the number of passes made by the players. This instruction focuses attention on the ball. As a result, events outside your attentional spotlight, even moonwalking bears, can be consciously missed. When people are told about the impending bear cameo, everyone sees it.[14]

---

* https://www.youtube.com/watch?v=Ahg6qcgoay4. If the link is dead, pop 'awareness test' and 'basketball' into YouTube or your preferred video provider.

Researchers found the same effect when they asked professional radiologists to look for abnormalities in CT scans of people's chests.[15] The radiologists were shown five scans, the last of which included a small matchbook-size picture of a gorilla in the lung. Unnervingly, eighty-three percent of radiologists didn't notice that their patient had inhaled a gorilla. Researchers did the study using eye-tracking technology, so they could see precisely where the radiologists were looking. Of the twenty radiologists who didn't report seeing the gorilla, twelve had looked directly at it.

Radiologists failing to spot gorillas is unlikely to have much impact on patient care. Yet missing other unexpected entities on CT scans can be profoundly impactful. A recent study asked fifty radiologists to check seven sets of chest CT scans for the presence of lung cancer. The final set contained a large breast mass. Yet sixty-six percent of the radiologists missed the mass and hence did not pick up on the breast cancer.[16] This phenomenon is called 'inattentional blindness'.

The moral of these stories is that whosoever controls attention controls thought. Furthermore, as the philosopher Thomas Metzinger argues, we can't engage in rational thought if we can't control our attention.[17] Controlling our attention is crucial to being able to think freely. We need to understand how attention works, how others can hack it and how we can control it ourselves.

There are two elements to attentional control: selective and sustained attention. Selective attention is our ability to attend to the relevant whilst filtering out the irrelevant.[18] Thinking becomes much harder if this ability becomes degraded.[19] In contrast, sustained attention refers to how long we can focus on an object before our minds drift off elsewhere.[20] To get back on track, we need to notice we have drifted off and pull ourselves back to the task. As mentioned in the last chapter, this skill is called 'attentional agency'.[21] You may have already utilised this ability multiple times in this chapter. Catching and recalling your mind to the matter at hand is a powerful ability. Sustained attention allows us to create a train of thought, thereby thinking better. Knowing how our selective and sustained attention work allows people to manipulate our decision making for both good and ill. In chapter 6, we will consider how this can threaten our freedom of thought and what we can do about it.

But to think freely, we need more than just the ability to control our attention. Metzinger argues that mental autonomy, which he defines as 'the ability to control the conscious contents of one's mind in a goal-directed way', also requires cognitive agency.[22] This term refers to the ability to control deliberate thought related to specific goals, which brings us to our abilities to reason and reflect.

Reason

Today, we tend to view freedom as the absence of constraints. We believe we are free if no one stops us from doing what we want. In the philosopher Isaiah Berlin's terminology, we focus on 'freedom from'. But thought freed from all constraints is like a toddler freed from its parent's hand near a busy road; it can lead only to disaster. Free thought is thought constrained by reason. This idea has echoed throughout the philosophic ages.

For Ancient Greeks such as Plato, we are free if we choose to 'follow the demands of reason rather than being coerced by external forces or by unruly desires'.[23] Plato believed the soul had three parts. In our head was *logos*, our ability to reason, which seeks to learn. In our chest was *thymos*, our spiritedness, which fights for honour and status. And in our stomach was *eros*, our appetites, which seek physical gratifications. The wise, said Plato, rule their souls through *logos*.

Plato was adamant that freedom wasn't the same as being able to choose to do whatever you wanted. Following your appetites didn't free you; it enslaved you to your passions. For Plato, someone who 'isn't capable of ruling the beasts in himself, but only of serving them' would be better off being a slave to someone who could rule themselves by reason.[24] A detailed knowledge of history is not needed to see the problems that can result from such an idea. Nevertheless, for Plato, wise people are free not because they can choose to do or think whatever they like, but because they can recognise truth and guide themselves towards the Good. Freedom is submitting to reason in the service of the Good.

Many thinkers after Plato echoed the idea that free thought is reasoned thought. For Immanuel Kant, free thought was 'the subjection of reason to no laws except those which it gives itself'. If there is 'lawlessness in thinking', said Kant, 'the freedom to think will ultimately be forfeited'. Arrogance, he claims, will be to blame for this. The key example Kant gives of lawless thinking is what was once known as 'enthusiasm'. This was the phenomenon, common in the seventeenth and eighteenth centuries, of people claiming to receive knowledge directly from God. Enthusiasm wasn't enlightenment in Kant's eyes; thinking for oneself was enlightenment.[25] Similarly, for the Young Hegelians, nineteenth-century German followers of the ideas of Hegel, unexamined thoughts were the true chains of humanity.[26]

Despite this philosophical concordance, there is a significant problem with the view that reasoned thought is a truth-seeking endeavour that sets us free. A strong case has been made that reason evolved not to help us reach truth, but rather to let us argue and win. To personify this problem, we can again travel back to Ancient Greece. There we find Socrates as the embodiment of the idea that reason is a means of attaining truth. For Socrates, reason's job is to serve knowledge by showing others the problems or contradictions in what they believe to be true. In contrast, Protagoras believed all opinions were true. For him, reason's job was to make one opinion prevail by persuading others using rhetoric.[27]

Plato labelled Protagoras a sophist and warned against this sect's 'universal art of enchanting the mind by arguments'.[28] Once Athens had expelled its tyrannical rulers, the creation of democracy forced citizens to learn the art of argumentation to win political discussions in the Assembly.[29] As the Italian sociologist Robert Michels would later note, 'the essential characteristic of democracy is found in the readiness with which it succumbs to the magic of words, written as well as spoken'.[30] Reason can be used to both reach truth and win arguments, but truth is not needed to win arguments.

Mercier and Sperber have convincingly argued that reason functions less to reach truth and more to help us win arguments.[31] They point out that most human traits have a primary function whose benefit to our survival explains their existence. For example, feet primarily evolved because they allowed us to walk (yet they also allow us to run, and to pick up remote controls without leaving our chairs). Mercier and Sperber propose that the primary function of reason, that for which it evolved, is to win arguments by convincing and persuading others. The ability of reason to find truth was an unexpected bonus.

Mercier and Sperber's argument starts from the fact that we all face the problem of getting truthful information from others. We must evaluate others, and what they say, to decide whether to trust them. We judge others' trustworthiness based on how competent we think they are ('trust me, I'm a doctor') and whether we think they have our interests at heart (are they a family member, have they sworn an oath to help?). But how can you convince someone to listen if you want to give them information but can't signal your competence or altruism? Well, you can give them reasons why they should believe you. The other person can then evaluate the soundness of your argument, opening the door to the receipt of information from someone they would otherwise not trust. Reasoning boosts our ability to communicate information to others.[32]

This 'argumentative theory of reasoning' leads to a range of testable predictions, which all stand up to scrutiny. As the theory predicts, we are better at producing and evaluating arguments than assessing abstract logical claims.[33] Similarly, the theory suggests that reasoning should work best in an argumentative context. If someone asks us to justify a position, such as whether everyone should serve in the army, we generally produce terrible arguments. This is because we create these arguments outside the amphitheatre of argumentation. We haven't had anyone trying to knock down our justifications, so we lazily accept whatever arguments pop into our mind. If we know we will be forced to present our ideas to others, we come up with better arguments. Another reason we come up with poor arguments in such situations is that we are trying to convince other people of our position rather than to get to the truth. If asked to evaluate someone else's ideas (rather than defend our own), we will be more open to getting to the truth. Truth is more likely to result from assessing others' claims than from creating our own.[34] We find ourselves back at the conclusion that we think better, and hence more freely, with others than we do on our own.

The argumentation theory also does a great job of explaining why we seem to possess baffling biases and errors in our thinking. If we believe the purpose of reason is to find truth, much of our thinking seems to display grotesque biases. But, if we assume that reason is for winning arguments, then what once looked like biases no longer appear this way. For example, when we want to convince someone, we should look for an argument favouring our view. This leads us to expect that people will look for evidence that supports their views rather than seek evidence that undermines them. That is, we should expect people to display a 'confirmation bias'. From the argumentation theory's perspective, this bias is a feature, not a bug. It helps us create arguments supporting our position, making us more likely to persuade others. This is not to say that we always show a confirmation bias. We are very good at disconfirming ideas *if* we want to disagree with them. It is just that our confirmation bias helps us produce arguments in our favour, making us more persuasive.[35] In short, what looks like a bug if we think reason is for reaching truth, becomes a sensible feature when we see reason's purpose as being to win arguments.

Knowing this can help us design environments that help us think better. Reason's best shot at reaching truth will be when truth-seeking people who disagree can debate together.[36] In this situation, individual truth-seeking weaknesses, such as confirmation bias, now help the group reach the truth. This is because a division of cognitive labour occurs. Each person, driven by confirmation bias, pumps out arguments in favour of their position. As a result, the group gets a vast range of views to consider, which it can sift for truth.

If free thought is reasoned thought, and reason works better in groups, group deliberations may be the freest form of thought. Free thought may be better understood as a group phenomenon rather than an individual one. The right to freedom of thought should therefore focus on protecting our thinking aloud together, and not simply on defending a sacred realm inside our heads. Creating deliberative groups may be more effective when trying to secure freedom of thought than teaching individual critical thinking skills. Governments trying to support citizens' freedom of thought should take note.

Indeed, promoting individual critical thinking skills is not all it is cracked up to be. Teaching people critical thinking skills is often ineffective in making them better thinkers. As psychologist Daniel Willingham points out, decades of research into whether critical thinking can be taught 'point to a disappointing answer: not really.'[37] The reason for this, Willingham explains, is that critical thinking isn't a general-purpose skill you can easily transfer between tasks. Once you can drive one car, you can drive any car.

But being able to think critically about one issue does not automatically mean you can think critically about another.

Willingham gives the example of a student who can think critically about the American Revolution, viewing it from both the American and British sides. Yet, when thinking about the Second World War, the student completely fails to consider the perspectives of the Germans. Willingham proposes that this failure occurs because you must know something about a subject to think critically. You can't look at a problem, such as the best form of government, from multiple angles if you don't know all the potential forms of government.[38]

Yet there is a strategy we can utilise to mitigate this problem. Willingham calls this a meta-cognitive strategy for solving problems. This helps us apply general critical thinking to a specific situation. One such meta-cognitive strategy, Willingham explains, is the ability to see a problem's 'deep structure'. Across our lives, we might have to escape out of caves, navigate through mazes, or find our way through forests. These problems, which have different surface structures, share a deep structure: 'how to avoid getting lost in a situation without bearings'. All these problems are, at the root, the same, and we need to recognise this deep similarity to apply our general critical thinking to the situation. Other meta-cognitive strategies include seeing the relevance to a problem of strategies such as 'I must consider the perspectives of both parties in this conflict', or 'don't settle on your initial answer' or 'when I design experiments, I should try to control variables.'[39] If we can see a specific problem as an example of a more general one, we can then apply our critical thinking skills.

Even if you can use these strategies, individual critical thinking skills have their limits. If you don't have problem-specific knowledge, you can't know how the Germans viewed the Second World War or the benefits of authoritarian regimes. Similarly, when designing an experiment, you need to know the critical variables to control for in the specific case you face. For example, as Willingham points out, when comparing the fuel consumption of cars, you might not think that you should account for the car's colour. However, if you know that people with red cars tend to drive faster and brake harder, it becomes clear you need to consider car colour. To be good thinkers, we must combine critical thinking skills with knowledge of specific areas.[40] But how many of us have the time and resources to think deep and wide? Looking ahead, we may worry about the implications of this for decision making in democracies.

Importantly, it turns out that not all forms of critical awareness are equal. When we try to help people better detect high-quality information and avoid 'fake news', some skills are more effective than others. For example, one study examined how good different types of literacy were in helping people detect fake news (such as the Pope endorsing Donald Trump, and Hillary Clinton selling weapons to ISIS). The researchers examined the effects of four types of literacy skills. These were media literacy (knowing about using multiple media sources, holding a critical view of the media), information literacy (knowing which sources of knowledge are more reliable and being able to differentiate opinion from fact), news literacy (understanding how the news media work), and digital literacy (understanding about spyware, phishing, search techniques, etc.). The researchers found that only information literacy helped people detect fake news. It appears that the ability to know what is reliable and verified information is the key to helping people think better online.

We can boost our information literacy by using lateral reading. This involves looking around for other sources that can help verify an article's claim. Lateral reading is a crucial skill for fact-checkers. So, for example, one study looked at the ability of students, historians and professional fact-checkers to evaluate the credibility of online information. Students and historians spent most of their time reading the website that made the claim they were evaluating ('vertical reading'). In contrast, fact-checkers quickly left the site and opened other websites in new browser tabs ('lateral reading'). This led the fact-checkers to reach better conclusions in less time. They learnt more about the site by leaving it.[41]

More generally, people need a plan of attack when reading an online article. Many people lack this and aimlessly 'flutter' around the screen, not knowing what pieces of information are and are not valuable.[42] Studying how fact-checkers work gives us three simple questions for boosting our reasoning:[43]

1. Who is behind the information?
2. What is the evidence?
3. What do other sources say?

Such skills can be quickly learned and make a difference. One study found that two 75-minute lessons on evaluating the credibility of online sources led to (albeit modest) increases in students' online reasoning skills.[44] For example, before the training, one student accepted at face value a tweet of a picture of a child sleeping between graves in Syria as evidence of conditions there. They did not ask who took the picture or consider whether it was from Syria. After the training, they saw a tweet containing a picture said to bear on conditions today in Fukushima. Now the student wondered about the credibility of the person posting the image. They also queried whether the photograph came from Fukushima. We can learn to reason better online. We *must* learn to reason better online.

Reflection

We can read a hundred books, listen to a thousand lectures, watch a million YouTube videos and still not think a single thought. To think, we must separate ourselves from the information we receive and reflect upon it critically. Indeed, the word 'critical' comes from the Greek '*krinein*', meaning to separate or judge.

When we explored the nature of thought in chapter 1, I emphasised its dialogic nature. Thought involves taking alternative perspectives on something, and then putting these perspectives in conversation with each other. Think of Gollum's conversations with himself in *The Lord of the Rings*. We reflect when an idea bounces back and forwards in the mirror of our mind. The ability to reflect on one's thoughts, rather than just accept them, is crucial to freethinking. Freethinking puts an idea up on a ramp and walks around it to examine it from all angles. It kicks the tyres and sounds the horn.

Throughout history, thinkers have emphasised the centrality of reflection to free thought. 'Why should we not', asked Socrates, 'calmly and patiently review our own thoughts, and thoroughly examine and see what these appearances in us really are?'[45] Likewise, propagandists have always emphasised that preventing reflection is central to effective propaganda. Propagandists don't necessarily want to block our access to information, but they do want us not to reflect on the message we are given.[46] Propaganda is treacherous because it gives us ready-made thoughts it encourages us to take no further.[47]

A good propagandist does not allow time for thought. One German who lived through the Nazi era described how the regime bombarded people with activities and things to do. This new 'rigmarole', he explained, 'consumed all one's energies, coming on top of the work one really wanted to do. You can see how easy it was, then, not to think about fundamental things. One had no time.'[48] Similarly, as one of Andrew Carnegie's steelworkers put it during the Homestead Steel Strike of 1892, 'what good is a book to a man who works 12 hours a day, six days a week?'[49]

Even if one has time to read, one may still not reflect. Taken in isolation, a right to read is merely a right to be propagandised.

Lenin knew this. He saw that newspapers were a vital tool of propaganda. Helping everyone to read, Lenin realised, would not be enough to combat capitalism. He pointed out that capitalists 'call freedom of the press that situation in which censorship is abolished and all parties freely publish any paper they please'. But, said Lenin, 'this is not freedom of the press, but freedom for the rich, for the bourgeoisie to mislead the oppressed and exploited masses.'[50] If the masses could not reflect on what they were reading, seeing it through the lens of class consciousness, they would be misled. A reader who cannot reflect on what they read will end up in a place well described by the German polymath Oswald Spengler. This is a place where the reader 'neither knows, nor is allowed to know, the purposes for which he is used, nor even the role that he is to play'. As Spengler concluded, a 'more appalling caricature of freedom of thought cannot be imagined.'[51]

This caricature of free thought arises partly because the topics that the news media choose to cover influence what the public thinks are important political issues.[52] When researchers reviewed the effects of the news media on public perceptions between 1972 and 2015, they found that the media had a strong agenda-setting effect.[53] The strength of this effect did not vary over the years, suggesting that the news media retains a strong ability to set the public agenda even in our internet age. In this way, the media discourages us from thinking about certain topics.

To increase reflective thought, we need to know what factors promote it. For John Dewey, there were two steps to reflective thought. The first was uncertainty. 'The origin of thinking', said Dewey, 'is some perplexity, confusion or doubt.' As he put it, we think when we face a 'forked road situation'. When two roads diverge in a yellow wood, we must think about which one to take. Problems generate thought. As Dewey wrote, 'As long as our activity glides smoothly from one thing to another . . . there is no call for reflection.'[54]

Propagandists and other enemies of free thought need to remove all obstacles from the road they wish us to take, greasing the path to profit. A contemporary example of this would be the endless-scroll function on many social media sites. The endless stream of new videos or tweets leads to the lack of a trigger that would cause us to pause and reflect on what we are doing. Similarly, advertising encourages us to purchase, rather than reflect on our decision to make a purchase, by giving us an obstacle-free road to run down. It first sells us a problem, then offers a simple solution: buying their product. As David Foster Wallace once wrote, 'It did what all ads are supposed to do: create an anxiety relievable by purchase'.[55] Adverts neither want you to pause to consider whether you need the proffered solution nor to wonder if the problem is even real.

We must be able to tolerate uncertainty to reflect in the face of anxiety or unease. Otherwise, we will flee these emotions by accepting a message, making an impulsive decision, or jumping to conclusions. Sitting with our emotions gives us the time and space to test, investigate and rectify the problem in the best way we see fit. A freethinker bears emotions where others bolt. Flawed thinking, said Dewey, involves grabbing any plausible answer to a problem to stop our mental uneasiness. In contrast, good thinking results from enduring uncertainty, persevering in the painful state of suspending judgement, giving us the best chance to come to the correct answer. We cannot hope to control our thoughts if we cannot first control our feelings.

Once we can control our emotions, opening a space for thought, reflection gets to work interrogating rather than accepting what pops into our head. Philosophy explains this using the concept of second-order thought. This idea emphasises the difference between the passive process of having thoughts and the active process of reflecting on them. To understand this, imagine you're pottering around your house when a sack of cocaine falls from a plane overhead. The sack crashes through your roof and lands next to you. You call the police, who pop round and promptly arrest you for possession. This is ridiculous. You are not responsible for things that fall into your house. Now imagine a parallel situation in which someone is minding their own business when murderous, violently sexual, or racist thoughts pop into their head. These thoughts are unbidden, unwanted and in complete opposition to their values. Should the person be held responsible for them? Should they be punished?

If they make these troubling thoughts public, perhaps tweeting in horror about them, they will quickly discover that they are held responsible and that they will be punished. The Twittersphere will descend, claiming the thinker is unfit to work and live in decent human society. But what if having such troubling thoughts turns out to be a common human experience? Because of the inability to voice such thoughts, the human condition would become literally unspeakable.

Across the world, people have troubling unwanted thoughts. You can ask people in Africa, Asia, Australia, Europe, North America or South America about such experiences. If you do, you will find that around ninety percent have recently experienced unwanted intrusive thoughts.[56] Many of these thoughts have content that goes against the person's values. One study found people were having 500 unintentional and unwanted thoughts a day, of which eighteen percent were unacceptable and uncomfortable. A further thirteen percent were 'out of character' or downright shocking.[57]

Research into the specific content of these thoughts does not paint a pretty picture. Early research found that unwanted intrusive thoughts involved 'unnatural' sexual acts, hurting or killing others (e.g. throwing a child out of a bus) and hurting or killing oneself (e.g. jumping in front of a train).[58] Most people experienced this weekly and had done for many years. However, they could easily dismiss these unwelcome visitors. Later research found that sixty percent of people admitted having intrusive, unwanted thoughts about running a car off the road and forty-six percent had them about hurting family members. Furthermore, thirty-three percent of people had such thoughts about wrecking something, twenty-six percent had them about fatally pushing a stranger and nine percent had them about stabbing a family member.[59] A study of college students found that six percent of people had such thoughts about sex with animals or non-human objects, nineteen percent of men and seven percent of women had them about a sexual act with a child or minor, and thirty-eight percent of men and twenty-two percent of women had them about forcing another adult to have sex with them.[60]

To be clear, we are talking about such thoughts occurring in people who find them abhorrent. If the person is not appalled by these thoughts, does not try to suppress or avoid them, does not try to avoid situations that trigger them, or is aroused by or acts on them, there is cause for concern.[61] Someone who felt this way about intrusive sexual thoughts they had about children is a potential sexual offender.[62] This brings us to a crucial distinction. You are not responsible for the thoughts that pop into your head, but you are responsible for how you react. As the theologian Martin Luther once counselled a person bothered by evil thoughts, 'you cannot prevent the birds from flying in the air over your head, but you can prevent them from building a nest in your hair'.[63]

This reflective process is at the heart of 'hierarchical' accounts of autonomy. These begin with the thoughts, desires and impulses that well up within us, which are called 'first-order' mental actions. We need to dissociate ourselves from these thoughts, which pop up from the basement of our brain without invitation. Such thoughts, the philosopher Harry Frankfurt suggests, are not truly ours.[64] Likewise, the media scholar David Edwards introduces the idea of 'Trojan thoughts', which we 'perceive as our own but which have in fact been pre-filtered by the society around us'.[65] We need to take a critical stance towards these first-order thoughts and interrogate their origins. As the Italian philosopher Julius Evola advised, we must be able to recognise the 'subtle influences that try to suggest ideas and reactions to us'.[66] To do so, we must take a critical stance toward these uninvited and often unwelcome guests of our mind.

To act on first-order thoughts without reflection is to fail to display mental autonomy. As the bumper sticker says, don't believe everything you think. Frankfurt called people who simply acted based on what came into their head 'wanton'. In his view, they were 'not persons' (clearly a very dangerous stance to take).[67] A slightly less offensive approach would be to view someone who felt free because they followed whatever first-order thought popped into their head as being, to adapt Orwell, only a rebel from the neck downwards.[68]

The alternative to simply following whatever thought pops into our head is to perform more mental work – to think about our thoughts. In doing so, to use Karl Marx's phrase, we 'revolt against the rule of thoughts'.[69] Thinking about our thoughts is called a 'second-order' mental action. Here we try to determine if our first-order thoughts, desires and impulses are authentic. Are they consistent with our own freely chosen values and goals? Here we get closer to traditional philosophical views of freedom, namely the ability to achieve our chosen goals. Frankfurt argues that by performing second-order mental actions, we make thoughts and desires more truly our own. We recognise that our first-order thoughts may represent 'a force other than' our own.[70] Reflective second-order thought lets us structure our initial thoughts, undertake logical trains of thought, terminate violent fantasies and guide our behaviour. Such reflective thought is central to our thought being free.

Notably, the conversational, to-and-fro nature of thought, discussed in the last chapter, seems crucial for reflective second-order thought. If we switch from talking to ourselves in the first person to using non-first-person pronouns (such as our name), then we think better.[71] The use of the third person by toddlers, Elmo and Donald Trump are exceptions that prove the rule.

Courage

'Nothing is easier than opting for autonomy', wrote the German author Ernst Junger, before adding that 'nothing is harder than bringing it about.'[72] Achieving mental autonomy, the ability to think freely, is just as challenging. This is not just because of the intellectual mindware it requires. Free thought also requires courage. It is for this reason that US Supreme Court Justice Brandeis spoke not simply of the power of reasoning, but of 'free and fearless reasoning'.[73] Courage, said Brandeis, was the secret of liberty.[74]

The freethinker must be willing and able to bear the reaction of others to hearing hard truths and have the courage to follow reasoned thoughts wherever they lead. As the former French Prime Minister, André Léon Blum, put it, 'the free man is he who does not fear to go to the end of his thought.' John Stuart Mill said much the same. For Mill, no one could be a great thinker if they were not prepared to follow their train of thought to whatever logical conclusion it may reach.[75]

A helpful concept here appears at the heart of the late writings of the French philosopher Michel Foucault, which focused on courage and truth.[76] The concept, which Foucault explores in depth, briefly mentioned earlier, is *parrhesia*. To understand this, we must once again voyage back to Ancient Greece. The Greeks were instructed to 'know thyself'. Yet they knew self-knowledge could not be achieved alone. They needed help in the form of someone who would tell them challenging truths. They needed a parrhesiast.

'Tell the truth and run', an old Yugoslavian proverb wisely counsels.[77] Hearing the truth can irritate, offend and anger us. We may lash out at the speaker. In Ancient Greece, a parrhesiast was someone prepared to suffer a backlash to speak a truth they genuinely believed. The philosopher Leo Strauss would have nodded at this idea. He observed that 'truth demands that we prefer her to all human friendship'.[78] But the failure of friendship to co-exist with truth is precisely the problem. Propaganda suppresses freethinking by making our fellow human 'no longer an interlocutor but an enemy'.[79] When seeking truth, we must suspend the friend–enemy distinction, which means suspending the political.[80] This is, of course, easier said than done.

The concept of parrhesia emphasises that free thought is not simply the result of individual freethinkers. Free thought only flourishes in a supportive society. To benefit from acts of parrhesia by freethinkers, fellow Ancient Greeks had to be open to hearing potentially painful truths. They had to be willing to bear what Foucault called 'the injuries of truth'.[81] Greatness of soul lay in both the courageous speaking and acceptance of truth. The Greeks referred to this as the 'parrhesiastic game'. To work, it required a parrhesiastic contract. The speaker agreed to risk telling the truth, and the listener promised not to punish them. In this way, Classical Athens created a safe space for truth.

Foucault was also clear that the parrhesiast saw truth-telling as a moral duty.[82] They were not forced to speak but felt a duty to do so. This duty was to help other people improve themselves. The other person could be a friend, or a ruler who needed help to make a decision. Of course, in this latter case, the parrhesiast faced a political quandary. They needed to be near power to have their view heard, yet distant from power to preserve their independence – a tricky balance to strike.[83]

The right to freedom of thought sits in close conjunction with the duty to tell inconvenient truths. Malcolm X, for example, was described by Cornel West as 'the great figure of revolutionary parrhesia in the Black Prophetic Tradition'.[84] After Malcolm's home was firebombed and his family threatened, he still felt a duty to travel to another city to speak truths others did not wish to hear.[85] There is also a strong tradition of feminist truth-telling inserting 'plurality into the singularizing regime of organized lying'.[86]

Whether contemporary democratic societies encourage parrhesia is debatable. Foucault was sceptical. He believed that people living in democracies only wanted to hear people with whom they already agreed. This meant that 'real parrhesia, parrhesia in its positive, critical sense, does not exist where democracy exists'.[87] This may explain why sixty-two percent of Americans say that the current political climate prevents them from speaking their mind because others may find it offensive.[88] The majority of Americans are self-censoring. As the philosopher and neuroscientist Sam Harris has put it, 'the first order of business, and the next one, [is] to figure out how we can have successful conversations'.[89] This will involve a renewed commitment to truth over tribe.[90]

Courage is needed not only to tell uncomfortable truths but also to bear the pain of inevitably erring. As the writer Douglas Murray observes, 'a great amount of thought often necessitates trying things out (including making inevitable errors)'.[91] You could object that the privacy of our heads would be the best place to make such errors. But this is not the case. First, we are unlikely to recognise our errors. And second, as we saw John Stuart Mill arguing earlier, society can benefit from these errors. A freethinking society would recognise this. It would tolerate and forgive honest mistakes rather than immolate people. There are few things more dangerous to free thought than self-righteousness. What we need, to adapt Orwell, is the right to think 'what one believes to be true, without having to fear bullying or blackmail from any side'. What this looks like, we will come to later.

Yet the courage to bear other people's reactions to challenging truths is not enough. We also need the courage to face our *own* resistance to truth. We must, as Kant put it, 'dare to know'. Just as we possess an immune system to attack foreign entities that invade our body, we also have a mental immune system. Ideas that threaten to harm our cherished beliefs cause our brains to try to fight them off. For example, one neuroimaging study explored what happened to the brains of liberals when researchers presented them with arguments that contradicted their firmly held views.[92] The study found that neural structures involved in creating negative emotions, such as the amygdala, were strongly activated. As the authors observed, this probably represented people experiencing negative emotions, causing them to try to downplay or discount the evidence. We naturally fight off, rather than embrace, challenges to our beliefs.

Finally, we must have the courage to bear the simple discomfort of thinking. For many of us, thinking is just so much damn effort. 'Thinking troubles us; thinking tires us', notes the writer Alan Jacobs.[93] For this reason, Jacobs argues, 'Relatively few people want to think.' If people make us think, we may resent them. As Nietzsche observed, 'when we have to change an opinion about anyone, we charge heavily to his account the inconvenience he thereby causes us.'[94]

The lengths we will go to avoid sitting with our thoughts are disturbing. Take a recent study led by Harvard psychologist Timothy Wilson.[95] The researchers had come across the results of a survey showing that eighty-three percent of American adults spent no time at all 'relaxing or thinking'. Could it be, Wilson and colleagues asked, that people don't enjoy thinking? To test this, they left college students in a room alone for fifteen minutes, without their belongings, and asked them 'to spend the time entertaining themselves with their thoughts'. Each student was allowed one potential activity: they could press a button and get an electric shock. Two-thirds of men and a quarter of women gave themselves at least one electric shock in the fifteen minutes.

Now there are clearly multiple ways to interpret the meaning of this study. We could see it as assessing curiosity or risk aversion. However, a plausible interpretation is that those who shocked themselves preferred this to sitting with their thoughts. As the comedian Mark Normand puts it, 'Thoughts are not good. My god, this whole time I thought I loved music. Turns out I just hate my brain.'[96] If true, how far have we come from the centrality of thought to human life when we would rather torture ourselves than think? Would a large percentage of us like governments and corporations to think for us, serving up predictions and nudges for us to follow? Would we prefer an auto-complete life?

We can find thinking an aversive experience because when we face a problem, our brain's preferred approach is to avoid getting involved. As noted earlier, we are cognitive misers who try to think as little as possible. There are good reasons for this. Evolution wants us to expend as little effort as possible to make decisions. As has been observed, 'all animals are under stringent selection pressure to be as stupid as they can get away with'.[97] Whilst we could set our brain's fearsome processing abilities to work on any problem, this comes at the cost of using a lot of mental juice (attention and time). This interferes with our ability to do other things. Our default state is to use a less effortful approach. Although this was probably good enough in the world we evolved in, it may not help us make the best choices today.[98]

Our brains push us away from thought by making concentrating and 'mental effort' feel unpleasant.[99] This seems odd. Working hard leads to later success and reward. Resisting junk food makes us healthier in the long term. Shouldn't trying to do these things feel good? The short answer is that when we decide whether to think, our brain signals that using effortful thought is expensive. It makes us feel bad, entering a cost into our cost–benefit analysis of this course of action.[100] To have free thought means learning to cope with and enjoy the effort of thought. To do this, we must realise that our brain is looking for cheap rewards. And we need to get such things the hell away from us. We are unlikely to want to think if our brain knows that something low-effort and high-reward is nearby. A common culprit is a smartphone with access to games and social media designed to reward us.[101] Removing barriers to thought may be a much better way to promote thinking than trying to sell people on the joys of thought.

So far, we have focused on what factors make thought free. Yet there is an alternative approach. The philosopher Ludwig Wittgenstein famously argued that when we want to understand a word, we should ask not for its meaning, but for its use.[102] Following this approach, we should try to determine what free thought and the right to freedom of thought are by looking at how people use these terms.

At its most basic level, we use the idea of a right to freedom of thought to legitimise our actions and to delegitimise those of others. It is a way to exert power. As we move forward, we need to be aware of what people seek to achieve by wielding this right. Are they using it to try to stumble towards truth, or are they using it for partisan political ends? People will try to wrap their political agenda in the garb of free thought. Those who are successful in doing this will laud the right to free thought, whilst those who are unsuccessful will damn it.[103] Ultimately, we are all safer when we are forced to justify ourselves based on clearly enunciated principles, ideals and arguments, rather than simply claiming what we do is justified by the right to freedom of thought. Rights-talk has a troubling way of ending conversations rather than starting them. We must avoid this.

In this chapter, we have seen how we can boost our ability to think freely by developing our abilities to attend, reason and reflect, and by having courage. These are key tools for free thought. In chapter 6 we will look at how these abilities are threatened and how the right to freedom of thought could protect them. Yet focusing on ourselves is insufficient. The world is full of governments, corporations and individuals who don't want us to think freely. If we do not understand these threats, we will fall victim to them in the most insidious way, unknowingly.

3

THE THREATS
TO OUR MINDS

Writing in 1913, the historian J. B. Bury believed freedom of thought had been secured.[1] In his view, a hopeful person could view this victory as permanent, with intellectual freedom 'assured to mankind as a possession for ever.' But Bury was trained in history, not hope. Was there the possibility, he queried, of 'a great set-back?' He had a previous example in mind. For him, Christianity had 'laid chains upon the human mind'. Bury worried that a new force emerging from the unknown could cause something similar to happen again. A century later, we see his fears realised. The cause of this set-back is corporate power.

In the real world, no one really wants you to have freedom of thought. Or, if they do, it is on the condition that you reach the conclusions they want you to. Everyone wants you to think something. Governments want you to think they are legitimate. If you do, you will let them set rules you will obey. You will grant them power over your life, death, body, mind, child and god(s). Corporations and financial elites want you to think that capitalism is legitimate. If you do, you will let them set rules you will obey. You will treat property rights as sacred, sell your labour to corporations for less than your work is worth, agree that owners not workers should control corporations, accept psychological warfare in the name of marketing, believe elected officials are chosen by you to enact policies you want rather than being creations and servants of corporate money, and be bound by the rules of the free market whilst corporations receive state subsidies and bail-outs. Our fellow citizens want us to think they deserve esteem, attention and tribute from our taxes. Every waking second sees a forest of hands reaching out to shape the clay of our spinning minds. Yet at the root of this forest are corporations. Corporations target our minds directly through advertising and corporate propaganda. They also indirectly target us through their influence on government and by responding to citizens' calls for employees to be fired for wrongthink.

These pressures on our minds mean many of our thoughts are not strictly our own. For example, write down five things you think are important in life, along with their opposite (e.g. democracy and monarchy, religion and atheism). Now, consider *why* you think these are important. Did you decide that your preferred pole was valuable because you independently sat down and figured it out? If religion is important to you, did you survey all the world's religions and carefully think about which was correct? Or did you adopt your parents' religion, which, by happy coincidence, happens to be the 'correct' one? And democracy? Can you explain why this is the best system of government beyond a series of stale platitudes? And even if you changed your mind, would this necessarily be due to thinking? Or does 'thinking for yourself' emerge as being 'switching from one set of culturally conditioned ideas to another equally preformulated set'?[2] If free thought is reasoned thought, then any thoughts you have that you have not reached through reason are not free. Someone has shown you what to think, told you what to think, or rewarded you for thinking what they want you to think.

Governmental and corporate/financial power is established over our thoughts by their control of key information channels, including the education system and the media. Information sent through these channels is designed according to the laws of human psychology to have maximal persuasive power. Our fellow citizens influence our thoughts by modelling desired actions. They also reward us for agreeing, punish us for disagreeing, and increasingly use proxies (such as the state, employers and mobs) to crush dissent. An invisible matrix of pervasive propaganda and conformity holds our minds rigidly in place. The real question is not how to protect freedom of thought but how we ever believed we had it in the first place.

Thought control in autocracies

A tyrant may try to rule by force alone. The population may not like or want their ruler, yet they cannot overthrow them. Such tyrannies are inherently unstable because the ruled always outnumber their rulers. If the ruled unite and organise, they can sweep away their rulers. As the Scottish philosopher David Hume observed, rulers have 'nothing to support them but opinion.'[3] Napoleon, who knew a thing or two about power and ruling, agreed with Hume. 'Power is based on opinion', said Napoleon, before adding 'what is a government not supported by opinion? Nothing.'[4]

The fragility of force in the face of numbers means that regimes which can freely utilise violence to control citizens' bodies must still attend to citizens' minds. Hannah Arendt saw this when the Soviet Union crushed the Hungarian Revolution in 1956. The Soviets viewed the Hungarians' 'freedom of action' as being most dangerous to total Soviet domination, she noted, 'closely followed by freedom of thought.'[5]

As a result, all human societies are to greater or lesser degrees ruled by the power of ideas. 'The ideas and images in men's minds', John Locke wrote, 'are the invisible powers that constantly govern them.'[6] The seventeenth-century Dutch philosopher Baruch Spinoza agreed. As he memorably put it, 'firmest dominion belongs to the sovereign who has most influence over the minds of his subjects.'[7]

Of course, control of the public mind is not easy. Rulers can silence speech, but thoughts are much harder to control. As Voltaire wrote, 'it would be easier to subdue the whole world by arms than to subdue all the minds in a single city.'[8] A ruler who could achieve this would gain unlimited power. 'Were it as easy to control people's minds as to restrain their tongues', wrote Spinoza, 'every sovereign would rule securely'.[9]

Tyrants work to make the population adopt ideas that support their legitimacy. An absolute monarch may have the force of arms at their disposal. Yet without the accompanying power of a legitimising myth, all the swords and soldiers in the world may not keep them on the throne. The Italian elite theorist Gaetano Mosca called such myths 'political formula'.[10] If the people stop believing that a monarch has a divinely delivered right to rule, crowned heads can roll. As a result, every good monarch is a great propagandist. As propaganda scholars have noted, 'every time a medieval king or queen donned an ermine robe and picked up a sceptre, they were propagandizing themselves.'[11]

Suppose an autocracy can remove or restrict its citizens' freedom of thought. In that case, the population becomes more vulnerable to ideas the government injects into society. José Ortega y Gasset argued that the 'majority of men have no opinions'. This meant ideas had to be 'pumped into them from outside, like lubricants into machinery'.[12] In Gasset's view, a ruler must exercise authority to let the people begin to have opinions. Having opinions gives the populace a feeling of power, which is why they are keen to adopt ideas. And, if people feel their team is in power, they will adopt their ideas as a badge of identity. A culture of free thought would work against the injection of such ideas. As the Soviet dissident Andrei Sakharov wrote in 1968 while Soviet tanks were rolling into Czechoslovakia, 'freedom of thought is the only guarantee against an infection of people by mass myths, which, in the hands of treacherous hypocrites and demagogues, can be transformed into bloody dictatorship'.[13]

In addition to encouraging people to adopt chosen ideas, tyrants also control thought by stopping other ideas from occurring. In *The Screwtape Letters*, C. S. Lewis has the senior demon, Screwtape, observe that 'It is funny how mortals always picture us as putting things into their minds: in reality our best work is done by keeping things out.' Orwell uses a related idea in *1984*, where Big Brother introduces the idea of Crimestop. This is a faculty through which people can immediately rid themselves of dangerous thoughts. Actual totalitarian rulers also appreciate the need to disarm the population mentally. As Stalin apocryphally said, 'ideas are more dangerous than guns, we took away their guns, why should we let them have their ideas?' Liberal societies also recognise this hierarchy. The US Constitution protects citizens' speech (First Amendment) before protecting their guns (Second Amendment).

Consistent with all this, even contemporary autocracies do not rule by force alone. They try to get their population to buy into their ideas. The North Korean leader Kim Jong-un rules a country imbued with the concept of *Juche*. This is the idea that the people should seek a spirit of self-reliance. Such self-reliance can only be achieved by following the guidance of a Supreme Leader (*Suryong*) whose genius is hereditary.[14]

In China, the CCP's tanks are a formidable way to control the bodies of their population. But the minds of the Chinese people must also be bound. The CCP must convince the Chinese people of their ruling ideology. The CCP needs the population to buy into the idea that the governmental leaders 'reflect people's interests, serve the people, and submit to supervision by the people'.[15] They also need the people to believe the CCP's claim that China is democratic, as many of its thinkers argue is the case.[16] In addition to being exposed to the CCP's ideas, the Chinese people must be shielded from competing ideas. Such ideas could be profoundly destabilising.

The CCP has put an enormous amount of work into influencing the thoughts of the Chinese population. They have promoted the idea that thought should be centralised rather than free.[17] Traditional Chinese thought emphasises 'right thinking' rather than a right to free thought.[18] To this end, the Propaganda Department of the CCP undertakes 'thought work' to ensure compliance with party ideology.[19] Since 2018, China's Propaganda Department has been given the power to regulate news, films, books and the internet.[20] A 'Great Firewall' restricts the Chinese population's access to information available to the rest of the world, blocking sites. To achieve 'cyberspace sovereignty', domestic internet service providers and foreign businesses must store their data on Chinese servers.[21] Consequently, the CCP has access to all user data,[22] which cannot but chill users' thought. Such policies are conceived and implemented in the name of controlling order and securing the nation's well-being. This well-being is said to stem from 'unity of thought'.[23] Such unity involves the populace subscribing to the CCP's ideology and rejecting 'spiritual pollution' from liberal Western ideas.[24]

Freedom of thought is essential to combat tyranny. The lies that underlie a tyrant's reign must be unmasked. Once the 'coercive force' of truth is released, the regime is in trouble.[25] A regime may be able to continue in power after its ideology has been undermined. As the American theologian Reinhold Niebuhr observed, 'when power is robbed of the shining armor of political, moral and philosophical theories, by which it defends itself, it will fight on without armor'.[26] But once legitimacy is lost, force can only rule for so long. And all this is as true for liberal democracies as it is for authoritarian states.

Thought control in liberal democracies

In liberal democracies, using force to control the public is largely off the table. Governments can send in tanks and militarised police forces to break up protests, but this is not a good look. Of course, when capital's interests are threatened, a capitalist-controlled state can show its teeth. When coal miners in Colorado went on strike in 1913–14, Colorado's governor sent in the National Guard to break the strike. This resulted in the Ludlow Massacre on 20 April 1914, when the National Guard (whose salaries were paid by the mine owners, the Rockefeller family) killed men, women and children. Today, incarceration rates in the United States do not speak of a society where force has no place in ruling. Liberal democratic governments certainly do not subscribe to the philosophy of Huxley's World State in *Brave New World*, namely that 'Government's an affair of sitting, not hitting. You rule with the brains and the buttocks, never with the fists.' Democracies make widespread use of what have been called Repressive State Apparatuses, which we will come back to later in this chapter.

We should not overemphasise the importance of ideas to the rule of democracies. In *The Neverending Story*, the werewolf Gmork argues with the warrior-hero Atreyu. Gmork claims that 'the power to manipulate beliefs is the only thing that matters'. Of course, it is not. Force needs to be wielded too. Shifts from tyranny to democracy are not always associated with a lesser role for force. As de Jouvenel argues, 'there goes with the movement away from monarchy to democracy an amazing development of the apparatus of coercion. No absolute monarch ever had at his disposal a police force comparable to those of modern democracies.'[27] And this is before we consider the force of economic oppression.

But let us turn back to the control of thoughts. Those of us who live in democracies have little difficulty seeing how autocracies control the minds of their populations. The mistake many of us make is thinking that thought control only happens 'over there'. There is good reason to suspect that thought control may be *more* widespread in democracies than autocracies.[28] If force is mainly off the table, then controlling thought becomes, as Noam Chomsky argues, '*more* important for governments that are free and popular than for despotic and military states'.[29]

To understand this, we need to see democracy for what it is. Until the nineteenth century, democracy was understood to mean rule by the poor.[30] A democratic government was expected to implement redistributive, egalitarian policies. Indeed, in the United States in the early 1780s (before the creation of the US Constitution), state governments began interfering with private property and helping people escape their debts (through schemes such as issuing paper money).[31] This was in addition to taking a range of other short-sighted and unethical actions. This is why people like the US Founding Father James Madison did not like democracy.

As Madison put it, democracies 'have ever been incompatible with personal security or the rights of property'.[32] Madison didn't want popular movements of working people anywhere near the levers of power. Such movements would support narrow 'localist' solutions and not take into account the wider good of society. A wider, long-term view could only be taken by natural aristocrats, as they were called. Madison's solution was a national government with representatives elected by such large groups of people that specific local interests could not grab the reins of power. This would lead to the 'right kind of people' being elected.[33] As an extra layer of security, the decision making of these representatives would be limited by a constitution. This constitution would aim 'first to obtain for rulers men who possess most wisdom to discern, and most virtue to pursue, the common good of the society'.[34] For Madison, the key factor that made the US Constitution distinct from democracies in the ancient world was its 'total exclusion of the people in their collective capacity' from any share in government, with reliance instead placed on representatives.[35] And ultimately, as the historian Edmund Morgan argues, 'representation had always been a fiction designed to secure popular consent to a governing aristocracy'.[36]

As more and more people were able to struggle successfully for the right to vote, wealthy elites faced the challenge of how to legitimise, and hence retain, their money, power and position. The obvious answer was to control the public mind. Elites would do this by influencing the education system and media to control the flow of information. In particular, claims Chomsky, the twenty percent of the population who are educated and pay attention to politics, such as managers, teachers and writers, needed to be deeply indoctrinated.[37] The remaining eighty percent, whose primary function, says Chomsky, is to 'follow orders and not to think and not to pay attention to anything', need less indoctrination. Democracy, as we understand it today, is not rule by the poor. It is a system of governance where all adults can vote, yet which is ruled by an oligarchy through their control of the public mind. Naturally, at least some of the oligarchy may believe this is all done for the greater good.[38] But as Socrates said, it is helpful if the rulers believe the noble lie too.

Everything we know points to democracies being managed from above, not below. Take theories of how organisations and groups work, which emphasise that small groups of well-organised people get things done.[39] When we apply these principles to the idea of self-government in a mass society, we conclude, as the American political scientist James Burnham did, that pure democracy 'is impossible'.[40] As Robert Michels put it, most of us become, 'by tragic necessity', merely the 'pedestal of an oligarchy'.[41]

Similarly, consider the practicalities of power. To think we live in a real democracy is to believe that the minority of people with much power will voluntarily cede this to the majority who have little. To think this would be beyond naive. Hans-Hermann Hoppe, a former student of Murray Rothbard, puts forward a more brutally realistic account. Hoppe's words nicely predicted the outcome of the Global Financial Crisis of 2007–9:

> It is not very likely that dullards, even if they make up a majority, will systematically outsmart and enrich themselves at the expense of a minority of bright and energetic individuals. Rather, most redistribution will take place within the group of the 'non-poor,' and frequently it will actually be the better-off who succeed in having themselves subsidized by the worse-off.[42]

The desires of elites strongly inform government policy. As sociologist Martin Gilens concludes, the more you earn, the more your views are linked to government policy. Conversely, there is a 'complete lack of government responsiveness to the preferences of the poor'. When the views of low- and middle-income Americans differ from the views of the rich, government policy ends up being 'virtually unrelated' to the poor's desires.[43] Elsewhere, Gilens notes that 'economic elites and organized groups representing business interests have substantial independent impacts on U.S. government policy, while average citizens and mass-based interest groups have little or no independent influence.'[44]

Likewise, scholars have argued that US foreign policy is most influenced by business leaders, then experts, followed some distance behind by organised labour. The general public lag much further behind.[45] Naturally, there has been a healthy critical response to these claims,[46] but Gilens appears to be on to something.[47] It would be amazing if it were to be any other way.

The idea of a pure democracy, as Burnham notes, is a myth that a ruling minority uses to legitimise itself. The idea of democratic government representing the will of the people is just another legitimising myth or political formula. In reality, as Lippmann saw, we live in a society where the people neither express their opinion nor govern. Instead, the people 'support or oppose the individuals who actually govern'.[48] As Strauss put it, freedom of thought today effectively amounts to 'the ability to choose between two or more different views presented by the small minority of people who are public speakers or writers'.[49] We select between the views of competing elites. Voters have the same vocabulary as a twelve-month-old child, two words: yes and no.[50] The idea that democracy is rule by the people is just as delusional as was the belief in the divine right of kings.[51] And, tellingly, it is just as effective. The democracy delusion survives because, like Fox Mulder in *The X-Files*, we want to believe.

The reason the political formula of democracy creates a stable society can be found in one of Machiavelli's greatest insights. Machiavelli observed that 'if we examine the aims which the nobles and the commons respectively set before them, we shall find in the former a great desire to dominate, in the latter merely a desire not to be dominated over'.[52] A propagandised, managed democracy offers a stable solution to this problem. The elites are happy because they can dominate. The masses are happy because they live under the illusion they are free. As the political scientists Achen and Bartels put it, democracy allows people to 'feel like they are thinking.'[53] In a society that runs on feelings, not facts, this is good enough for most people. The illusory feeling that we are thinking, in conjunction with our satisfaction with this state of affairs, is a major threat to free thought.

Today we live in societies of illusion. The French philosopher Gilles Deleuze proposes that we moved from a disciplinary society, where we spent defined periods of time in institutions such as schools, factories, the army or hospitals, into a new society of control.[54] But, for me, this doesn't capture the society we live in. It is a society of control, but the crucial innovation is that we do not feel as if we are controlled. Control has put on the clothes of freedom. Freedom has become the ignorance of what controls us. We live under a series of illusions, which we embrace because they give us a sense of power. Chief among these is the illusion that we live in a democracy where the people are sovereign. In reality, we live in an aristocracy and, to pull the curtain right back, it may be a good thing we do.

In a disciplinary society, according to Foucault's formulation, we were kept in place by a process he characterised as surveil, examine, normalise.[55] Yet such discipline offended against the feeling of freedom. Today, in societies of illusion, 'surveil, examine, normalise' has become 'surveil, suggest, monetise'. We can do what we please because others are pleased that we do. The challenge we face is to identify the hidden sources of our behaviour and to consider reasserting control over them. Free thought is central to this task.

Returning to the matter in hand, government and corporations cannot do the heavy lifting of thought control alone. Liberal democracies also outsource control of thought to the population. As C. Dale Walton observes, 'Chinese citizens worry about the knock on the door, while Americans fear that a random comment – even a sentence taken entirely out of context – will result in the destruction of their careers and personal lives'. China under Mao saw brother denouncing brother, husbands denouncing wives, and children denouncing parents, often with fatal consequences.[56] Yet denouncement culture is prominent in the West today. As journalist Matt Taibbi recently pointed out, 'people everywhere today are being encouraged to snitch out schoolmates, parents, and colleagues for thoughtcrime'.[57] A new Red Guard has arisen. As a result, the West now suffers from what Walton calls 'crowd-sourced totalitarianism'.[58]

The danger we pose to each other's freedom of thought has long been recognised. John Stuart Mill warned that the greatest threat to free thought in democracies was not the state but the 'social tyranny' of one's fellow citizens.[59] Similarly, Orwell argued that the main danger to free thought was not the government but public opinion. As he put it, 'If publishers and editors exert themselves to keep certain topics out of print, it is not because they are frightened of prosecution but because they are frightened of public opinion.'[60] To protect free thought means protecting us from the tyranny of each other.

Let's dive deeper into the mechanics of thought control in liberal democratic societies. There are two ways to view the relationship between such governments and the minds of their people. An optimistic view is that government reflects the thoughts of the majority. In such a society, whosoever best *reads* the public mind will rule. Government is then of the people, by the people, for the people. Unfortunately, the public mind is often contradictory, transitory and confused. Public opinion can become, as the philosopher Jacques Derrida put it, 'the silhouette of a phantom'.[61] This means that the government can neither follow public opinion, nor escape it.[62] Even if the public mind were consistent, governments typically serve those who pay them paper bills, not those who hand them paper ballots.

If the government cannot or will not follow public opinion, public opinion must follow the government.[63] And given the above, the only way a government can follow public opinion is if it makes it. In this situation, as the American political blogger Curtis Yarvin observes, public opinion becomes an effect, not a cause.[64] We now reach the more pessimistic, and more realistic view that those in power *write* the thoughts of the people to match the ruling interests. As Thomas Hobbes put it long ago in *Leviathan*, 'the common people's minds, unless they be tainted with dependence on the potent, or scribbled over with the opinions of their doctors, are like clean paper, fit to receive whatsoever by public authority shall be imprinted in them.'

Elite theorists argue that elites create public opinion. In this view, if public opinion takes control, then 'for an instant, the engine becomes a *real* democracy – then it turns into something else, or just catches fire and explodes. Think Germany in 1933.'[65] We enter a situation where governments need to manage the thoughts of the masses.[66] To do this, they use propaganda.

Democratic governments actively use propaganda to shape the public mind. For example, in 1916, Woodrow Wilson was re-elected as President, emphasising how he had kept the US out of the First World War. Indeed, one of his campaign slogans was 'He kept us out of war'. Then in 1917, he decided the US would enter the war. Now Wilson had to engineer public opinion to support this decision. America needed to issue its soldiers with beliefs as well as bayonets.

To do this, Wilson issued an Executive Order that set up the Committee on Public Information. The Committee's aims were as Orwellian as the name suggests. Chaired by a muckraking journalist, George Creel, this body was commonly referred to as the Creel Committee. Journalists, academics and public relations specialists were all employed to get the public behind the US's entry into the war. Creel was clear that he was in a fight 'for the minds of men' and for 'the conquest of their convictions'.[67]

Creel claimed that the Committee only gave 'people the facts from which conclusions may be drawn' and changed minds through 'unanswerable arguments'.[68] Yet this was highly disingenuous. Creel really believed that Americans needed to be 'thrilled into unity and projectile force.'[69] As scholars of this era have noted, Creel 'wanted the public to cheer, not think'.[70] Many progressives at the time believed that 'the public mind should not only be made up, but made up right'.[71] As a result, information, such as the number of war deaths, which the public could have used to demand peace, was not made available.[72]

The Committee's work involved censorship, banning books and producing misinformation. As one of its senior officials would admit after the war, 'we never told the truth – not by any manner of means . . . We told that part which served our national purpose.' Creel's name led to the creation of the term 'creeling', which meant to twist or suppress news. When the Committee 'shot propaganda through every capillary of the American bloodstream', this created information clots that clogged the arteries of free thought.[73]

The work of the Creel Committee during the First World War, expertly documented by John Maxwell Hamilton, was a blatant infringement of the American people's right to free thought.[74] As it stands, the right to freedom of thought is an absolute right which can never be violated, even if heavily armed Germans (or Russians) are marching towards you. The question of whether the right to freedom of thought should be able to be permissibly violated in the interests of national security is something to which we will return. Yet the impact of Creel's work was not limited to the war years. Techniques honed by his committee were subsequently used to crush the US labour movement in the 1920s.[75]

Another way that the public mind is made consonant with elite interests is by promoting thoughts desired by elites and suppressing competing ones. Many inhabitants of liberal democracies haughtily claim that 'no one tells me what to think!' The depressingly obvious retort to this, as Chomsky has pointed out, is that no one has to.[76] Anyone with beliefs that threaten elite interests will not be employed in a position of power. Notably, in the UK, journalism is dominated by people from advantaged backgrounds. Sixty-six percent of journalists come from families with professional and managerial backgrounds, with only twelve percent having working-class parents.[77] People whose beliefs already match elite interests are those who go places. As the American political scientist Michael Parenti put it, 'you say what you like, because they like what you say.'[78]

Members of the press face a range of pressures. Given their dependence on advertisers, stories that put certain businesses in a bad light are incentivised not to appear. The owners of the press can also be influential. David Yelland, a former editor of the *Sun* tabloid newspaper in the UK, described it thus:

> All Murdoch editors, what they do is this: they go on a journey where they end up agreeing with everything Rupert says. But you don't admit to yourself that you're being influenced. Most Murdoch editors wake up in the morning, switch on the radio, hear that something has happened and think, What would Rupert think about this? It's like a mantra inside your head, it's like a prism. You look at the world through Rupert's eyes.[79]

The corporate media, propaganda scholars argue, are a system of mass thought control.[80] One of the most transparent corporate media thought control campaigns in recent years was undertaken in response to Jeremy Corbyn, leader of the British Labour Party, running for Prime Minister in the 2017 UK parliamentary elections. Corbyn was an odd thing for modern Britain, a socialist leader of a socialist party. When he took over the leadership of the Labour Party, a genuine alternative to neo-liberalism emerged in the UK for the first time in almost half a century. Corbyn announced plans to nationalise public utilities, raise taxes on the rich and scrap university tuition fees. These policies won the support of many voters, especially the young. In the 2017 election, Corbyn achieved the biggest increase in the vote share for the Labour Party between two election cycles since 1945: rising from 30.4% in 2015 to 40% in 2017. In 2017, 12.9 million people voted for Corbyn's party, compared to a previous high from 2001 to 2015 of 10.7 million votes.[81]

Despite Corbyn's message resonating with many voters, the media, dropping any pretence of neutrality, attacked Corbyn relentlessly. A report by the Department of Media and Communications at the London School of Economics found 'an overall picture of most newspapers systematically vilifying the leader of the biggest opposition party, assassinating his character, ridiculing his personality and delegitimising his ideas and politics'.[82] The report added that 'with the vast majority of the British newspapers situated moderately to firmly on the right of the political spectrum, the analysis of our data also points to a strong ideological bias'. The report's authors concluded by rhetorically asking whether it is 'acceptable that the majority of the British newspapers uses its mediated power to attack and delegitimise the leader of the largest opposition party against a rightwing government to such an extent and with such vigour?'

More generally, ruling powers control the public mind by getting the population to adopt their ideas. Antonio Gramsci made this point, identifying two methods of controlling the people of a country.[83] The first method was direct domination. This was the force-based authority exercised by the government and legal system. The second method was hegemony. A ruling class achieves hegemony when the population spontaneously adopts the worldview of its rulers.[84] The ruled are infected by ideas that support the interests of the rulers. Once a ruling class achieves hegemony, the population are bound by chains into which they have willingly walked.[85] For this reason, as the murdered South African activist Steve Biko put it, 'the most potent weapon in the hands of the oppressor is the mind of the oppressed.'[86]

The French philosopher Louis Althusser elaborated on Gramsci's ideas. The ruling class, said Althusser, aimed to make the masses reproduce the existing form of society that served the ruling class's interests. He proposed that the ruling class influenced the masses in two ways: violence and ideology.[87] By ideology, Althusser meant the background ideas we possess about how the world works and how we should behave.[88]

Through what Althusser called Repressive State Apparatuses, the ruling class had a body of institutions at its disposal that controlled people, at their root, by violence. These bodies included the government, army, police, courts and prisons. Althusser also identified another body of institutions that primarily controlled people through ideology. He called these Ideological State Apparatuses. The primary such institution was school, with other bodies including churches, the family, political parties, the media, literature, the arts and sports. Althusser believed the ruling class could only stay in power if they controlled Ideological State Apparatuses.[89] Although Ideological State Apparatuses were often privately owned, and hence had some autonomy, the ruling class could still use them to control the population. Althusser explains that this is because this diverse range of bodies was unified by their realisation of the ruling class's ideology. But he doesn't spell out the details of this. For these details, we can turn to Curtis Yarvin.

Yarvin introduces us to a concept closely related to an Ideological State Apparatus, the 'cathedral', which he defines as 'journalism plus academia'.[90] Yarvin was puzzled why a diverse range of newspapers and universities all held the same basic view of the world. Why was there a 'herd of independent minds'?[91] What, he wondered, was the 'mysterious force' that drives them all in the same direction? Yarvin's short answer is power. His longer answer will need its own paragraphs.

Yarvin argues that truth is the only selective pressure on the ideas of dissidents in a totalitarian regime. Ideas are selected based on their truth value and the truest win out. However, professors and journalists in a liberal democracy experience another selection pressure: the promotion of power. The state will adopt ideas that validate the ruling class's use of power. It will ignore ideas that invalidate such use of power.

A concrete example that Yarvin gives involves climate change. The idea that governments must do something about climate change validates the government's use of power – namely taxation and spending. The idea that governments shouldn't do anything, leaving it to private individuals to make changes, invalidates the government's use of power. The idea validating state use of power wins out and is what the media and academia promote. Thus, a diverse range of journalists and professors align in their views, not necessarily because an idea is true, but because it validates the state's use of power. Obviously, this example is problematised by the fact business tries to push responsibility for climate change onto the consumer, but you see Yarvin's essential point.

Yarvin then traces the roots of this problem deeper. The ultimate problem, he argues, is that governments leak power. Governments want to outsource responsibility to outside actors to depoliticise matters. This allows governments to place blame for things going wrong onto others. But in doing so, state power leaks out into the hands of non-governmental actors. The only form of government that can escape this problem, claims Yarvin, is a monarchy. In such a system, the buck rests with the sovereign, meaning power does not leak. No power leakage means the 'cathedral' can pursue the truth rather than being corrupted by incentives to serve power. In short, ideological conformity is achieved in a liberal democracy through the widespread adoption of ideas that serve the state.

Not only does thought control become the primary tool of control in democracies, but liberal democracy also makes individuals increasingly susceptible to mental manipulation. As Jacques Ellul notes, the more individualist a society is, the less power an individual has. Individuals become broken off from community, guilds and even family. They become isolated, weak and alone, yet still responsible for their choices and actions. This makes them ideal targets for propaganda. These isolated yet responsible people don't read newspapers to help them think, says Ellul, but rather to be told what to think. They read not to create their own opinion but to find and adopt the opinion of their tribe.[92]

Our desire to appear knowledgeable and to participate in the public conversation means we are happy to swallow and regurgitate received wisdom. As Ellul puts it, 'the majority prefers expressing stupidities to not expressing any opinion . . . they are hence ready to accept propaganda if it allows them to participate in the conversation.' Readers, says Ellul, offer their throats to the knife of the propaganda they choose. Spengler called this 'mind training'.[93]

What we have discussed so far assumes that if a regime can rule with force, it will. If it can't use force, then it must rely on ideas. But maybe we have things backwards. All regimes may begin by trying to rule with ideas. The extent to which they are successful determines the amount of force needed. In a totalitarian state, there is only one narrative. This means citizens are likely to object, making the need for force high. In contrast, in a democratic state, people are typically offered a choice of two stories (a left and a right narrative). The feeling of choice makes them more accepting of these ideas, and therefore less force is needed.[94] Yet is the leap from one party to two parties so big?

Technologies of mind control

Given the importance to ruling powers of controlling the public mind, any new technological developments that might support their quest should scare the living daylights out of us. We need to spot these technological developments and prevent elites from using them to control our minds. The right to freedom of thought will be an important shield in such a battle.[95] There are two main threats to look out for.

New technologies allow old persuasion methods to work in a new medium. We now swim in a new digital world. 'Minds', as US Supreme Court Justice Anthony Kennedy observed, 'are not changed in streets and parks as they once were. To an increasing degree, the more significant interchanges of ideas and shaping of public consciousness occur in mass and electronic media.'[96]

The development of new technologies also allows the development of new persuasion methods. Recent decades have seen significant advances in the psychological understanding of how our minds work. This knowledge can be used to create new ways to manipulate our minds. Society has also seen the development of behaviour-reading and brain-reading technologies that can see ever more clearly into our minds. Today our skulls are no longer opaque but translucent, with future technology threatening to make them transparent. These technologies promise to achieve something that Big Brother could never do. 'With all their cleverness', wrote Orwell, Big Brother 'had never mastered the secret of finding out what another human being was thinking.' New technologies may be the holy grail of rulers; a way to know and control what the ruled are thinking.

The law saw this threat to freedom of thought coming long ago. 'Advances in the psychic and related sciences', warned US Supreme Court Justice Brandeis in 1928, 'may bring means of exploring unexpressed beliefs, thoughts and emotions.'[97] We must understand these developments before considering how these new technologies may create new mechanisms for the ancient desire to control minds.

Developments in psychological science

Throughout the Middle Ages and Renaissance, writers published books termed *specula principum* or 'mirrors for princes'.[98] These books advised princes on how to rule their kingdoms most effectively. The most famous is Machiavelli's *The Prince*. Today's pop psychology books on how we think are mirrors for the modern prince. They give the Machiavellians among us the information they need to manipulate voters, consumers or whomever their target may be. After all, as Machiavelli himself wrote, 'Princes who have accomplished great deeds are those who have . . . known how to manipulate the minds of men.'[99] By providing understanding, psychology offers control.

In this light, psychology is a satanic project that follows the dark prince's strategy in Milton's *Paradise Lost*. After his defeat by the forces of heaven, Satan suggests that he and his band of fallen angels go to study Man in the Garden of Eden:

> Thither let us bend all our thoughts, to learn
> What creatures there inhabit, of what mould,
> Or substance, how endu'd, and what their Power,
> And where their weakness, how attempted best,
> By force or subtlety.[100]

For centuries, psychology has sought knowledge about the workings of the human mind. Those in (or seeking) power were always going to use such knowledge to manipulate the populace. To think otherwise is like a Manhattan Project worker believing nuclear fission would only be used to keep the lights on.

In the golden age of advertising, advertisers used numerous psychological tricks to encourage people to buy products. Books such as Vance Packard's *The Hidden Persuaders* tried to expose such tricks. Since then, scientific research has given us a much better idea of how the mind works, leading to many new techniques being discovered and deployed for profit. Every insight from research into biases in human decision making has been utilised by business to enhance the probability of people buying products.

Such techniques can be found in any contemporary marketing textbook. These techniques centre on encouraging people to engage rule-of-thumb rather than rule-of-reason thinking. Specific techniques include loss/gain framing (emphasising a choice in terms of either a loss or a gain, knowing that each will have a different effect on the customer), reciprocity (offering a customer something free, knowing this will trigger an automatic bias in them to reciprocate and buy your product), scarcity (making a product seem scarce to panic customers into valuing it more) and social proof (pretending that lots of other people have bought the product to make it seem more desirable). Such techniques take advantage of quirks of the human mind, rather than engaging with people's rational faculties, to encourage purchasing behaviour.

The term 'dark pattern' refers to tricks used by websites and apps to manipulate you into doing something you didn't intend.[101] Such tricks often interfere with your rule-of-reason thinking and try to trigger rule-of-thumb thought. There is a range of named types of dark pattern.[102] In the 'roach motel' dark pattern, named after a trap for cockroaches, something is easy to get into but hard to exit. You may be able to subscribe to a service with a single click, but then have to go through a long series of steps to unsubscribe. In the 'nagging' dark pattern, you are given two options, neither of which lets you give a final 'no' response. Instead, you are forced to choose between 'yes' or 'not now', with the latter option leading you to be asked again at a later point. In the 'bait and switch' dark pattern, you get what you asked for but are then offered something you never inquired about. There is also the 'false demand' dark pattern. In this, you are falsely told that someone like 'Bill in Buffalo' has just purchased a product or service. Such information encourages you to buy, either by triggering a sense of scarcity or setting a social norm of people buying the product by offering false social proof.

The most effective dark patterns involve hiding information (e.g. putting details in small print in a non-prominent location), obstruction (making people perform lots of steps to get out of a service), using trick questions (double negatives, etc.) and social proof.[103] Even the mild use of dark patterns has been found to double the rates at which people sign up for a questionable service. More aggressive use of such patterns makes people nearly four times as likely to subscribe.[104]

Social networking companies are clear manifestations of the knowledge of human psychology being deployed for profit. For many early users, the internet was a way to let thought roam free. When John Perry Barlow issued his *Declaration of Independence of Cyberspace* in 1996, he stated his aim to 'create a civilization of the Mind in Cyberspace' where 'no one can arrest our thoughts'.[105] This was, at best, optimistic; at worst, catastrophically naive. Today, social networking companies are motivated to maximise the time users spend using their product. By doing so, they maximise users' exposure to adverts, thereby enhancing advertising revenue. Guillaume Chaslot, a computer programmer who worked with YouTube engineers on their recommendation system, describes how 'watch time was the priority ... Everything else was considered a distraction.'[106] This goal is achieved through the accumulated knowledge of the human mind gained by psychology.

All this is enabled by a knowledge gap, which is itself a manifestation of a power gap. Internet companies know more about how your mind works than you do. They know how to exploit your rule-of-thumb thinking and create dark patterns. Indeed, with the increased use of gamification to draw us in, we are often literally 'being played'.[107]

In addition to corporations manipulating people's thoughts to serve corporate interests, governments may use similar techniques to try to serve citizens' interests. In recent years, governments have used behavioural science research findings to 'nudge' citizens into individually and socially beneficial actions. These include joining organ donor registers, increasing fruit and vegetable consumption, and boosting tax collection.[108] Nudges work by designing the environment in which people choose (the 'choice architecture') to make specific options more accessible. For example, the option you want people to choose should be put at the front of a physical display, made a pre-selected option, or require an explicit opt-out to avoid. This means the consumer has to do more work to select alternative options. In essence, nudging takes advantage of human laziness.

Nudging has been referred to as 'libertarian paternalism'. Its ethical justification is that it is acceptable for policymakers to manipulate the choice architecture individuals face to promote 'better' choices, as long as people are still free to make a choice.[109] Of course, not everyone is happy with this justification. Nevertheless, every environment must be designed in some manner. Given our knowledge of nudges, we may as well design them to promote healthy options.

A less controversial approach involves 'educative nudges'. Here, rather than designing the environment to nudge people unknowingly into a certain choice, people are explicitly given information to aid decision making. People may receive nutritional or safety information about a product to support their ability to make a healthy choice.[110] Furthermore, once people understand the idea behind nudges, they can design their online environments to 'self-nudge'. That is, they can set up their social media interfaces to help them act in the way they desire. We can all become 'citizen choice architects'.[111]

Not everyone is equally nudge-able. For example, one way to nudge people is to tell them what others are doing. Due to our tendency to conform our behaviour to social norms, this information nudges us to act like everyone else. For example, providing households with information about how much energy they use compared to their typical and efficient neighbours reduces energy consumption.[112] However, some people react to such information more than others. One study found that whilst information on relative energy usage led liberals to reduce their usage by 2.4%, conservatives only reduced theirs by 1.7%. Conservatives were also more likely than liberals to opt out of receiving energy reports and say they disliked them (the reports, not the liberals, though probably both). Would-be nudgers need to account for the politics of the nudged.

The historian Yuval Noah Harari once claimed that 'Freedom depends to a large extent on how much you know yourself, and you need to know yourself better than, say, the government or the corporations that try to manipulate you.'[113] If this is true, we are all utterly screwed. We will never know ourselves better than large organisations armed with masses of data on us and the advanced technology to analyse this. Yes, we can know ourselves better than we do now, and this will offer us some defence against manipulation. But ultimately some form of regulation will be required. Otherwise the only freedom that will remain will be the freedom to exploit. How we can design the right to freedom of thought to address this situation is a key question we will return to later.

The threat from brain-reading

Have you ever felt guilty about a dream you had? If so, you may have taken comfort from the fact that no one else could ever know what you dreamt. Or could they? In 2013 three volunteers came into a lab at Kyoto University in Japan and were asked to sleep in an MRI scanner.[114] While carbon slept, silicon calculated. The scanner recorded the activity in the brains of the sleeping volunteers. The scientists then roused the volunteers and asked what, if anything, they had seen in their dreams. Having previously shown the volunteers a range of pictures whilst they were awake, the scientists knew what kind of brain activity was associated with seeing things such as streets, houses, men or food. But could they use these neural dictionaries to decode the brain activity of the dreamers? They could. The accuracy wasn't great, but it was better than a random guess. There is no doubt this technology will rapidly improve over time. Not even your dreams are safe.

Much more work has been done looking at decoding our neural activity whilst we are awake. Brain-reading analyses our brain activity to try to decode what we are thinking. Early studies were able to take people's brain activity and accurately predict what type of pre-defined object (e.g. a face, house, cat, etc.) they were looking at.[115] But technology can now use people's neural activity to predict what novel images people are viewing.[116] Once, neuroimaging could only create rudimentary reconstructions of what an individual was seeing.[117] Today it can create 'remarkable' reconstructions.[118] Brain imaging data can also be used to decode what people are hearing.[119] Research has even attempted to decode individuals' verbal thoughts from the activity of both their brains[120] and their throat muscles,[121] with some success. The brain activity when people mime speech can even be directly converted into recognisable speech.[122]

A great leap forward occurred when machines became able to decode thoughts they had not previously been trained to recognise in someone's brain.[123] This technology used the idea that the brain represents words as 'semantic vectors'. Any word can be coded by the extent it is associated with a finite number of specific features.[124] This means that any word in a basic vocabulary can be represented by a unique pattern of scores on these vectors. Knowing the brain activity associated with each of these features allows you to decode any word someone thinks.

We are still a long way from the Precrime police of Philip K. Dick's *The Minority Report*. We will not soon be arresting people for crimes they have been foreseen to commit in the future. Nevertheless, neuroscience can now decode people's basic intentions.[125] It is possible to predict what participants will do (for very basic tasks) before they know.[126] However, the ability to do this in real time is still limited. Our ability to predict what someone intends to do when they are playing a first-person shooter computer game is limited to the movements they intend to make. Someone's plan to fire their weapon can't be detected because this subtle neural signal is swamped by the neural activity associated with the emotions of firing or being fired upon.[127]

Much of this work is preliminary, with accurate predictions only possible under highly controlled laboratory conditions. Yet there is enormous research interest in this area, with money pouring in, suggesting progress could be rapid. Work in this area is often framed as being motivated to help paralysed people communicate again. This is a noble aim, obviously. However, whilst one would like to think the deluge of money pouring into this area is due to a concern for the paralysed, this is clearly not the case. Brain-reading technology will be a huge money spinner for tech firms, explaining why companies are throwing money at such research.

Back in 2017, annual spending on neurotechnology by for-profit industry was estimated to be $100 million per year and growing.[128] Facebook is funding research into decoding our thoughts from our brain activity.[129] Such work appears to be part of Facebook's goal to develop a brain–computer interface that can decode users' thoughts and transmit them to Facebook. Likewise, Microsoft has patented brain-reading technology and Elon Musk's Neuralink is developing brain–computer interfaces. Taxpayer dollars are also being spent. State funding for research in the US has come from the Army, Air Force, Intelligence Advanced Research Projects Activity (IARPA) and Defense Advanced Research Projects Agency (DARPA), suggesting a national secu-rity interest in brain-reading. We should not exaggerate the current abilities of these brain-reading technologies. Nor should we encourage a moral panic. Yet it is clear that powerful interests are driving the field rapidly forward.

At present, no one is going to scan your brain and read your thoughts without you knowing and consenting. As one philoso-pher dryly notes, echoing a *Lord of the Rings* meme, 'one does not just end up in an MRI-scanner without knowing this'.[130] Yet it is not hard to imagine a future where brain-reading technology becomes effectively mandatory. When firms develop wearable brain-reading technology, this will be enthusiastically embraced by society due to the convenience it will offer. This technology will be discreet and not like the prominent (and decidedly gooey) EEG technology I and others use in laboratories today.

In a world where working your computer is significantly quicker with brain-reading technologies, the competitive advantage this yields may force everyone to adopt this technology. People will line up to have their brains read. Society will be designed around this technology, just as planners design cities with car drivers in mind. All of this will push us into using such technology, which has huge potential for abuse. Concerns over the misuse of our Facebook data[131] would pale in comparison compared to the abuse or hacking of our thoughts. For this reason, as the barrister Susie Alegre points out, we may wish to rigorously apply an approach to managing risks that is known as the precautionary principle here and take action to diminish the plausible yet uncertain morally unacceptable harm such technologies may cause.[132]

Yet applying the precautionary principle creates new problems. As psychologist Thomas Hills points out, this principle plays into our tendency to focus on the negative.[133] Evolution has biased our attention towards threats and problems ('if it bleeds, it leads'). As a result, we share more information about perceived risks than benefits, termed 'social risk amplification'.[134] For example, one study examined how people passed on information about an antibacterial chemical. As the message passed from one person to the next, facts about the benefits of the antibacterial stopped being mentioned. Not only did the risks and downsides of the antibacterial continue to be mentioned and passed on, but new risks were made up out of thin air and passed on.[135] All this contributes to creating a risk-society[136] in which we base decisions about new technologies on their potential downsides, without considering their potential benefits.[137]

Impairing the development of brain–computer interfaces would have downsides. Technological progress is already behind where we hoped it would be. As the American entrepreneur Peter Thiel puts it, we were promised flying cars and instead we got 140 characters.[138] Thiel argues that less regulation, not more, is needed to kickstart innovation again. More innovation means more economic growth. More growth means we won't have to engage in a zero-sum game where we are all trying to grab a piece of the same pie, promoting peace and security.[139] Brain-reading technology could be an important driver of growth and productivity.

There are many specific benefits that brain-reading could bring. Slowing the development of brain-reading technologies would undoubtedly impede our ability to help people suffering from various forms of paralysis. Furthermore, using the right to freedom of thought to justify slowing technological progress could damage the very things freedom of thought is meant to protect. The discovery of new truths could be delayed or even prevented if we cannot obtain and analyse big data relating to our minds. Freedom of thought is meant to promote autonomy. Brain-reading would be a boon to the development of immersive virtual reality technology. In the meta-verse, we will all be able to achieve the fantasy of complete autonomy. Brain-reading will also help address a key limiting factor on thought – our mental bandwidth.[140] We are incredibly slow at getting our thoughts into a phone or computer. Brain–computer interfaces could rapidly increase our speed of communication, in turn increasing our ability to think. We need to reap the rewards of this technology whilst managing its risks.

If brain-reading troubles you, then you're really not going to like brain-writing. This involves stimulating someone's brain to make them think specific thoughts. In 2019, neuroscientists at Columbia University were able to read the thoughts of a mouse and then insert thoughts back into its brain.[141] The researchers first trained mice to lick at a waterspout when they saw a striped pattern. The researchers then injected viruses into the regions of these mice's brains involved in vision. Now, when the mice saw striped patterns on a screen, the researchers could see which areas of the mice's brains were active. In the crucial next step, the researchers used a laser to activate these visual areas of the mice's brains. This caused the mice to lick. The researchers' laser had caused the mice to hallucinate striped lines, triggering them to lick. Soon, we may not need glasses or goggles to experience a virtual reality world. A new world could be beamed directly into our brain. If unsure whether to be excited or afraid about this, be both.

The threat from behaviour-reading

In the 2018 film, *The Great Hack*, Professor David Carroll asks a room of students about the adverts they see on their smartphones and computers. Had anyone seen adverts that uncannily coincided with products they had been looking for? Most had. Had any concluded that their device must have been listening to their conversations? Again, most raised their hands. That many of us feel Big Tech has been bugging us shows how much we underestimate their ability to use our personal data to infer what is happening in our heads. We can use the term 'behaviour-reading' to refer to the process of analysing our observable actions to gain insights into our thoughts.

Behaviour-reading is possible because we bleed data. For millennia, this data seeped into the earth where it fell or was washed away by the tides of time. Now, technology can capture and store our data indefinitely. Tech giants, like Facebook and Google, possess unimaginable amounts of data on their users. Data-analytic companies possess thousands of pieces of data on millions of people. It seems likely that national security agencies possess even more data on us,[142] with the exact size of their troves being unknown.

People can analyse this data using advanced machine-learning algorithms to crowbar their way into your inner world. Facial analysis technology can work out more than just what emotions you are feeling. Deep learning neural networks are twenty percent better than average people at guessing someone's sexuality from their face alone.[143] Things such as your musical preferences or the words you use on Facebook can be used to predict your personality and political orientation.[144] When researchers reported they could accurately predict detailed personal information about people simply from the pages on Facebook they had 'liked', many people were shocked.[145]

When Big Tech combines its powerful machine-learning algorithms with the vast amount of data we have put into the internet, it can see into our souls. As the CEO of Google, Eric Schmidt, put it, 'you give us more information about you, about your friends, and we can improve the quality of our searches. We don't need you to type at all. We know where you are. We know where you've been. We can more or less know what you're thinking about.'[146] Artificial intelligence will soon tell us truths about ourselves that we were not even aware of.[147]

A century ago, US Supreme Court Justice Brandeis crusaded against big banks' use of 'other people's money'.[148] Today the challenge is Big Tech's use of other people's data. The ease with which our inner states can be inferred is a significant departure from the past. These abilities are already being utilised for both financial and political ends. We need to consider if, and why, this may threaten freedom of thought.

Is this a moral panic?

Yet there is an objection to the argument that brain- and behaviour-reading threaten freedom of thought, which new legislation needs to address. This objection is that we are amidst a moral panic over these technologies; or, more specifically, a techno-panic.[149] If so, then other people's anxieties should not determine laws that govern your behaviour.

We love a good moral panic. Every era has them. They are characterised by a societal reaction to an alleged threat that is completely out of proportion to the actual harm done. Moral panics are often based on the idea that 'people are motivated to act by mysterious and unrealistically powerful forces'.[150] Sound familiar? In the 1950s, a moral panic emerged over the idea that reading comic books produced juvenile delinquency.[151] In the 1960s, it was the effects of LSD. In the 1980s, it was satanic child abuse in day care centres. In the first decade of the 2000s, it was the idea that violent video games lead to school shootings.[152] The target of the moral panic is typically seen as threatening children. As parodied in *The Simpsons*, the cry goes up: 'Won't somebody think of the children!' Action is demanded.

Action may be taken. Unsurprisingly, this often doesn't work out well. Laws passed because of a moral panic can be ill-conceived, unsuitable, or wildly disproportionate to the problem.[153] As one scholar notes, 'driven as it is by irrationality, the reforms usually miss the point of the original problem and suffer from disproportionality'.[154]

Moral panics do not simply identify a problem; they actively construct one. They begin by identifying people who are felt to threaten societal values. This typically happens through the mass media. Hostility arises towards these deviant individuals who become 'folk devils'. The literati then man the moral barricades to help create a consensus that there is a problem. Crucially, claims of the size of the problem are disproportional to the actual threat or harm caused. Statistics may be exaggerated or fabricated. Experts then pronounce their diagnoses and solutions. Ways of coping emerge. The panic may then evaporate, but the new legal or social policies remain.[155] So, does anything we discussed above look like a modern moral panic?

In the Cambridge Analytica scandal, we got to see a social problem being constructed in a way that looked much like a moral panic. This scandal involved a data-analytic company collecting personality data on people and using this information to create targeted political advertisements to try to influence these individuals' voting behaviour. The use of personal information about voters to tailor adverts to them, deemed likely to result in desired behaviours, is called microtargeting.[156] This practice raised two concerns. The first was that microtargeting was so powerful that people were powerless to resist it, effectively becoming brainwashed. The second was that because only certain people saw these adverts, debate about the content of these adverts was being prevented in the public square.[157]

In this scandal, clear folk devils were identified. They included both groups (Big Tech in general) and specific individuals, ranging from the CEO of Cambridge Analytica, Alexander Nix, to its billionaire US backers and individuals involved in data analyses. As one headline put it, 'Cambridge Analytica's Shady Professor "Dr. Spectre" Is a Perfect Trump Villain'.[158] A consensus emerged in the media that there was a problem.

A key part of moral panic analysis is testing whether an issue is being distorted and exaggerated, prompting massive overreaction. In the case of Cambridge Analytica, the problem was blown out of all proportion. The media did not offer a sober, evidence-based assessment as to whether the machinations of Cambridge Analytica could, as the headlines proclaimed, brainwash voters.

Such an analysis would have undermined the sensational headlines. At the time of the scandal, there was no clear evidence to support Cambridge Analytica's alleged ability to influence voters. At the time, the ability of microtargeting to change political views looked likely to be small.[159] However, there were suggestions that microtargeting might notably affect turnout, which could have a decisive effect in close political contests.[160] Indeed, the political strategist Steve Bannon emphasises that the priority is not to persuade people, as they already largely know what they believe, but to mobilise them.[161] This fits with what the scholar of propaganda, Jacques Ellul, stressed long ago, namely that the aim of modern propaganda is 'no longer to modify ideas, but to provoke action.'[162]

Subsequent work has helped clarify what microtargeting may and may not be able to do. Microtargeting does appear, in specific contexts, to be able to change levels of voter turnout. A study published by Katherine Haenschen in 2022 examined the effects of microtargeted issue-focused Facebook adverts on the turnout of 870,000 users in Texas in the 2018 US midterm elections.[163] Individuals deemed unlikely to vote were targeted with adverts relating to the issues of abortion rights, healthcare, immigration and gun control. Overall, showing users these microtargeted ads did not significantly alter turnout rates. However, some of the specific adverts did increase turnout. Specifically, turnout increased in competitive congressional districts that received adverts about abortion rights, women's healthcare and immigration. In particular, in these competitive districts, female voters shown abortion rights adverts demonstrated an increase in predicted turnout.

Haenschen concluded that 'these findings speak to the power of microtargeting: Facebook's platform enables advertisers to target a specific list of individual voters, and its algorithm seemingly exacerbates that effect by exposing individuals it predicts to be most likely to respond.' Yet giving people political information on issues they are concerned about is a long way from brainwashing. So, what was the Cambridge Analytica scandal really about?

The Cambridge Analytica scandal was an attempt to create a narrative about how the political right should be viewed. It did what all moral panics do, namely it prescribed 'who has the right to speak, on what terms and to which ends'.[164] Moral panics are a form of moral regulation. People called 'moral regulators' object to the behaviour of others which they 'seek to control, by legal or other means'. Moral panics are an extreme case of a story being created about a new risk, which regulates moral behaviour.[165] Arguably, the Cambridge Analytica phenomenon represented an attempt by the political left to morally regulate the behaviour of the political right. Yet, as Susie Alegre has pointed out, 'if you feel uneasy about political behavioural microtargeting when it delivers Trump, you need to question the ethics of using this technique at all, regardless of the politics of the candidate involved.'[166]

A similar analysis can be applied to claims about social media, such as that they create echo chambers and make our society dangerously polarised. You can find a piece of research that supports almost any perspective you wish to be the case. However, one study does not make a summer. When the basket of empirical evidence for such claims is level-headedly analysed, we reach markedly different conclusions from those shouted from media headlines. Take, for example, an open-sourced review of the evidence of the impact of social media curated by the psychologists Jonathan Haidt and Chris Bail.[167]

Haidt and Bail's review found evidence both for and against the idea that social media increases the polarisation of people's political views. Studies have found that watching videos recommended by the YouTube algorithm[168] and seeing negative political tweets[169] does make people's political views more polarised. However, an examination of people's use of Facebook News during the 2016 US presidential election found that people who used Facebook News were more likely to view material that challenged their views and came to hold slightly less polarised beliefs.[170] Another study's findings suggested that social media use did not increase users' dislike and distrust of people who held different political views. Instead, this study found that the more people disliked others with opposing political views, the more they used social media.[171]

Haidt and Bail's review also showed evidence both for and against the idea that social media creates echo chambers. A study of the 2012 US presidential election found that public exchanges on Twitter mostly took place between people with similar political views.[172] Similarly, Twitter users have been found to be three times more likely to follow back accounts whose political views matched their own.[173] The YouTube recommendation algorithm has been found to push real users into mild ideological echo chambers. Yet it does not appear that many users go down rabbit holes that take them to ideologically extreme content.[174] Indeed, one study found no strong evidence for the existence of echo chambers.[175] A further study found that only around five percent of internet news users inhabited a politically partisan left or right online news echo chamber.[176] This finding was consistent with a body of work that challenges the impact of echo chambers.[177] In short, anyone wishing to cherry-pick the literature can find data to support their pre-existing prejudices. Yet the reality is more complicated, less clear, and utterly unsuited to a catchy headline. We need more Athenian discussion of details and less Spartan shouting of claims.

Three causes of moral panic have been identified by Goode and Ben-Yehuda, so which may we be amidst?[178] First, this could be a grassroots phenomenon. Maybe the public's anxiety over new technologies has been mapped onto a scapegoat. Second, it could be an elite phenomenon. Maybe elites are strategically creating a moral panic to divert our attention from social problems. A third option is that it comes from neither the top nor the bottom of society, but results from a moral crusade by activist groups, social movements and moral entrepreneurs ('organizers, activists, do-gooders, movement advocates who push for a given cause').[179]

The current techno-panic over mental manipulation appears to stem from a mix of activist and elite concerns over the rise of the political right. Part of the debate over freedom of thought is really a contest over who should rightfully influence people. James Burnham once wrote that 'we imagine we are arguing over the moral and legal status of the principle of the freedom of the seas when the real question is who is to control the seas.'[180] Conceivably, the debate over freedom of thought is not about trying to free minds but about determining who gets to control them.

Another way to understand how a non-evidence-based moral panic may have emerged around Big Tech is through the concept of what Timur Kuran and Cass Sunstein have termed 'availability entrepreneurs'.[181] These are activists who manipulate the public discourse to serve their agendas, which may be selfish or altruistic in motivation. Availability entrepreneurs influence the public discourse by creating 'availability cascades'. Such cascades make the activist's idea more available to the population by boosting their presence in the public discourse. As we saw in our discussion of rule-of-thumb thinking earlier, the more available an idea is to our minds, the more plausible or believable we take the idea to be, all other things being equal.

An availability cascade draws its power from the merging of two smaller cascades. The first is an informational cascade. If people aren't sure what is true, they will turn to others and adopt what they believe. This increases the availability of such ideas in public discourse. The second is a reputational cascade. Here, people realise that holding a certain view will make them look good in the eyes of others, so they adopt it. Again, this increases the availability of the idea in public discourse.

If activists want to push on an open door, they need to emphasise issues that possess traits which make them likely to be deemed unacceptably risky by the public. Big Tech has a wide range of traits that Kuran and Sunstein highlight as increasing the chance that something is deemed an unacceptable risk. Specifically, Big Tech is: 1) new; 2) out of our personal control; 3) something which if we want to be an integrated member of society we may feel we have no choice but to use; 4) receives heavy media coverage; 5) puts children at special risk; 6) has the potential to impact future generations; 7) is hard to roll back; 8) is human-generated, rather than natural; 9) comes out of institutions/companies we don't trust; and 10) has immediate negative effects. These factors influence whether the public finds the risk of Big Tech acceptable, a risk which may have little basis in evidential fact and reality.

Competing discourses about the actual extent of risk, or the benefits of Big Tech, are not able to achieve an availability cascade. Perception, not reality, then informs policy decisions. Assessments of the risk of Big Tech to our thoughts must be guided by evidence, including weighing both costs and benefits, and not the extent to which activists have been able to create moral panics and availability cascades.

The partisan threat

The issue of partisan political concerns points to another potential threat to freedom of thought. The idea of a right to freedom of thought is likely to be undermined by people embracing it only when it serves their partisan political aims. This would lead the right to be viewed as a disposable tool rather than a basic right. This has been the fate of the right to freedom of speech.

In the 1960s and 70s, many in the left-wing 'free speech movement' did not value the right to free speech itself. There were obvious and noble exceptions. Lawyers at the American Civil Liberties Union (ACLU), including some who were themselves Jewish, stood up to defend the right of American Nazis to speak.[182] This principled stance led to the ACLU facing a severe backlash from many of its fair-weather supporters. Despite such committed actions, for many left-wing activists the concept of free speech proved merely a means to a variety of ends, albeit commendable anti-racist and anti-imperialist ones. The left shouted 'free speech' to get their voices heard. But when the political right wanted to exercise their right to free speech, the left rapidly moved to shut them down. One silencer-in-chief was the German-American philosopher Herbert Marcuse.

Marcuse believed the truth was out there and that free speech could be used to find it.[183] Unfortunately, the masses could only do this if someone silenced the elites. By dominating the media and the machinery of idea propagation, elites had fooled the masses into parroting the elite's own opinions. These elites were not using their free speech to find truth, argued Marcuse, they were using it to oppress 'the Damned of the Earth'. So, Marcuse made the classic tyrant move of claiming that the world was in a state of emergency. He decided exceptions needed to be made to the right of free speech. The rule of law needed to give way to the rule of Marcuse.

On some issues, Marcuse claimed, there was no 'other side'. As such, he demanded intolerance 'toward the self-styled conservatives, to the political Right'. This included 'intolerance even toward thought, opinion, and word'. To restore freedom of thought, said Marcuse, the right to free speech needed to be taken away from certain groups. The groups he had in mind were those which 'promote aggressive policies, armament, chauvinism, discrimination on the grounds of race and religion, or which oppose the extension of public services, social security, medical care, etc.' Marcuse also wanted to see extensive restrictions on teaching and research in schools and universities. His overall aim was 'shifting the balance between Right and Left by restraining the liberty of the Right'.

Marcuse's ideas helped propel what Antonio Gramsci called the left's long march through the institutions. Those with leftist political beliefs were installed in prominent teaching roles at universities and various other institutions. The origins of this march in the US can be traced back to 1962.[184] It was in this year that the Students for a Democratic Society (from which the Rodin-bombing Weather Underground would emerge) issued the Port Huron Statement. This argued that 'a new left must be distributed in significant social roles throughout the country. The universities are distributed in such a manner'.[185] This was to be a war of position. It would be fought 'in opposition to common sense and mainly situated at the level of the cultural and ideational whereby the educational system, the media and civil society should be used to construct and promote a counter-hegemony to capitalism and to the liberal democratic system which sustains and legitimates it.'[186]

Parts of the political right saw how much the left had achieved through these tactics and sensibly adopted them wholesale. The right attempted its own march through the institutions. In 1953, John Olin, an executive at an ammunition company, created the Olin Foundation. This invested in the development of conservative ideas, particularly at universities. The Foundation's president would write that 'ideas are weapons' and that 'capitalism has no duty to subsidize [the ideas of] its enemies.'[187] Grants from the Foundation supported conservative law schools and campus newsletters, promoting a 'counterintelligentsia' in academia.[188] The Foundation gave scholars 'grants, grants, and more grants, in exchange for books, books, and more books', whose ideas they hoped would then trickle down into society.[189] By supporting works that would become bestsellers, such as Allan Bloom's *The Closing of the American Mind*, the Foundation was successful in its aims.

In 1974, the billionaire industrialist Charles Koch funded the establishment of the Cato Institute. This libertarian think-tank funded research and idea dissemination.[190] It aimed to incrementally advance libertarian goals (promoting liberty and shrinking government, i.e. less business regulation and lower taxes) by influencing US government policy debates and legislation. When one of Cato's members, Murray Rothbard, was forced out, he co-founded the Ludwig von Mises Institute, which again aimed to spread libertarian ideas.[191] Mises believed that 'the masses of men do not create their own ideas, or indeed think through these ideas independently; they follow passively the ideas adopted and disseminated by the body of intellectuals'.[192] He called for intellectuals and writers to shape public opinion by creating sound theories and selling these to the masses. Likewise, Rothbard argued that intellectuals needed to take on the 'vital social task' of transforming society by becoming public 'opinion-molders'.[193] A new aristocracy was to be created. To achieve this, Rothbard focused not only on lobbying but also on the radical education of new libertarian thinkers.[194]

More recently, the billionaire DeVos family, including Betsy DeVos (Secretary of Education in the Trump administration), has spent more than $200 million funding conservative causes at universities and think thanks.[195] This included supporting the James Madison Center for Free Speech, a think tank that contributed successful arguments to the US Supreme Court case *Citizens United*.[196] This case ruled companies could give unlimited amounts of money to support political campaigns (as long as they are not formally co-ordinating with a candidate or party), on the basis that such contributions were 'speech'. This was a major victory for the political right. Although universities remained left-leaning,[197] the actions of the right secured a conservative majority on the US Supreme Court, the effects of which we see all around us.[198]

To wrap up this chapter, it is clear that the ruling powers of any society, whether totalitarian or democratic, governmental or corporate, have an interest in controlling the thoughts of its citizens. Scientific progress offers such powers new tools for accessing and controlling these thoughts. Psychological science, brain-reading and behaviour-reading all pose threats to freedom of thought.

For some, this news will vindicate the conclusion of Ted Kaczynski (better known as the 'Unabomber') that 'the Industrial Revolution and its consequences have been a disaster for the human race.'[199] Others will look to see how we can best develop and use these remarkable technologies whilst still safeguarding our right to freedom of thought. To address this issue, we need to know how our freedom of thought is currently secured, a question to which we now turn.

4

PROTECTIONS FOR
FREE THOUGHT

In the interregnum between the beheading of Charles I and the restoration of Charles II, a short pamphlet appeared in England. It was 1657, and Oliver Cromwell was taking on more and more power. 'Who made thee a prince and a judge over us?' asked the pamphlet. 'If God made thee, make it manifest to us. If the people, where did we meet to do it?'[1] Three hundred and fifty years later, the same question was flung at the European Union as part of the UK's Brexit referendum. Campaigners raised questions about the European Court of Human Rights' ability to overrule the sovereignty of the UK's Parliament,[2] an issue that had been rumbling on for some time.[3] So, where did human rights come from, and who made them prince and judge over us?

After the Second World War, the United Nations began drafting a statement on human rights. Responsibility for this fell to a Human Rights Commission headed by Eleanor Roosevelt. The Commission wanted members who were independent, knowledgeable human rights experts with the ability to challenge governments. It did not want government representatives primarily interested in maintaining national sovereignty. Yet member states insisted on governmental representatives being involved.[4] In the face of this, it is remarkable what the Commission members achieved. The process had an ambitious aim: to 'morally unite liberal and socialist by entrenching their common support of antifascist values'.[5] In 1948 the process bore fruit with the publication of the Universal Declaration of Human Rights. Article 19 of the Declaration stated that 'everyone has the right to freedom of thought, conscience and religion.'

The UN neither addressed what was to count as 'thought', nor what it meant for thought to be free, nor why thought needed to be free. The philosopher Alasdair MacIntyre argues that, in the Universal Declaration of Human Rights, 'what has since become the normal UN practice of not giving good reasons for any assertions whatsoever is followed with great rigor'.[6] To better understand what the UN was thinking, we must delve deeper.

The creation of the right to freedom of thought

The individual who was central to the creation of the right to freedom of thought in 1948 was Charles Malik. A Lebanese academic and diplomat with a PhD in philosophy, Malik worked as the Rapporteur of the Human Rights Drafting Committee. His life shaped his approach to freedom of thought. Malik had visited Germany in 1935 to study under the German philosopher Martin Heidegger. Malik was there on 29 March 1936 when the German people 'voted' on whether to elect an approved list of all-Nazi candidates to the government. This was an election in which the ballots did not contain an option to vote 'no' and any uncompleted ballots were deemed to represent a 'yes' vote. It may have been the only election in history where the winner was disappointed to only get ninety-nine percent of the vote.

Malik viewed Hitler's rise as a victory of 'systemic, controlled propaganda' and a remaking of 'human nature'. He saw how people in Germany had come to feel they were members of a group (the Aryan race) and that individualism was 'dead, wholly dead'. Malik witnessed individual reason succumbing to the power of the state.[7]

As a result, he would come to rail against states that determined 'all relations and ideas, thus supplanting all other sources of conviction.'[8] This led him to believe that civil society groups were needed to protect the individual from the overwhelming power of the state. Institutions such as family, friends, universities and the Church were, to Malik, 'intermediate sources of freedom', to which individuals could be loyal. These institutions represented the 'real sources of our freedom and our rights'.[9] In particular, Malik saw the family as 'the cradle of all human rights and liberties'. Indeed, Malik was the driving force behind the only part of the Universal Declaration of Human Rights that mentions non-individual rights. This is to be found in Article 16, paragraph 3, which states that 'the family is the natural and fundamental group unit of society and is entitled to protection by society and the state.'[10]

Views put forward by other delegates, in particular, the Yugoslavian delegate Vladislav Ribnikar, spurred Malik's approach. Ribnikar had voiced concerns that the human rights declaration would reify the social and political ideas of the middle classes. He argued that 'personal freedom could only be attained through perfect harmony between the individual and the community.'[11] Karl Marx had previously argued that 'the so-called rights of man' were only the 'rights of egoistic man, of man separated from other men and from the community . . . an individual withdrawn into himself, into the confines of his private interests and private caprice, and separated from the community'.[12] Marx believed that civil and political rights in capitalist societies were bourgeois rights that capitalists would exercise to protect themselves and oppress the working class. The pro-American, anti-communist Malik rejected these views, being deeply concerned about 'the tyranny of the masses and the State'.[13]

Malik's goals for the human rights declaration were overly ambitious. As he put it, 'The Bill of Rights must define the nature and essence of man . . . It will, in essence, be an answer to the question:

What is man? It will be the United Nations' answer to this question.'[14] Malik's answer to this question was essentially that humans
were thinking creatures. This concorded with philosophers from
Aristotle onwards, who had stressed that the capacity for reasoned
thought was uniquely human. For Descartes, reason was the 'only
thing that makes us men and distinguishes us from the beasts'.[15]
Similarly, for Hegel, 'what man has, as being nobler than a beast, he
has through thinking'.[16]

For these reasons, Malik proposed that the Universal Declaration
should adopt the principle that: 'The human person's most sacred
and inviolable possessions are his mind and his conscience,
enabling him to perceive truth, to choose freely, and to exist.' He
proposed text for the Declaration that described humans as
'endowed with reason and conscience'. This made some delegates
nervous. They felt that such abilities could end up being required
for people to be deemed rights holders.[17] Defining the human is a
perilous task because inevitably it is used to classify some people as
non-human. As such, it is a roadmap for genocide.

How universal was the Universal Declaration?

Despite attempting to create a *universal* right to freedom of thought,
Malik paid little attention to how thought could vary across
cultures. In 1947 the American Anthropological Association was
already asking how the Universal Declaration could be 'applicable
to all human beings, and not be a statement of rights conceived
only in terms of the values prevalent in the countries of Western
Europe and America?'[18] They warned against those who 'profess
equality and practice discrimination, who stress the virtue of
humility and are themselves arrogant in insistence on their beliefs'.
'There can be', the Association observed, 'no full development of
the individual personality as long as the individual is told, by men

who have the power to enforce their commands, that the way of life of his group is inferior to that of those who wield the power'.

Many of Malik's statements showed a Western bias. Take the following passage, arguably a victory of rhetoric over reason, in which he explains how committed he is to the right to inner freedom:

> What constitutes humanity more than anything else is this inward freedom which should therefore be absolutely inviolable ... Without the full and unimpaired right to think and believe freely, the value of these other rights pales into relative insignificance. One enjoys these other rights precisely in order to be free, and being free means nothing if it does not mean freedom to think and believe and change in your belief from the good to the better and better as the truth progressively reveals itself to you. The right to be free inwardly is the end and justification of all other rights.[19]

What stands out in this statement is its particularly individualist Western view. Many Asian societies would probably offer a very different view.

One way to understand this is to consider differences between Western and Asian societies, as the American psychologist Richard Nisbett does in his book *The Geography of Thought*.[20] Nisbett documents differences in thinking between Western and Asian societies, tracing these back to their roots in geography. The mountainous terrain of Ancient Greece, Nisbett argues, led to people developing occupations such as hunting and fishing. These required relatively little co-operation with others. In contrast, the fertile plains of China led to agricultural practices. These required much more co-operation between people, who had to work together to farm and irrigate the land. This meant social harmony was of crucial importance.

As a result, the West came to conceptualise the social world as consisting of atomised individuals, whereas Asian societies saw it as made up of groups. For the West, individual people and objects moved around with fixed identities, causing things to happen. The world should change to accommodate the individual. Combative debate and truth-seeking were possible because co-operation between individuals was not essential to the stability of society. Truth was more important than order. In contrast, in Asian societies, people existed in the context of relationships with their groups and the world. People should change to accommodate the harmonious function of the group. Combative debate and truth-seeking were problematic because they could destabilise social harmony. Order was more important than truth.

Remarkably, these different worldviews have led to measurable differences in how people in the East and the West view the world. For example, Nisbett reviews research showing that Asians remember more than Westerners about the background context of a picture. There are also differences in how we think about the world, with Asians being more likely to explain someone's behaviour as influenced by their environment and context than Westerners. Nisbett further highlights the more combative and truth-seeking nature of the West. He describes how the ratio of lawyers to engineers is forty times greater in the US than Japan. The US has many more lawyers because it is keen to engage in legal confrontations where the goal is to determine who is right and who is wrong. In Asian countries, the goal is less to work out who is right and wrong, and more to try to restore harmony through intermediaries.

Similarly, Nisbett points out that children's literature in the West focuses on the individual, e.g. 'This is Dick. See Dick run. See Dick play'. In contrast, Asian books for children focus more on relationships, e.g. 'Big brother loves little brother. Little brother helps big brother'. Given three words, such as 'cow', 'grass' and 'chicken', a Westerner is more likely to say cow and chicken go together, as they are both animals. In contrast, someone in Asia is more likely to say that cow and grass go together, because they think of relations between objects (cows eat grass).

Different countries are likely to disagree on what constitutes free thought. For example, the CCP rejects a one-size-fits-all approach to human rights and calls for 'a human rights development path with Chinese characteristics'.[21] The CCP's attempts to put society ahead of the individual, to promote harmony over conflict, and to use a social credit system to re-establish trust in society, are immediately written off in the West as the evil workings of totalitarianism. The West may be right. But if we want to promote free thought, we should consider, following Mill's advice, what truths, if any, the Chinese government may be onto. For example, how can the West deal with its crisis of trust?[22] If we do not invite global perspectives on the right to freedom of thought, claims of universal rights will mask a grinning colonialism.

Denazification and the Universal Declaration

A notable irony, given Malik's high-minded words about freedom of thought, was that during these deliberations the victorious Allies were undertaking one of the largest thought control operations in history. Germany could not be allowed to threaten world peace again, and this created a 'political imperative to eradicate Nazi thinking'.[23] The German people had to be denazified. The Allies wanted to do this through 're-education'. After the Potsdam Conference, the Allies agreed that 'German education shall be so controlled as completely to eliminate Nazi and militarist doctrines and to make possible the successful development of democratic ideas'.[24] Eisenhower initially estimated that this denazification process would take '50 years of hard work'.[25] It is unclear why Eisenhower believed this would take so long. Perhaps he thought the sound of John Stuart Mill rolling in his grave would be distracting.[26]

The Supreme Headquarters Allied Expeditionary Force originally planned to leave books promoting National Socialism on bookshelves. The idea was that these books were now discredited and would soon be replaced with new ones.[27] Indeed, the creation of new educational books was a priority. As Field Marshal Montgomery, military governor of the British Zone of Germany, put it, 'the text books with which the Nazis poisoned the minds of your children will not be used again in schools . . . New books, written by Germans in Germany and reflecting a wholesome spirit, are coming along.'[28]

However, guidelines were soon given for removing books from German public libraries. The Allied Control Council's Order No. 4 stated its aim to 'eradicate as soon as possible National Socialist, Fascist, Militarist and Anti-Democratic ideas in all forms in which they found expression throughout Germany'.[29] This was to be done by removing books, pamphlets, magazines, songs, slides and anything else containing pro-Nazi and anti-UN views from libraries, bookshops and publishers' offices. German library directors had to sign certificates stating that 'I fully understand that it is my responsibility to see that the library is completely denazified'.[30] However, when it came to purging private libraries of Nazi books, the Americans weren't so confident. The French proposal to remove books from private libraries was blocked by US General Lucius Clay. He worried this was 'an invasion of individual rights to which he could not agree.'[31]

The Allies kept tiptoeing around the issue. They claimed that a list of 1,000 books they issued (the Soviet list was fifteen times as long)[32] was not a 'blacklist' but just for illustrative purposes. However, they were clearly ordering the destruction of books. Thousands of volumes were removed and destroyed by German libraries. Burning the books would have been a bad look, so strict orders were given not to do this. They were pulped instead. When

President Roosevelt had claimed 'books never die', he hadn't been entirely correct.

The order to destroy these books made the front page of the *New York Times* and led to an editorial claiming the order imitated and exceeded Hitler. The point was also made that the destruction of books would signal to the Germans that Hitler had been on the right track when he burnt books. Many Germans thought the Allies' destruction of books was a bad idea too. One German librarian argued that 'ideas have to be killed by ideas, not by governmental orders.'[33] Yet there were some immediate benefits. Due to paper shortages, pulping of books did help print new books.

Overall, the book purge did not and could not achieve the re-educational aim of 'leading the Germans to think and feel as democratic individuals'.[34] None of the Allied military governments of Germany gave the German people the freedom to read. This made them subjects, not citizens. Despite the work that Malik and colleagues were doing to create a right to freedom of thought, it was not something the German people were granted at the time. It was an inauspicious start.

Beyond the Universal Declaration

After the Universal Declaration of Human Rights had been created, the next job was to create a binding covenant of human rights. Because this would have the status of international law, it took much longer for everyone to agree on this text. The resultant International Covenant on Civil and Political Rights would finally come into force in 1976. Most countries have ratified this covenant, meaning they are obligated to protect and preserve the human rights set out in this treaty. Germany ratified it in 1973, the UK in 1976, India in 1979, and the US in 1992, yet some countries, most notably China, have still not signed on.

Of course, there is a difference between ratifying the treaty and buying into its values. Anyone who believes that countries will respect the international human rights of citizens of other countries when national security is at stake is deluded. Take the following statement from Michael V. Hayden, CIA director from 2006 to 2009, written in the context of widespread praise for Edward Snowden for his disclosures of classified materials:

> Let me put this in the starkest possible terms . . . if you are not protected by the U.S. Constitution, and your communications contain information that would help keep America free and safe, information that would not otherwise be made available to the U.S. government, then game on. Your privacy is simply not the concern of the NSA director.[35]

So much for international human rights.

Today, the right to freedom of thought is found in most countries' national human rights legislation. In Europe, it appears in Article 9 of the European Convention on Human Rights. In the United States, constitutional law technically provides no direct protection for freedom of thought.[36] Nevertheless, the First Amendment is widely understood to protect an individual's thought from government interference.[37] Although the First Amendment does not mention freedom of thought explicitly, US courts have noted that a concern for freedom of thought was present at the time of its adoption.[38] Subsequently, US courts have explicitly referred to a 'First Amendment right to freedom of thought'.[39] Indeed, the US Supreme Court has stated that the 'purpose of the First Amendment is to foreclose public authority from assuming a guardianship of the public mind.'[40]

Reflecting the value put on the right to freedom of thought, both the International Covenant on Civil and Political Rights and the European Convention on Human Rights deem it an absolute right. This means it cannot be interfered with in any circumstances. Article 4(2) of the International Covenant on Civil and Political Rights sets out the 'fundamental character' of this right. This is 'reflected in the fact that this provision cannot be derogated from, even in time of public emergency'. In the United States, scholars argue that the right to freedom of thought is as close to an absolute right as any that exists in the Constitution.[41] Yet occasional statements by US courts have undermined the idea that the First Amendment offers absolute protection for thought.[42] For example, *Doe v. City of Lafayette, Indiana* (2003) raised the idea of an 'exercise of determining what state interests might outweigh a person's right to think'.[43] This is a case we will return to later.

Some countries have not made freedom of thought an absolute right, which should make their citizens nervous. The Canadian Charter of Rights and Freedoms guarantees freedom of thought but does not state this is an absolute right. Instead, it states it is 'subject only to such reasonable limits prescribed by law as can be demonstrably justified in a free and democratic society'.[44] Similarly, the South African Constitution also fails to categorise freedom of thought as an absolute right.[45]

Despite law emphasising the existence of the right to freedom of thought, the meaning of the right remained undefined and unde-cided. As the legal scholar Jan Christoph Bublitz puts it, in the more than fifty years of the European Court of Human Rights, it 'has only decided a handful of cases regarding freedom of thought . . . [it] is an almost empty declaration. There are no defi-nitions over its meaning, scope or possible violations.'[46]

Finally, in 2021, the content and scope of the right was finally discussed at the level of the UN in the Special Rapporteur's report on freedom of thought. The report pointed to four possible elements of the right:

1. Not being forced to reveal one's thoughts;
2. No punishment and/or sanctions for one's thoughts;
3. No impermissible alteration of one's thoughts; and
4. States fostering an enabling environment for freedom of thought.[47]

All these elements of the right, except the fourth, have featured in the courts over the years. Such cases give us a better feel for what the right to freedom of thought is and isn't understood to mean.

Not being forced to reveal one's thoughts

In the 1890s, photography's widespread availability led to people's lives being vividly splashed across newspapers. Sometimes the papers intruded on the wrong people, those who could fight back. One such person was Mabel Bayard. After her wedding, photos of the happy day began appearing in the American press. Mabel was not happy about this. Unfortunately for the press, she was the daughter of a US Senator. Even more unfortunately for the press, her new husband, Samuel Warren, had come second in his class at Harvard Law School. And the cherry on top of the press's misfortune was that the person who had come top of Warren's Harvard class, and who would later sit on the US Supreme Court, Louis Brandeis, was the best man.

Such intrusive press practices led Warren and Brandeis to introduce the idea of a right to privacy, also known as the 'right to be let alone'.[48] Their conception of privacy was broad and covered people's 'thoughts, sentiments and emotions'.[49] Ever since, US courts have signalled that an individual's Fourth Amendment right to privacy includes mental privacy. For example, in 1986 California's Supreme Court stated that 'if there is a quintessential zone of human privacy it is the mind'.[50] More generally, the US Supreme

Court has stated that 'the Constitution protects the right . . . to be generally free from governmental intrusions into one's privacy and control of one's thoughts'[51] and that the constitutional right to privacy 'is broad enough to include the right to protect one's mental processes from governmental interference'.[52] In the US, privacy laws shield thought.

Whereas the American right to privacy stemmed from a husband trying to uphold the dignity of his wife, the French took a slightly different route, although it also involved photography. The French came to a right to privacy via a man trying to get control over saucy photos he'd had taken of himself. The right to privacy in France stems from what is known as the Dumas case.[53] Alexandre Dumas, of *Three Musketeers* fame, had posed for photos with a twenty-three-year-old actress and, as it was France, her mother was there too. These were pretty risqué photos. In some shots Dumas was not even wearing his jacket. The photographer obtained the copyright for the images and then sold them to other people. Dumas was not happy with this, but as he didn't own the photos, there wasn't much he could do. However, a Parisian court decided that Dumas had a 'right to privacy' that he could invoke. Privacy rights, the court proclaimed, could sometimes trump property rights. This brings us to an important distinction between privacy in the US and Europe. For Americans, privacy is about freedom, but for Europeans, privacy is about dignity.[54]

Today, European human rights law recognises a right to mental privacy. Article 8 of the European Convention on Human Rights ('Right to respect for private and family life') protects a space, whether it be 'the head or the home', where people can be free from unwanted intrusions.[55] But this right to privacy has limits and can be violated. Yet if mental privacy were deemed to come under the right to freedom of thought, it would get absolute protection.

No punishment for thought

The idea that we should not be punished for our thoughts is ancient. The forerunners of the Seven Deadly Sins were the less marketable 'eight tempting thoughts' identified by the Christian monk Evagrius Ponticus (345–99). Evagrius claimed that eight tempting thoughts rode in on the wake of an initial evil thought, love of self. These eight were gluttony, fornication, love of money, sadness, anger, listlessness, vainglory and pride. Evagrius was clear that the thoughts themselves were not sins. They couldn't be because, for something to be a sin, you had to be responsible for it and consent to it. And, as Evagrius put it, 'it is not in our power to determine whether we are disturbed by these thoughts, but it is up to us to decide if they are to linger within us or not and whether or not they are to stir up our passions.'[56] His position was much like that of Martin Luther. Today, we would agree that we are not responsible for first-order thoughts of the type Evagrius discussed, but only our second-order thoughts about them.

Despite Evagrius's perspective on matters, the New Testament seemed to deem individuals morally responsible for their thoughts. As the Gospel of Matthew (5:28) put it, 'whosoever looketh on a woman to lust after her hath committed adultery with her already in his heart.' Yet a maxim of Roman law was *cogitationis poenam nemo patitur* (no one shall be punished for his thoughts alone).[57] This maxim was widely adopted, with only occasional deviations. Even in the sixteenth century, it was said that Queen Elizabeth I, 'not liking to make windows into men's hearts and secret thoughts, except the abundance of them did overflow into overt and express acts or affirmations, tempered her laws so as it restraineth only manifest disobedience'.[58]

Naturally, there were some occasional violations of the Roman maxim. The upheavals of seventeenth-century England saw the jurist Sir Edward Coke propose that, at least in the case of treason,

148

thoughts alone could be punished. As he put it, *voluntas reputatur pro facto* (the will is to be taken for the deed).[59] A similar aberration occurred in the 1683 trial of Algernon Sidney. Accused of treason against Charles II, Sidney could not be convicted by law unless two accusers could be found.[60] In the absence of a second accuser, his unpublished private writings were introduced as a second 'witness' to his crime. This was done using the novel principle that *scribere est agere* (to write is to act). Furthermore, the Chief Justice, George Jeffreys, argued that people should not curse the king, even 'in thy thoughts'.

Despite such periodic exuberances, today it is a truth almost universally acknowledged among lawyers that the law should not punish people for their private thoughts.[61] Yet, on the face of things, the law looks like it often punishes thoughts. A famous principle of criminal law states '*Actus reus non facit reum nisi mens sit rea*' (an act does not make a defendant guilty without a guilty mind). This means that every crime with a *mens rea* component technically involves people being punished for thoughts, specifically their intentions.

Hate crime laws also appear to punish people for happenings in their heads. As the legal scholar Robert Corry Jr puts it, because hate crime laws 'seek to punish certain unacceptable prejudiced thoughts, more accurate terms for these laws are "thought crimes"'.[62] Corry gives an example of this from the case of *Apprendi v. New Jersey* (1999).[63] Charles Apprendi was deemed to have had hateful thoughts whilst committing a crime. Due to this, he was given a sentence twenty percent higher than the maximum sentence he would otherwise have received.

Corry warns that punishment for thoughts may be a 'cure worse than the problem', if the cure involves 'an Orwellian police state where government agents probe into the deepest, darkest recesses of our minds, our thoughts, and our pasts, seeking any evidence, however circumstantial, that we have harbored improper thoughts or emotions'. As a result, he argues that governments 'should tread more cautiously in the area of our thoughts'.

The way the law tries to get around the idea that hate crime cases involve punishing people for their thoughts is by claiming that people are not being punished for 'mere thought'. For the person to be punished in such cases, they must have already taken substantial steps towards performing a criminal act. This represents punishment for thought plus action, not mere thought. Without such steps having been taken, US law recognises 'the fundamental constitutional principle that a person's thoughts are his own – however distasteful they may be to the state or to the populace'.[64] As US courts have noted, 'one would not punish a person with a history of bank robbery for simply standing in the parking lot of a bank thinking about robbing it.'[65]

A good example of a court deciding that people can't be punished for 'mere thought', even when the thoughts in question are 'offensive', 'repugnant and deplorable', comes from a series of cases involving *Doe v. City of Lafayette*.[66] Doe, a convicted sex offender, was driving home from work one day when he began to have sexual thoughts about children. He then drove to a park in the city of Lafayette where he:

> watched several youths in their early teens playing on a baseball diamond. Doe admits that, while observing them, he thought about having sexual contact with the children. After watching them for fifteen to thirty minutes, and without having any contact with them, Doe left the park. Because he was upset about the incident, Doe contacted his psychologist to report the incident. He also reported the incident to his self-help group.[67]

An anonymous source (potentially from Doe's Sexual Addicts Anonymous group)[68] reported Doe's thoughts to his former probation officer, who in turn contacted the Lafayette Police Department.

Eventually, a district court decided to ban Doe from entering any city parks, even though he was neither serving a sentence nor on probation.

Doe appealed against this ban to the US Seventh Circuit Court of Appeals in 2003. The court observed that:

> we are faced with a question not typically before a court: may a city constitutionally ban one of its citizens from public property based on its discovery of that individual's immoral thoughts? This scenario is quite unusual, as it is a rare case where thoughts, as separated from deeds, become known. Technology has not yet produced a mind-reader, and thus most thinking, unless purposefully revealed to others, remains one's own. Unlike other cases in which the state becomes aware of an individual's mental state because of his or her actions, here the City acknowledges that Doe's own revelation of his thoughts, not any outward indication of his thinking, is the basis for its actions.[69]

The court also pointed out that presumably many of the people of Lafayette wandered the city's parks thinking offensive or objectionable thoughts. The judges concluded that Doe's actions didn't reach a level justifying punishment and that being in the park, thinking sexual thoughts about children, did not qualify as a 'substantial step' towards an attempted sex offence. The court reversed Doe's ban from city parks, ruling that such a ban would punish Doe for 'thinking alone' and therefore violate his First Amendment rights.[70]

There was, however, a dissenting judge. This judge argued that the city had 'adopted a reasonable proscription designed to protect a vulnerable part of the population, its children, against the danger of a relapse by Mr. Doe'. However, this judge was still not advocating Doe be punished for his thoughts, but rather for engaging in 'a sort of psychiatric brinksmanship by placing himself in a situation that increased substantially the possibility of his acting on these impulses'.

When the case came back to the court the following year, the rest of the judges now agreed with this dissenting judge and reinstated Doe's ban.[71] As the court again explained, Doe was not being punished for 'mere thought'. The ban was not introduced because he 'admitted to having sexual fantasies about children in his home or even in a coffee shop'. He was banned because he 'did not simply entertain thoughts; he brought himself to the brink of committing child molestation'.

A similar conclusion about the inappropriateness of punishing thoughts was reached in the case of thirty-one-year-old Gilberto Valle.[72] Valle was arrested after his wife examined his computer. She found he had been discussing 'butchering her and raping and torturing other women whom they knew', as part of an internet sex community called Dark Fetish Network. He had also been having 'chats' about cannibalising various women. Valle was acquitted of conspiracy to kidnap. In response to an appeal (which upheld Valle's acquittal), Circuit Judge Barrington D. Parker stated that this was a case 'about the line between fantasy and criminal intent'. Judge Parker stressed that although this line was increasingly difficult to identify in the internet age, 'it still exists and it must be rationally discernible in order to ensure that "a person's inclinations and fantasies are his own and beyond the reach of the government."' The judge clarified that 'Fantasizing about committing a crime, even a crime of violence against a real person whom you know, is not a crime.' However, interestingly, given the argument that thoughts should not be punished because they cannot harm others (which we will shortly discuss), Judge Parker added that 'this does not mean that fantasies are harmless. To the contrary, fantasies of violence against women are both a symptom of and a contributor to a culture of exploitation, a massive social harm that demeans women.' Nevertheless, he concluded that 'we must not forget that in a free and functioning society, not every harm is meant to be addressed with the federal criminal law'.

A recent example in the UK echoed Judge Parker's conclusion. A Scottish Working Group on Misogyny and Criminal Justice examined whether malign behaviour towards women should be considered a hate crime.[73] The report they produced stated that:

> It is the conduct that flows from hatred that can be criminal, but what goes on in our heads cannot and must not be criminalised . . . We were very clear from the outset that misogyny itself must not be criminalised as it is a way of thinking and freedom of thought must remain sacrosanct. 'Thought crime' is the stuff of totalitarianism and people in a free society have to be free to think the unthinkable.

Despite this forthright defence of freedom of thought, one cannot help but look back to Judge Parker's warning about the real-world harms that stem from fantasies of violence against women. It is one thing to put freedom of thought on a pedestal, but another for that pedestal to be made of pain.

The law is hence clear that, however despicable one's thoughts, one should not be punished for them. The extent to which the public is happy with this, as I will explore later, will probably determine whether such a position remains tenable.

No impermissible alteration of one's thoughts

In the wonderful words of the author Kathleen Taylor, 'Human beings have been trying to change each other's minds since they first discovered they had them'.[74] One of the most important questions we face when developing the right to freedom of thought is where to draw the line between what we deem the permissible and impermissible alteration of people's thoughts.

Governments continually influence our thoughts in ways that we accept as fair. This may be through the education system or public information campaigns. But the law is clear that the government cannot simply influence our thoughts in any way it chooses. The US Supreme Court has repeatedly ruled that the government cannot interfere with people's thoughts. A classic example is the 1969 case of *Stanley v. Georgia*.[75] Here the US Supreme Court ruled that the defendant could not be forbidden to own a pornographic film (involving consenting adults) because to do so would be for the government to claim the right to 'control the moral content of a person's thoughts'. The Court made clear the depths of its aversion to regulating thoughts. As it famously stated, 'our whole constitutional heritage rebels at the thought of giving government the power to control men's minds'.

Whilst most of us would agree with this ruling, we may feel more uneasy about the government being forbidden to interfere with thoughts we find abhorrent. In such cases, we may be more open to the idea that some thoughts should be regulated. For example, in 2002 the US Supreme Court ruled that a person could not be punished for possessing visual depictions of an actor who appeared to be a minor engaging in actual or simulated sexual intercourse.[76] Simulated child pornography was deemed not to involve direct harm to a child. As a result, the Court felt that the only reason to ban simulated child pornography was because it negatively affected viewers' minds. Any ban would be made because the material 'whets the appetites of pedophiles and encourages them to engage in illegal conduct'. The Court decided the material couldn't be criminalised for this reason and concluded that 'the government cannot constitutionally premise legislation on the desirability of controlling a person's private thoughts.' As we will see later, such cases could be used to move the public in favour of regulating minds.

Outside the US, the European Union's 2009 Charter of Fundamental Rights introduced the idea of a 'right to mental integrity'. Article 3.1 of the Charter states that 'Everyone has the right to respect for his or her physical and mental integrity.' But the EU left it unclear what exactly they meant by 'mental integrity'.[77] One scholar has since defined mental integrity as 'the right to be free from unwanted mental interference or manipulation of a direct and forcible sort'.[78] But how exactly are we going to define 'manipulation'?

What courts determine to fall into the category of 'impermissible alteration' of people's thoughts will be hugely consequential. To get a sense of what sort of things are likely to be put into this category, we can look at what the legal scholar Manfred Nowak has given as examples of what he terms 'impermissible external influence'.[79] These include indoctrination, brainwashing, psychoactive drugs or other means of manipulation. Nowak also claims influencing others is impermissible 'when it is performed by way of coercion, threat or some other prohibited means against the will of the person concerned or without at least his or her implicit approval'. The European Court of Human Rights has also waded in on this matter. They suggest that improper means of influencing thoughts include exerting improper pressure on people in distress or need, violence, brainwashing and unreasonable propaganda.[80] However, the Court has stayed clear of defining 'unreasonable propaganda'. It is profoundly disconcerting to know that a court could deem any communication to be 'unreasonable propaganda' and thereby restrict free speech in the name of free thought.

One factor a court has already ruled to be an illegitimate form of thought manipulation is using a power imbalance to influence other people's thoughts. In a 1998 case, the European Court of Human Rights agreed with the conviction of high-ranking army officers who tried to convert lower-rank soldiers to Jehovah's

Witnesses.[81] The hierarchical structure of the army meant that these senior officers could be deemed to have 'improper' undue influence. This may yet turn out to be a very important ruling. Consider, for example, the power imbalance between you and Big Tech. Do you think their influence upon you could be deemed improper undue influence? If so, then Big Tech could be legally forced to redesign both its interfaces and its algorithms.

To get a handle on what should count as an illegitimate alteration of thoughts, we need to distinguish between coercion and manipulation. Coercion is using a power disparity to narrow someone's choice to a single rational option that accords with your desire. Someone uses a disparity in physical power by putting a gun to your head and asking you for money. This leaves you with one rational choice – give them the money. Someone uses a disparity in economic power by offering you a job at less than a standard wage because they know you have a family to feed. You have one rational choice – take the job.[82] This is wrong because it damages the autonomy of the person who is the target of this coercion.

In contrast, manipulation degrades someone's ability to reason in order to increase the likelihood they will act the way you want. Again, this is wrong because it diminishes the autonomy and capacity for self-government of the target. Degrading someone's ability to reason can be done by corrupting the input to their mind. We will reason to erroneous conclusions if we have bad or incomplete information. As an ancient Chinese saying puts it, 'the meal you cook can never be better than the rice you cook it with'.[83] Or, as scientists more succinctly put it, garbage in, garbage out. Degrading someone's reasoning ability can also be done by inhibiting their rational processing capacity. Manipulation can hence target your inputs and your processing. We can link these ideas to our earlier discussion of the ARRC of free thought to create four elements of impermissible alteration of thought.

Impermissible manipulation of input

1. The presentation of corrupt information (lying, misrepresenting, encouraging false assumptions).[84]
2. The unacknowledged use of techniques known to substantially bias our attention.

Impermissible manipulation of processing

3. The disabling or circumvention of, or failure to engage or appeal to, our capacity for reflection and reason sufficiently.[85]
4. The use of inappropriate pressure to discourage us from pursuing a line of inquiry (including guilt, fear, greed, self-esteem, belonging and gratitude).[86]

Clearly, the devil is in the detail of how we interpret the terms 'substantially', 'sufficiently' and 'inappropriate', which each open a significant grey area.

At the far impermissible end of the thought-alteration spectrum lies subliminal advertising. We cannot attend to such information, and our lack of consciousness of the stimuli means we cannot reflect on and reason about the message.[87] If information does not enter consciousness, the brain cannot broadcast it to all neural areas, reducing its ability to process the material. Beyond subliminal messaging, the permissibility of many other forms of communication is more debatable. Let's consider potential violations of principles 2, 3 and 4, listed above. I will not address the first item here, the presentation of corrupt information, as we will explore this in chapter 5.

First, let's take a relatively trivial example of techniques that bias attention, other than the well-known machinations of social media. People can shape our eating habits using the knowledge that our decision making is more impacted by that to which we first attend. The American legal scholar Cass Sunstein gives an example of this from a study that examined the effects of how restaurants present nutritional information on menus. One menu gave the calorie content first, then the food's description (e.g. 550 kcals, Hamburger). The other menu had the food's description first, then the calorie count (e.g. Hamburger, 550 kcals). The researchers found that, on average, diners with the calories-first menu ordered food with sixteen percent fewer calories than those with the calories-last menu.[88]

What seems like a trivial bias is not so trivial when candidates for political office are on the menu. Voters, it is said, are not fools. But you don't need to be a fool to be fooled. If we attend more to the first thing on a list, there is a conceivable advantage to being the first candidate named on a ballot paper. Some countries, such as Ireland, require candidates to be listed in alphabetical order on a voting ballot. However, US courts have ruled that an alphabetical ordering of candidates is unconstitutional because being named first on the ballot secures a slight advantage.[89] As a result, many states in the US require the order of names on ballots to be chosen at random, with the order of names then being rotated across voting districts.[90]

Depending on the type of election, being listed at the top of a ballot may be able to boost your share of the vote by between one and five percent.[91] As one study concluded, 'a number of Danish local and regional councilors [sic] owe their council seat to their position on the ballot paper'.[92] This may be a small effect, but as we noted in the case of microtargeting, small effects can still swing elections. A party in power should not be allowed to place their candidate first on a ballot, as some countries currently do.[93] Requiring that each ballot in an election has a randomised order of candidates' names, easily done in this digital age, would respect voters' right to freedom of thought.

Let's now take an example of a failure to engage or appeal to reflection and reason sufficiently. Earlier, we saw how suggesting a low minimum repayment on a credit card can encourage us to pay off a lower amount than we otherwise would. This works by activating a specific rule-of-thumb within us, namely the anchoring heuristic. Is this an impermissible alteration of thought? Arguably, when presented with such a credit card bill, we still have the capacity to engage rule-of-reason thought about how much to pay off. If we were given the statement and told we had fifteen seconds to decide, our capacity to reflect and reason would have been impaired. Yet we typically have weeks to decide how much to pay off. However, if the credit card statement disclosed that its presentation of a low minimum repayment could bias us to pay off a smaller amount, our reflection and reasoning would be more engaged. If a company makes a deliberate design choice known to engage rule-of-thumb thinking, they should be mandated to disclose this to consumers.

Moving away from consequential financial decisions, what about attempts to engage rule-of-thumb thinking to encourage us to make seemingly more trivial purchases? Take the use of celebrities in adverts for sugary drinks. The use of celebrities encourages us to view the products they advertise as markers of status, thereby making the product more attractive to us. This attempts to sidestep our use of rule-of-reason thought to assess the health or social benefits of the drink to us. Our capacity to reason is not diminished, but nor are we encouraged to employ it.

So, is the use of celebrities in adverts an impermissible alteration of thought that should be regulated? We may worry that we are reaching the point where the costs of onerous regulation exceed the benefits gained.[94] We may also worry that we are violating people's right to free (commercial) speech, or simply being killjoys. I'm going to come down on the side of this clearly being manipulative and therefore requiring regulation of some form. But what about the slippery slope this could lead us down, one may ask? Sometimes worrying about where we may end is merely a distraction from the more pressing question of where we should start.

Finally, let's consider inappropriate pressure to discourage us from pursuing a line of inquiry. Take the case of an employer firing a person because of a political conversation the employee had with another person, in which they were thinking aloud together. The employee's dismissal has used inappropriate pressure to chill thought in both the staff member who was fired and the staff that remain. It has raised an unnecessarily high barrier of courage to thought. This should be deemed impermissible alteration of thought. Employees, even those under at-will employment, should not be able to be fired for thought. Thinking should be a protected characteristic.

There are a range of reasons it is hard to assess whether thought is being impermissibly altered. As Cass Sunstein asks, what if someone's thoughts are manipulated for their own good?[95] For example, let's say that Andrew has reasoned and reflected and has come to an erroneous conclusion. Perhaps he has concluded that smoking won't make that much difference to his health in the long term. Now let's say that Barry, a government-employed health official, comes along, realises that there is a high probability that smoking will damage Andrew's long-term health, and knows that Andrew wants to be healthy. Barry first tries to reason with Andrew and fails. Barry then switches to appealing to Andrew's rule-of-thumb thinking by mentioning high-profile people who have died of lung cancer, playing into Andrew's availability heuristic, and manages to convince him to stop smoking. Has Barry now impermissibly altered Andrew's thoughts and violated his right to think freely?

The traditional objection here is that Barry does not know what is in Andrew's interests better than Andrew does. Maybe Barry fails to account for how much Andrew enjoys smoking. But Barry cannot always be wrong. Ultimately, to protect the dignity of

Andrew, we may be forced to forbid the use of any manipulative attempts at influence. The promotion of human dignity may do more for humanity in the long run than reducing lung cancer rates would do in the short term. One exception to this rule could be if Andrew asks to be manipulated to help him stop smoking, as he recognises his weakness of will.[96] In this case, manipulative actions would support rather than undermine his autonomy.

Yet there are plenty of other issues swirling around. Was Andrew manipulated by cigarette companies into smoking in the first place? Would two manipulative wrongs make a right or violate one? Maybe we should be focusing our attention further upstream. If Andrew is addicted to nicotine, is he capable of thinking freely about smoking? Would manipulating him out of smoking *restore* rather than restrict his freedom of thought? Do we need to balance people's right to freedom of thought against their right to life? Such issues highlight how hard it is to judge if thought has been impermissibly altered.

Recent years have seen new legislation that has the effect of protecting our right to freedom of thought in the digital world. As my legal colleagues have pointed out to me,[97] in 2022, the European Union passed a new regulation, the Digital Services Act. Article 25(1) of this Act states that:

> Providers of online platforms shall not design, organise or oper-ate their online interfaces in a way that deceives or manipulates the recipients of their service or in a way that otherwise materi-ally distorts or impairs the ability of the recipients of their service to make free and informed decisions.[98]

The EU introduced this provision due to its concerns about the effects of dark patterns. Although the Act does not explicitly refer to the right to freedom of thought, it nonetheless protects it. Yet what counts as deceiving and manipulating users, or impairing their ability to make free and informed choices, still lacks a coherent underpinning. As my colleagues and I have argued, a well-elaborated account of the right to freedom of thought will give us a firmer basis for identifying and justifying what counts as a violation of this right.[99]

Who does the right to freedom of thought protect us against?

In the last chapter, I talked about how governments, corporations and our fellow citizens threaten freedom of thought. Yet the US Constitution only protects people's freedom of thought from government interference. We must think about how the law can or should protect us from the remaining two threats.

The United States' Founding Fathers were well aware of the threats that corporations could pose. As Thomas Jefferson put it, 'I hope that we shall crush in its birth the aristocracy of our monied corporations, which dare already to challenge our government to a trial of strength, and bid defiance to the laws of our country.'[100] Unfortunately, whilst the US Supreme Court sees demons lurking in government, it has a more challenging time sensing them in corporations.[101] As one legal scholar has warned, we need to take corporations as seriously as government, recognising that they are our institutions, under our control.[102]

International human rights law puts direct obligations on states but not corporations. States must not violate human rights themselves. But states must also protect the rights of their citizens from interference by non-state actors.[103] As the United Nations Human Rights Committee explains, states will only live up to their obligations under international human rights law if 'individuals are protected by the state, not just against violations of Covenant rights by its agents, but also against acts committed by private persons or entities.'[104] This means that both corporations and our fellow citizens have an indirect duty not to violate our right to freedom of thought, which states need to enforce.

Although calls for human rights law to apply directly to corporations have not had much success, there is evidence of change.[105] In 2011 the United Nations Human Rights Council endorsed the Guiding Principles on Business and Human Rights. Guiding Principle 11 states that 'business enterprises should respect human rights. This means that they should avoid infringing on the human rights of others and should address adverse human rights impacts with which they are involved.'[106] It is hardly a mystery why states have largely failed to enforce this Principle. If money was taken out of politics, and states did enforce this Principle, our world would look very different.

In summary, the right to freedom of thought emerged as part of a unifying rejection of fascist ideology. However, its development reflected a Western conception of thought, which we need to rectify. The UN sees the right to freedom of thought as involving mental privacy and neither punishing nor manipulating thoughts. Yet such concepts have primarily been discussed through the lens of the law. We need to integrate this legal conception with a psychological understanding of thought. We can do this by using what we have learnt about thought, what makes it free and what threatens this freedom. Creating a clear and comprehensive account of what the right to freedom of thought involves will help governments protect this right from interference by citizens, companies and itself.

5

CREATING FREE
THOUGHT I: CLAY

On the evening of Saturday, 25 September 2010, engineers at JSTOR, a digital library providing online access to scholarly works, noticed something unusual. Someone at the world-leading Massachusetts Institute of Technology (MIT) was pulling an all-nighter that was beyond heroic. Nearly half a million academic papers were downloaded between 5 p.m. on Saturday night and 4 a.m. on Sunday morning.[1] As even good Will Hunting couldn't read that much in a night, JSTOR were suspicious. They contacted MIT, who promptly blocked the unidentified individual's access.

Shortly before Christmas, the downloading frenzy started again. By January, the mystery individual had downloaded nearly five million academic papers.[2] JSTOR realised someone was trying to download every paper in their archive. But MIT now had a lead. They knew from where these documents were being downloaded. On 4 January 2011, one of MIT's engineers began searching the pinpointed building. They found a suspicious cable, which they followed into a closet in the basement. Inside the closet was a box which, when lifted, revealed a whirring laptop underneath. It was 8 a.m.

By 11 a.m., a detective from the Boston Police Department, three members of the New England Electronic Crimes Task Force and a US Secret Service special agent were peering into the now slightly overwhelmed closet. The authorities decided to set a trap.

They left the laptop sucking data, rigged a camera facing the closet, and waited. They did not have to wait long. Thirty minutes later, the camera picked up someone entering the closet. The figure changed the laptop's bulging hard drive and began to head out again. But law enforcement was myriad, caffeinated, and ready. They pounced. And caught precisely no one.

Two days later, the camera again picked up a figure entering the closet. This time the law enforcement officers swiftly swooped in. And missed again. Frankly, it was amazing that the closet hadn't escaped to Mexico by this point. Later that afternoon, an MIT police cruiser would fortuitously spot a cyclist matching the suspect's description. The officer approached the cyclist, showed his badge, and asked to speak with him. 'I don't speak with strangers', replied the stranger-danger-aware cyclist, before dismounting and running away. The officer gave chase but, in keeping with the progress of the investigation, couldn't catch up. After calling for backup, other officers arrived, chased the cyclist through parked cars, and finally caught and handcuffed him.

'Information is power', the cyclist had written three years before his arrest, 'but like all power, there are those who want to keep it for themselves.' These words formed the opening lines of his Guerrilla Open Access Manifesto.[3] Two years later, the cyclist would hang himself in his Brooklyn apartment. He was only twenty-six. His name was Aaron Swartz. He left no note, but the world would take note.

Swartz was a computing prodigy. In his early teens, he helped develop the RSS web-feed system. In his late teens, he co-founded the website Reddit. Swartz would dedicate his life to fighting the institutional control of information.[4] He opposed academic publishing houses requiring universities and the public to pay to access scientific research papers. Much of this research had already been paid for by the taxpayer. As Swartz put it, 'forcing academics to pay money to read the work of their colleagues? . . . Providing scientific articles to those at elite universities in the First World, but not to children in the Global South? It's outrageous and unacceptable.'[5]

Swartz felt that the availability of scientific research was a human rights issue.[6] In 2013, he was posthumously awarded the American Library Association's James Madison Award for his work. This prize honours people who have 'championed, protected and promoted public access to government information and the public's right to know'.[7] Somewhere in the next life, Swartz is probably smiling at the irony of John Podesta having been the 2001 recipient of this award.

Swartz wasn't the only person believing that access to scientific research was a human rights issue and prepared to do something about it. The same year he was arrested, a Kazakhstani post-graduate student, Alexandra Elbakyan, was struggling to freely access academic research in Kazakhstan.[8] She created the website Sci-Hub. Elbakyan placed millions of pirated research papers and books on this site, allowing anyone to access this research for free.

Elbakyan's actions have been described as 'an awe-inspiring act of altruism or a massive criminal enterprise, depending on whom you ask'.[9] Clearly, her website violated copyright law. In 2015, a New York court awarded one of the largest academic publishing houses $15 million in damages. Elbakyan's actions, the publisher's lawyers said, had caused 'irreparable injury' to, amongst others, the public.[10] 'This ruling should stand as a warning', said a spokesperson for the International Association of Scientific, Technical and Medical Publishers, 'to those who knowingly violate others' rights'.[11] Yet Elbakyan defended her actions on human rights grounds. Article 27 of the Universal Declaration of Human Rights, she pointed out, gives us the right 'to share in scientific advancement and its benefits'.[12]

The activities of Swartz and Elbakyan fed into a larger movement calling for free access to scientific research. This Open Access movement aimed to make the research that 'scholars give to the world without expectation of payment' freely available to every-one.[13] Today, Open Access publishing involves scientists' funders paying publishing houses to make research freely available. Public repositories such as bioRxiv and arXiv also make early drafts of scientists' research papers freely available. This makes it less likely that we will click on a link in an interesting tweet only to find the relevant evidence hidden behind a paywall. Patients can access research into their diseases. Activists can access research to support their cause. Anyone can check anyone else's claims. We can think.

Swartz, Elbakyan and the Open Access movement echo a call made by the science-fiction writer H. G. Wells more than seventy years ago. In 1940, Wells proposed his own Declaration of Rights. His fourth article proposed a 'right to knowledge'. Wells argued that it was the duty of the community to equip everyone with suffi-cient education to enable them to be 'as useful and interested a citi-zen' as they could be. As part of this, he argued that people should have 'easy and prompt access to all information necessary . . . to form a judgment upon current events and issues.'[14]

Although no such 'right to knowledge' exists today, it seems clear that such a right should be embedded in the right to freedom of thought. Without clay, sculptors cannot sculpt. Without infor-mation, humans cannot think. Those who block, conceal or restrict our access to information impede our ability to develop our mind. A right to freedom of thought that doesn't address our ability to access information would be like a right to car ownership that doesn't mention fuel.

A right to information

A democracy that fails to give citizens access to the information necessary to cast an informed vote is not a democracy. Without such information, citizens may feel they are making autonomous choices. If an oligarchy controls the flow of information, then they are steering voters' choices. An invisible hand lies on our tiller. The Founding Fathers of the United States recognised this. As James Madison put it, 'people who mean to be their own governors, must arm themselves with the power which knowledge gives ... A popular Government, without popular information, or the means of acquiring it, is but a prologue to a Farce or a Tragedy; or perhaps both.'[15] All that remains to be debated about a society that restricts information is whether it is a tragic farce or a farcical tragedy. It is certainly not a society in which the people govern themselves.

Hindering access to information also stymies human creativity. As the co-founder of Apple, Steve Jobs, pointed out, creativity is just connecting things. Indeed, the Latin verb 'to think' (*cogitare*) comes from another Latin term meaning 'to shake together' (*coagitare*).[16] But to connect things, you must first know things. Access to information is essential for the innovation that allows individuals and societies to flourish.

Legal scholars recognise the link between free thought and free access to information. Judge Loukis Loucaides, a former member of the European Court of Human Rights, claims that 'it cannot be emphasised enough, that a prerequisite to the exercise of freedom of thought, is the effective exercise of the right to freedom of information'.[17] Yet a right to information is only one element in the broader class of rights that the scholar Lani Watson calls epistemic rights. Watson defines epistemic rights as 'concerning goods such as information, knowledge, and truth'.[18] An example of an epistemic right is a consumer's right not to be misled by adverts.[19] Such rights are effectively 'knowing' rights. As Michael Bolton thankfully didn't put it, how can we be thinkers if we can't be knowers? We must build the right to freedom of thought on a foundation of subsidiary rights to information, knowledge and truth. Although this sounds straightforward enough, working out the practicalities is like tap dancing in a tar pit.

The first problem with a right to information is that the state's provision of information may hide ulterior motives. These include desires to pacify and fragment. We see this when we consider the dark side of public libraries. We may not all agree with the writer Jorge Luis Borges' vision that 'paradise will be a kind of library', but it is hard not to love public libraries.[20] Many of us will have fallen in love with libraries as children. As the actor John Goodman has recalled, 'when I was young, we couldn't afford much. But, my library card was my key to the world.'[21] The British Library has been a home from home for writers from Karl Marx to Colin Wilson. These libraries are trusted. A Pew Research poll found that seventy-eight percent of adults in the US felt that public libraries helped them find trustworthy and reliable information.[22] By supporting public libraries, governments fulfil part of their duty under the positive aspect of a right to information and support our right to freedom of thought.

Yet when we scratch below the surface, worries emerge about how much public libraries support freedom of thought. In the second half of the twentieth century, most people viewed public libraries as drivers of democracy. In 1960, President John F. Kennedy stated that if the US was to be both wise and strong, 'then we need more new ideas for more wise men reading more good books in more public libraries'.[23] Forty years later, the American Library Association still linked public libraries to democracy. As they put it, 'the public library is the only institution in American society whose purpose is to guard against the tyrannies of ignorance and conformity'.[24] It is unclear whether this was an intentional jibe at the US public school system.[25]

Yet governments did not create public libraries simply to help the public inform themselves. Governments introduced public libraries as a barrier to democracy, not a support. In the mid-nineteenth century the UK passed the Public Libraries Act 1850, which paved the way for free public libraries. At the time, only one in five men (and zero in five women) were allowed to vote. As the working-class Chartist movement was campaigning for all adult men to have the right to vote, the elites wanted to pacify the working classes. Libraries were one tool to do this, being thought to calm people, keep them out of pubs, and promote morality and order.

William Ewart sponsored the library bill in Parliament. He believed that public education could teach the lower classes 'the value of property and the benefits of order'. He also felt education was a 'great preventive measure' against crime.[26] As was said at the time, 'nothing would tend more to the preservation of order than the diffusion of intelligence'.[27] Another observation made at the time, which gives insight into the thinking behind public libraries, came from George Alexander Hamilton, the Tory member for Dublin University. Hamilton opined that the people of Ireland had not been given libraries 'from which proper books could be got' and that this was 'the reason why they were driven to read inflammatory publications'.[28] The government meant public libraries to put out fires, not start them. Public libraries were to be the camomile of the people.

Many in the United States shared this pacifying conception of public libraries. As an American librarian wrote in 1902, 'free corn in old Rome bribed a mob and kept it passive. By free books and what goes with them in modern America we mean to erase the mob from existence'.[29] Public libraries would, it was hoped, keep 'boys at home in the evening by giving them well written stories of adventure', civilise 'the conduct of men and . . . [suffer] them not to remain barbarians' and 'keep people out of bad company . . . [and] divert working-men from the street corner and the low, corrupting dram-shop'.[30] Some librarians explicitly discussed how libraries should aim to keep ideas of revolution away from the working class.[31] Yet other early librarians had more revolutionary visions. In 1897 Frederick Crunden wrote that the wisdom to revolutionise people's minds 'is not to be obtained from schools or colleges, but from the higher education of mature minds – the masses of people – which the public library alone can give.'[32]

So are public libraries 'cauldrons of democracy or the tools of social control?'[33] They are clearly repositories of information. Yet we often visit alone, read alone, and become atomised processors. As Abigail Van Slyck argues, the Carnegie programme of library building led to a narrowing of what libraries were. As a result, 'the efficiency-driven public library of the twentieth century defined reading as a solitary activity'.[34] Public libraries lost their potential to be a place for public discussion and debate.[35] Consider what public libraries partially replaced in 1850, aside from smaller, informal lending libraries. Before this, working-class men and women set up institutes where educated workers and middle-class reformers taught classes, gave lectures and discussed matters of the day.[36] In comparison, public libraries contributed to the diffusion of the masses.

The point here is not to denigrate our great public libraries but to show how providing information can be both a boon and a barrier to thought. We should approach the information orgy engendered by the internet with caution. When creating an internet ecology that supports freedom of thought, we need to take two tacks. The first is to ensure that government regulates internet companies that provide us with information to ensure that users' right to freedom of thought is respected. The second is to boost users' abilities to think freely on the internet. To achieve these goals, we need to understand how problems arise from the way in which information on the internet is presented by others and used by us.

Too much information?

If someone else controls the window through which we see the world, we become as dependent upon them as a prisoner is on a guard to see the sky. Yet in a world where Google indexes more than 130 trillion web pages,[37] we need people to act as curators and gatekeepers of information. Without this, the wave of information coming at us overwhelms rather than enlightens. Imagine visiting the Louvre only to find they had increased their collection to include every picture ever drawn or painted. Would this enhance or diminish your experience? The problem of how to deal with vast amounts of information is not new. Over a century ago, psychologist William James pointed out that there was so much information in books that 'the art of being wise is the art of knowing what to overlook'.[38] Nevertheless, in the internet era, this problem has reached new heights.

Search engines have become our window to the world, and Google our guard. 'The perfect search engine', said Google co-founder Sergey Brin, 'would be like the mind of God'.[39] Until someone develops such an engine, we are stuck with human, all-too-human search engines. Today, around ninety percent of internet searches are performed with Google, resulting in trillions of Google searches annually. With its stated mission to 'organize the world's information and make it universally accessible and useful', Google has become a digital library at our fingertips.[40] Thinking and googling have become intimately linked. How we think impacts how we google and how Google works impacts how we think. Each stage of the googling process has important implications for the information we obtain and our thinking.

It starts with our choice of terms to google. Here we immediately fall victim to one of the classic biases, confirmation bias. We tend to type searches into Google that confirm our beliefs rather than disconfirm them.[41] Suppose we believe that the death penalty reduces crime. In this case, we are more likely to google 'death penalty crime reduces' than 'death penalty crime ineffective'. The problem is that the phrase we choose impacts dramatically the results we get.[42] If you want to know if the TV ratings for football are up or down, googling 'football ratings up' will give you very different results from 'football ratings down'. In turn, this will affect what you think about the issue.[43]

Knowing how to use a search engine effectively improves your ability to think. Most of our queries contain only a couple of terms. This makes it harder for Google to find the most relevant answers.[44] We can search better if we are precise with our search terms, put critical phrases in quotation marks, and use a minus sign in front of terms we want to specify as not relevant. Many other tips are available.[45] Such search literacy is an essential support for thought.

Once we have entered our query, the decisions that Google's algorithms make about what information to present powerfully impact our thinking. Imagine going into a library, seeing shelving stacks heading off to infinity, and then simply taking the first book on the front shelf. How could you not think that you might be missing something? Yet this is what we do with Google. Around a third of all clicks on Google search results are made on the first result listed. More than ninety percent of clicks are made on the first page of results.[46] We eat what we are served. We do so because we trust that what Google shows us is most valuable to us, even though we typically have no idea how Google makes these decisions.[47] We rarely, if ever, go in against Google when truth is on the line.

If Google's algorithms became politically biased, they would change governments. Changing search engine results to favour one political candidate over another can lead to significant changes in people's views of candidates. Research suggests this could increase the vote for a candidate by just over one percent.[48] In many elections, this would be enough to swing the outcome. Perhaps most troublingly, voters would not even be aware it had happened. If it were to be the case, as President Donald Trump claimed in 2018, that Google had 'RIGGED' its search results against him, this would be concerning indeed.[49]

There does seem to be political bias in Google's search results, yet this is not Google's fault. Although some studies have reported political biases in Google search data,[50] most researchers argue that Google's search algorithm is not politically biased.[51] Instead, Google seems to reflect mainstream media sources. This means that if the mainstream media are politically biased, then so too will be Google's results.[52] In this context, the liberal bias of the mainstream press becomes even more problematic.[53]

Nevertheless, decisions made by Google can politically bias what information users can or cannot find. For example, in 2017, Google announced plans to keep users from accessing 'fake news'. As a result, traffic to many left-wing, progressive and anti-war organisations fell sharply. This included sites such as *WikiLeaks*, *Alternet*, *Counterpunch*, *Global Research*, *Consortium News* and *Truthout*.[54] Google's increased use of mainstream media in its search results made many valuable left-wing websites harder for readers to find. Readers would now be insulated from particular views. These included the view of *Consortium News*' Robert Parry, that the mainstream press 'believe in a "guided democracy" in which "approved" opinions are elevated – regardless of their absence of factual basis – and "unapproved" evidence is brushed aside or disparaged regardless of its quality'.[55]

In a way, Google's lack of inherent political bias makes sense. Google would lose much in bad publicity from having systematically biased results relating to social and political issues. Any personal differences in search results can be seen as 'collateral damage' rather than as a deliberate goal.[56] Furthermore, the idea that Google results will lead you into your own filter bubble is also not well-supported.[57] Neither does there appear to be a significant impact of personalisation on your search results, apart from your location.[58]

All this is rather fortunate as Google's ability to display search results in any way it wants to is effectively protected by the company's own right to free speech.[59] As one court ruled, Google's rankings are 'opinions of the significance of particular web sites as they correspond to a search query'.[60] As there was no way, the court noted, to determine if the significance of a website was falsely estimated by this method, Google's ranking system had 'full constitutional protection.' We may have a right to think, but we don't have a right to force Google to push people into thinking the way we believe they should.

If Google is transparent about its methods and other search engines are available,[61] this would not seem too objectionable. Unfortunately, we can never know what search results we aren't shown. These unseen results are Donald Rumsfeld's famous unknown unknowns.[62] Yet this is more of a problem if we think of an isolated individual at their computer, knowing little of the world and being entirely guided by Google. In reality, we know people. They suggest websites to us. We suggest websites to them. These websites then lead to further websites. Google is not (yet) the world; if it is, you need to make a change. Perhaps the best change you can make is to meet people who know an area in depth and can offer you the signposting you want or need. Equally, there is a duty for those who have specialist knowledge to make it available for the public to think with.

The internet makes vast amounts of information available and social media and discussion forums connect people. It seems like this should empower people and encourage their participation in the political process. There is some evidence that using the internet encourages people to participate in politics, such as trying to influence government policy or their local community. However, the effect is weak. Studies of whether social media impacts levels of political participation have found conflicting results.[63] What does seem clear is that simply scanning information you come across online has no real impact on your political activities.[64] Mere information is unlikely to change the world.

That said, under certain circumstances the information the internet provides can help people accurately think about the world around them and cause major social change. The most-cited example of this is the Tunisian uprising of 2011. The internet's ability to provide the Tunisian people with information helped them overthrow the authoritarian regime of President Zine el-Abidine Ben Ali.[65] A key barrier to social change is a regime's ability to censor and isolate citizens. Such actions prevent citizens from knowing how each other feels about the regime and how likely others are to revolt.[66] In Tunisia, digital activists shared previously censored information about the state's human rights violations. They were also able to rapidly inform people about just how big the anti-regime protests were. This information allowed Tunisians to see that many of their fellow citizens were also outraged and not afraid to do something about it.[67] The internet enabled people to think together. When it does so, the internet becomes reminiscent of the vigorous working-class movements of the late nineteenth and early twentieth century. The internet's most significant contribution may not be the information it provides but the new ways it creates for us to think together.

Restricting information

Despite the importance of information to thought, there are valid questions to be asked about when governments can keep information from us. Whilst secrecy and censorship are sometimes essential to security, they are always anathema to free thought. Governmental control over information is simultaneously necessary and problematic. As the economist Joseph Stiglitz has observed, 'secrecy gives those in government exclusive control over certain areas of knowledge, and thereby increases their power, making it more difficult for even a free press to check that power.'[68] Radical transparency would reduce the corruption and abuse of power made possible by governmental information control. As best-man extraordinaire Louis Brandeis put it, 'sunlight is the best disinfectant'. Yet there are valid reasons for governmental secrecy. Sunlight can cause cancer and disinfectant can be lethal, especially if injected into the lungs.

Governments may conceal information due to national security concerns.[69] Although some information requires governmental protection for national security purposes, governments classify swathes of information unnecessarily. As the American activist Steven Aftergood has observed, 'entire shelves of commission reports, congressional hearings, and independent critiques since then have blasted official secrecy for improperly shielding government operations from the public, impeding oversight, covering up malfeasance, and, ultimately, undermining national security itself.'[70] Aftergood points out that the US Under Secretary of Defense for Intelligence told a House subcommittee in 2004 that as much as fifty percent of classified materials may not need this designation.

In addition to national security concerns, governments block our access to information due to bureaucratic and political secrecy. As sociologist Max Weber noted, 'every bureaucracy seeks to increase the superiority of the professionally informed by keeping their knowledge and intentions secret.'[71] Political secrecy stems from the desire to keep information private that, if known by the public, could damage the government or its friends. Examples of this are depressingly familiar. Pressure from the Chinese government in 2017 prompted several Western academic publishing houses to remove the access of China-based users to scholarly articles on human rights, Taiwan, Tibet and Tiananmen Square.[72] US examples include governmental attempts to expose, disrupt, misdirect and discredit the Black Panther Party in the 1960s,[73] and secret spying programmes revealed by Edward Snowden in 2013. The problem becomes how to ensure appropriate independent oversight of what is legitimate for governments to keep secret and what is not.[74]

'Vigilante transparency' has tried to address the problem of government secrecy.[75] Organisations like WikiLeaks have gained access to and released large amounts of non-publicly available government information. WikiLeaks views itself as acting to uphold human rights, specifically the rights to freedom of opinion and expression. They could add the right to freedom of thought to this list.

Another way that the problem of governmental secrecy has been addressed, and some form of the 'right to know' introduced, comes from official freedom of information laws. The US's Freedom of Information Act has led to millions of requests for information each year. Despite the intuitive benefits such laws have for promoting free thought, they have significant limitations. Some of those in positions of power have chosen not to put information about decisions down on paper to avoid disclosing such material.[76] Even if a government does record information, it may still block its release. In 2009 the UK's Minister for Justice was able to veto a request for the minutes of the Cabinet meeting that discussed starting the Iraq War. The Minister was also able to prevent the release of the Attorney General's legal advice.[77] Although the government argued this was necessary to protect the 'integrity of the Cabinet', it did nothing to protect the integrity of voters' future decision making.

More information will not automatically help citizens better understand government. In the UK, freedom of information laws have had 'little effect' on public understanding of how government makes decisions.[78] Indeed, such laws can thwart public understanding. Studies of journalists' experiences of freedom of information laws have found that this legislation can provide 'ammunition for political control' and give governments new tools to control the news agenda.[79] The invocation of our absolute right to freedom of thought offers a way to justify strengthening freedom of information laws. This could provide access to previously withheld documents and allow citizens to insist that government records all information relevant to decision making.

Yet a paradox arises from our earlier observation that mental privacy matters because it supports the discovery of truth. Making public the thinking aloud of politicians in Cabinet meetings could lead to worse decision making. Such revelations could lead to short-termism and Ministers adopting policies based on 'crude majoritarian, intolerant or prejudiced policies'.[80] As a former UK Attorney General has put it:

> If there cannot be frank discussion of the most important matters of government policy at Cabinet, it may not occur at all. Cabinet decision-making could increasingly be driven into more informal channels, with attendant dangers of a lack of rigour, lack of proper accountability, and lack of proper recording of decisions.[81]

The same reasons that justify individual mental privacy may also justify governmental decisional privacy. Increasing public involvement in decision making through the institutions of deliberative democracy[82] may be a better route forward than potentially impairing existing decision-making processes.

There are several other reasons why the idea of a 'right to know' turns out to be problematic. First, there is the misguided assumption that access to more information makes us better thinkers. Philosophers have long pointed out that being surrounded by information can hinder rather than help thought. Nietzsche wrote of the reader who, 'when he has not a book between his fingers . . . cannot think.'[83] Nietzsche's compatriot Arthur Schopenhauer echoed this sentiment: 'when we read, another person thinks for us ... the person who reads a great deal ... almost the whole day ... gradually loses the ability to think for himself ... many men of learning: they have read themselves stupid.'[84] And, yes, I am aware of the irony of my use of quotations to make this point.

We can mistakenly think that encountering, possessing and owning information means we have absorbed it. To quote Schopenhauer again, 'to buy books would be a good thing if we could also buy the time to read them; but the purchase of books is often mistaken for the assimilation and mastering of their contents.'[85] What student hasn't deluded themselves into thinking they were learning simply by photocopying, printing or downloading articles? As the French poet, explorer, *enfant terrible*, trader, brawler and amputee Arthur Rimbaud pointed out, 'you've got a ball between your shoulders that ought to take the place of books. When you put books on your shelves, the only thing they do is cover up the leprosies of the old walls.'[86]

When others flood us with information, we don't know what matters or to what we should attend. This is 'reverse censorship'. Whereas normal censorship works by withholding information, reverse censorship works by bombarding people with information.[87] In such a situation, we find ourselves crying out for the trusted gatekeepers and curators that the internet attempted to circumvent. To think, we must trust. Medieval monks were given guidelines on how to discern what was and what was not God's voice. Today, we need to practise digital discernment.

Cass Sunstein gives an excellent exposition of the problems of a flood of information in his book *Too Much Information*. In this, he shows that simply throwing information at people does not automatically increase their freedom and autonomy.[88] Sunstein argues that we should determine what information improves people's well-being and prioritise promoting this information. He calls this approach 'welfarism'.

Indeed, giving people information can decrease their well-being. We instinctively know this and turn away from some information. For example, would you want to know when your partner will die or what they will die of? Ninety percent of people do not want to know this information. What about your cause and date of death? Eighty-seven percent of people do not want to know.[89] And how much do you really want to know about your romantic partner? A 2021 study found that thirty-four percent of participants didn't want to know if their partner was being unfaithful. Twenty-six percent didn't want to know how their partner felt they compared to previous partners, and fourteen percent didn't want to know about their partner's sexual history.[90] We even turn away from profound theological issues. Sunstein found that only fifty-three percent of people wanted to know whether heaven existed.[91] Our right to freedom of thought needs to include a right not to think, a right not to know, and a right to ignorance, which, after all, can be bliss.

Given the problem of too much information, research has explored the benefits of deliberate ignorance. Not knowing can help shield us from our biases. For example, in the television show *The Voice*, judges make their decision as to whether to support a singer before they have seen them. This removes a host of potential biases that someone's physical appearance can activate. Such an approach can have real-world benefits. Having applicants for orchestras audition behind a screen increased the number of women in orchestras in the late twentieth century.[92]

Not knowing can also help us by keeping our attention away from low-quality, unreliable data sources. Of course, this requires us to know from where reliable information will probably come. This again reinforces the importance of digital discernment in the form of information literacy.

A right to the truth?

The great hope of the internet was that it would empower us by granting us access to information kept hidden by traditional gate-keepers. After all, knowledge is power. Of course, things didn't work out that way. Part of the problem was that people were often presented with incomplete, misleading or simply wrong information. We face the triple threat of disinformation (false information spread by someone who knows it is false), misinformation (false information spread by someone who doesn't know it is false) and malinformation (correct information spread to harm or presented out of context to intentionally mislead). In the internet age, you are what you spread. Could we respond to this by claiming a right to truth?

To some extent, we already possess such a right. Governments can't make laws that hide the truth from us for our own good. If governments fear we will make bad decisions or harm ourselves if we get the truth, tough.[93] Corporations also have a legal duty not to mislead us. It's illegal for businesses to mislead customers by using false or deceptive messages.[94] When the American public was not given correct information about the dangers of OxyContin by its manufacturer, legal action ensued.[95]

Despite this, politicians have no legal requirement to tell us the truth. Governments can regulate adverts made by businesses to prevent lying, but they cannot do the same for adverts made by politicians. The UK doesn't allow politicians to lie about the character or conduct of their opponents during an election.[96] However, other false statements of fact are not prohibited, including those relating to publicly available statistics.[97] Politicians can even put them on the side of a bus. US courts have also rejected the idea that the government can make it illegal to lie in political adverts.[98] American politicians must not defame each other, but they can say pretty much anything else. Their false and reckless claims, as long as they aren't defamatory, cannot be punished by the government.[99] This is, as one judge has noted, 'an invitation to lie with impunity'.[100]

The law does not want to pass judgement on political truths. It wants politicians' statements to be scrutinised, analysed and judged by their political opponents and the public. But should this be the case? And what about private citizens? Do we have the right to lie? What, if anything, should the right to freedom of thought have to say about truth-telling and lies? As we will see, the problem we face here is that lies can both threaten and support free thought. For this reason, laws against lying also both threaten and support free thought.

Lies pose a major problem for humans. They are everywhere. We lie daily.[101] When we inevitably encounter lies, our minds are not well-equipped to manage them. Our ability to detect when people are lying to us is pretty appalling. If we relied on a coin toss to determine if someone was lying, we wouldn't be much worse off.[102] Even as adults, we are only slightly better than chance at detecting the lies children tell.[103] Despite being bad at detecting lies, we are great at spreading them. This is a terrible combination. As we are better at telling than detecting lies, we are continually muddying the waters of truth.

Lies are particularly pernicious because we tend to believe that what people say is true. This 'truth bias' makes us dangerously gullible. Hitler knew this and made appalling use of it. In *Mein Kampf*, he discusses the use of lies so colossal that people would 'not believe that others could have the impudence to distort the truth so infamously'.[104] Contemporary politicians make use of this bias too. If we hear a lie about a politician and have it corrected immediately, we can discount it. But six months later, we may have forgotten the information was corrected and remember the lie as being true. Lies have a sleeper effect.[105] This is why, even when judges instruct jurors to ignore inadmissible evidence, jurors still end up being influenced by this evidence.[106]

Furthermore, lies can evict truths from your mind. If you repeatedly tell people something they clearly know is wrong, they will nevertheless start to believe the lie. Repeatedly telling people that 'the Atlantic Ocean is the largest ocean on Earth' or that 'the sari is the short pleated skirt worn by Scotsmen' can make them believe these falsities.[107] This happens because the new 'alternative fact' you have been repeatedly given now easily pops into your mind. You use this ease of access as rule-of-thumb evidence of the idea's truth (the so-called 'availability heuristic' mentioned earlier). In fact, you don't even have to lie to people to evict truths. If you selectively recount an event, this causes people to forget about other important aspects of the events.[108] This isn't to do with people being stupid. It doesn't matter how smart you are. If you hear something repeatedly, you tend to believe it more than things you only hear once.[109]

In 1710, Jonathan Swift noted how 'Falsehood flies, and Truth comes limping after it'.[110] He was half right. Falsehood certainly flies. A study that examined how 126,000 true and false news stories spread on Twitter found a troubling pattern. Lies 'diffused significantly farther, faster, deeper, and more broadly than the truth in all categories of information'.[111] This was particularly the case for false political news. Although Swift was right about falsity flying, he was optimistic to think the truth would ever follow.

Consider a study of media coverage of the 2003 Iraq War.[112] During the war, the media reported all sorts of factually inaccurate claims. UK Prime Minister Tony Blair claimed that the Iraqis had executed prisoners of war who surrendered. This claim was effectively retracted by the government the next day. The media reported that a whole Iraqi division had surrendered to coalition forces, a statement that was also later corrected. Such flip-flopping led one research team to examine whether retractions changed people's memories of events.

The researchers found that Americans were unaffected by retractions. They continued to believe that the events had happened as initially reported. In contrast, Germans and Australians used retractions to remember better what happened. It appeared that the greater suspicion the Germans and Australians had about the official narrative surrounding the war led them to be more able to correct misinformation. But for those who wanted to believe the official account, lies stuck.

Swift also observed that by the time truth catches up with falsity, it may be too late to matter. Advertisers and public relations firms know this too. Take one of the most shocking lies of the 1991 Gulf War campaign. In 1990 a tearful fifteen-year-old Kuwaiti girl gave testimony to the United States Congressional Human Rights Caucus. She described seeing Iraqi soldiers entering a hospital in Kuwait City, taking babies out of incubators and then leaving them to die on the cold floor. The story made the US nightly news, reaching tens of millions of Americans. President George H.W. Bush repeated the story. It would later emerge the girl was the Kuwaiti ambassador's daughter and had been coached by an American public relations firm.[113] The story of these murdered babies, which had a powerful effect on Western sentiments, was a straight lie.[114] When the truth caught up, the war was over. Abraham Lincoln stressed that you can't fool all of the people all of the time. Unfortunately, you don't need to. Often a week or so will do the job.

Such wartime lies can have profoundly damaging consequences. During the First World War, the British press falsely reported that the Germans were boiling down the corpses of dead German soldiers to make pig food, candles and soap. This story would be officially repudiated in 1925. Unfortunately, when stories started circulating about Hitler and the Holocaust in the Second World War, people remembered this lie from the First World War. Unwilling to be fooled again, many nations were sceptical about the need to take in Jewish refugees. This slowed the international response and cost more Jewish lives.[115]

The stickiness of lies and their ability to evict truthful beliefs poses problems for the idea that true speech can correct false speech. Permitting lies on the basis that truth can mend them is like allowing violent hands to smash Fabergé eggs on the basis that sensitive hands can put them back together. The lie and the truth are not equal in power. This gives us a strong rationale for keeping lies out of the public discourse. Lies are an offence against freedom of thought. They also threaten many goods that freedom of thought supports. When other people lie to us, they commit an offence against our autonomy.[116] Consistent with this argument, some forms of lying are already illegal. Fraudulent statements, defamatory statements, pretending to be a government official and perjury can all get you in trouble with the law.[117]

A rationale for making other types of lying illegal comes from considering why courts made perjury illegal. The US Supreme Court has stated that perjured testimony 'is at war with justice' because it can cause a court to pass a 'judgment not resting on truth'.[118] If it is not OK for courts to make consequential judgements that don't rest on truth, then it should not be OK for voters to make choices based on lies. Similarly, why would we make lies that part us from our money illegal but legalise lies that part us from our vote? Votes are more valuable than money. To protect freedom of thought, and the ends to which it aspires, we must take steps to protect the integrity of circulating information.

Yet a case can also be made that lies promote free thought. In this view, legal control of lies will hurt, not help, freethinking. There are good reasons why much lying is legal. The US Supreme Court has discussed how the government has a duty to protect people from others who are after their money. Yet the government has no right to protect the public against 'false doctrine'.[119] The First Amendment, the Court notes, stops the state from 'assuming a guardianship of the public mind', meaning that 'every person must be his own watchman for truth'. The reason for this, says the Court, is because the US's forefathers 'did not trust any government to separate the true from the false for us.'[120]

This idea that we can't trust governments to separate fact from fiction correctly is central to the right to free speech. Governments given the power to arbitrate on the truth would end up punishing people who promoted ideas that threatened either governmental power or the corporate interests behind the government. As US Supreme Court Justice Robert Jackson put it, 'Those who begin coercive elimination of dissent soon find themselves exterminating dissenters. Compulsory unification of opinion achieves only the unanimity of the graveyard.'[121] This is a powerful reason why the public, not the government, should decide on what is and is not a lie.

An excellent example of how governments banning lies can threaten truth-seeking and free thought comes from 'ag-gag' laws in the US. The public should be able to know if businesses are harming animals. This means knowing the truth about the conditions under which the meat they buy is produced. Without such information, the public cannot think in the way they would need to make informed purchasing decisions. To bring this information to us, activists and journalists have attempted to film what goes on in factory farms. In response, in the late 1980s, agricultural business owners began to lobby state legislatures and Congress to pass laws to protect them against such filming.[122]

Money talks, and both Congress and individual states passed laws to protect business owners. New laws made it illegal to film in factory farms without the owner's permission. Getting a job at a factory farm was often the only way for activists to get into such private facilities. As a result, laws were also passed to stop people from applying for jobs at such factories without revealing they intended to get evidence of animal abuse.[123] Such laws were passed in many US states, including Kansas (1990), Montana (1991), Iowa (2012), Wyoming (2015) and Arkansas (2017).[124]

Ag-gag laws criminalise 'the dissemination of truthful, unprivileged information to the public'.[125] The public has a right to know what is happening in these factory farms. We can't trust the government to protect our right to know because of its dependence on corporate money. We can get on the right side of history by arguing that our absolute right to freedom of thought means we should know about such activities. If this means that activists must lie to get us the truth, so be it.

Legal punishment for lies would also inadvertently make truth-telling less likely. If you are open to prosecution for erring, you will be more cautious about venturing any opinion. If other people say less, we learn less. This hurts free thought. For this reason, the US Supreme Court has recognised that 'some false statements are inevitable if there is to be an open and vigorous expression of views in public and private conversation, expression the First Amendment seeks to guarantee'.[126] Only those unafraid to be wrong will ever get things right.

Permitting lies also promotes truth in other ways. The US Supreme Court agrees with John Stuart Mill's idea that false statements can lead to the 'clearer perception and livelier impression of truth, produced by its collision with error'.[127] In the scientific context, stating falsehoods can help promote thinking that leads to the truth, which, as the Court notes, Socrates sometimes did.[128]

Banning false statements would drive them underground and potentially make them more appealing.[129] A better approach, argues Cass Sunstein, is to employ counter-speech, labels and warnings. In the Supreme Court's view, 'the response to the unreasoned is the rational; to the uninformed, the enlightened; to the straight-out lie, the simple truth'. This is the (problematic) idea that truth will win out in the marketplace of ideas. For now, we can simply note Hannah Arendt's criticism of this idea: 'the chances of factual truth surviving the onslaught of power are very slim indeed.'[130]

The dominant ideology of the West, liberalism, is also a central factor in Western reticence to legislate against falsehoods. Liberalism views anyone who insists they have the truth as a menace to peace, order and prosperity. This view dates back to the seventeenth-century religious wars in Europe. Such wars ended in 1648 with the signing of the two treaties known as the Peace of Westphalia. This Peace was based on the idea that the only way to stop people from laying waste to Europe and each other was by getting them to agree that no one could be sure which religion was correct. Stability was achieved by building on shifting sands. Liberalism's antidote to destabilising fanaticism was to inject society with the idea that no one knows what the hell life is about. Accordingly, everyone should be left to make their own choices and muddle through as they see fit. Liberalism turned society into a giant bouncy castle. It's hard to fight when you can't find firm footing. No one ever said 'Here I stand, I can do no other' whilst on a bouncy castle.

Liberalism is, therefore, highly reluctant to pass official judgement on truth and falsity, preferring toleration instead. In a live-and-let-live society, truth is a secondary concern to peace. In a liberal society, one must hold one's beliefs lightly. The idea of absolute truth, which one would act on – in short, the idea of anything being sacred – is anathema to liberalism. The only exception to this is human rights, where liberalism is keen to state what should be the case, categorically and universally. Liberalism is the tyranny of tolerance.

To better understand the judicial thinking behind not legislating against lies, we can look at a landmark 2012 case in the United States, *U.S. v. Alvarez*.[131] The defendant, Xavier Alvarez, had a habit of lying. He had falsely claimed to play hockey for the Detroit Red Wings and to have once married a starlet from Mexico. Such false claims are common to many a bar room. Where Alvarez got in trouble was in standing up at a public meeting in 2007 and falsely claiming he was a recipient of the Congressional Medal of Honor. This claim entered the realm of law because lying about possessing this medal violated a federal criminal statute, namely the Stolen Valor Act of 2005.

Given that 'Congress shall make no law . . . abridging the freedom of speech', the Stolen Valor Act smelled unconstitutional from the get-go. In 2012, this Act ended up in front of the Supreme Court. The Court had to decide whether one of the freedoms for which US service personnel had fought and died was the right to lie about having fought. Yet this case wasn't really about a right to lie. Instead, as subsequent cases have emphasised, it was about a right not to have the truth of statements judged by the government.[132]

The Court's plurality opinion was concerned about setting a problematic precedent. What would happen if they let the government make lying about receiving this medal illegal? In the Court's view, this would effectively 'endorse government authority to compile a list of subjects about which false statements are punishable'. Explicitly referencing Orwell's *1984*, the Court claimed that the US constitutional tradition opposed the idea that America needed a Ministry of Truth. The Court felt letting the government ban speech based on its perceived truth would give it far too much censorial power. It could deter people from thinking and speaking freely. As a result, the Court ruled that the government could not punish Alvarez for lying. This did not mean the government could never punish lies. Rather, it meant that the government could not punish lies which did not cause or intend to cause injury or 'legally cognizable harm'.[133] In this case, the Court deemed the harm to the military caused by lying insufficient to justify the Stolen Valor Act.

Yet a minority of the Court's judges disagreed. They felt it was OK for the government to ban lying about having received this medal. False factual statements, they said, had no value worth protecting. They also questioned what other speech this Act would discourage. 'None' appeared to be the answer. In the dissenting judges' view, in contrast to 'hypothetical laws prohibiting false statements about history, science, and similar matters, the Stolen Valor Act presents no risk at all that valuable speech will be suppressed'. For this reason, it was fair in their view to prosecute such lies.

The dissenting judges explored a hypothetical law that made it illegal to falsely claim to have been a high school valedictorian. In such a case, they said, the problem would not be suppression of speech. The problem would be 'the misuse of the criminal law, which should be reserved for conduct that inflicts or threatens truly serious societal harm'. Cass Sunstein picked up on this point.[134] He argued that in well-functioning societies, citizens socially punish each other for lying (such as through ostracism) rather than resorting to the law. Trying to get the law involved in regulating lies signals that something has gone wrong with our normal cultural means of ensuring truthfulness. Similarly, our focus on the right to freedom of thought suggests something has gone very wrong with our culture's ability to think.

As Sunstein points out, one factor that has damaged our culture's ability to regulate lying is the increasing opportunities we have to speak anonymously. This lets us evade the traditional social consequences of lying. Yet, at the same time, anonymous speech makes it easier to speak truth to power safely. Any attempt to increase the circulation of truths in society may come at the cost of a simultaneous increase in circulating lies. This is a price we need to be willing to pay for truth. But it then becomes essential that we create societal mechanisms that can sift the wheat of truth from the chaff of lies in the marketplace of ideas.

Our need to be actively involved in this sifting process is another reason not to ban lies. The idea of a ban hints that others should decide what is true and then spoon-feed it to us. This would not increase our ability to think. It would substitute for it. A crucial element of thought is weighing evidence and deciding on what we think is true. We should not be sucking truth through a straw. After all, would we prefer ourselves or the government to wrestle with the greased pig of truth? We need accurate information to be able to think, and we need to be able to think to find accurate information. Freedom of thought requires us to know how to get information, access information, and make judgements as to the accuracy of that information. This is what self-government is.

You could object that without fact-checking, citizens will 'end up believing those who are best at fooling them, or who have the most power'.[135] Yet this assumes that lone individuals are making isolated decisions about the truth. The solution may be to both promote individual critical thought *and* support the development of institutional structures that allow people to deliberate together. As we have seen, we become better at detecting falsehoods when we gather in well-designed groups. When you put people into groups to determine whether someone is lying, these groups are better than individuals at identifying lies.[136] Again this emphasises how thinking together is a crucial element of freedom of thought. We need to give people the power to think together and find the truth, rather than simply tell individuals what is true. Again, this is self-government.

This brings us to my view on the problem with the Alvarez ruling. The judges could have reflected further on the purpose of the First Amendment as viewed by Alexander Meiklejohn. Meiklejohn argued the First Amendment exists to prevent the 'mutilation of the thinking process of the community'.[137] To what extent did the Stolen Valor Act mutilate the community's thought? Arguably, not at all. Indeed, the Act supported the community's thought. This case could have been seen as being about what Meiklejohn referred to as 'liberty of speech', namely, the right to say anything you want. Meiklejohn argued that the First Amendment guarantees unlimited public discussion to facilitate self-government. In contrast, he claimed that Americans' private speech, not focused on grand self-government issues, is protected by the Fifth Amendment and *can* be regulated by the government. The Stolen Valor Act was not regulating speech relevant to self-government and therefore should have been let stand. Furthermore, if we are unwilling to let government legislate against claims that are undeniably and unarguably empirically false, then we are subscribing to a culture of relativism that does no service to the truth. Instead, we are fostering a culture of mental timidity that we earlier saw US Supreme Court Justice Jackson warn us against. It should be permissible to legislate against some lies.

Will information free us or control us?

Cass Sunstein once set a noble goal, to create an informational policy that makes 'people's lives happier, freer, longer, and better'.[138] But is a happier life a freer one? Is a freer life a longer one? And is a longer life a better one? Aldous Huxley introduces us to the pitfalls of such assumptions.

A central scene in Huxley's *Brave New World* involves 'John the Savage' being offered what the elite felt was the good life. This involved comfort, happiness and pleasure, not truth, not God and not freedom. 'But I don't want comfort', John protests, 'I want God, I want poetry, I want real danger, I want freedom, I want goodness, I want sin.' The World Controller, Mustapha Mond, observes that John is 'claiming the right to be unhappy'. 'All right,' says John, 'I'm claiming the right to be unhappy. Not to mention the right to grow old and ugly and impotent; the right to have syphilis and cancer; the right to have too little to eat; the right to be lousy; the right to live in constant apprehension of what may happen tomorrow; the right to catch typhoid; the right to be tortured by unspeakable pains of every kind.' There is a long silence. 'I claim them all', John reiterates. Mond shrugs his shoulders and utters a two-word reply:

'You're welcome.' Human desires are painfully complex and resist central planning.

Sunstein's approach focuses on what information helps promote our well-being. It is this information, he argues, that governments should make available to citizens. This pragmatic approach has benefits. And, to steelman Sunstein's case, he makes it clear that it is not simply information that the government *thinks* increases people's well-being that should be released. Rather, the government should give out information that people themselves say is beneficial to know. The first problem with this idea is that the population's views are not independent of the government's. The government and elites can engineer public opinion. Ruling powers could engineer it so that people request information it is in the interests of the ruling powers for them to possess.

A deeper problem is that if we concede experts can calculate what is good for us to know, we may also concede they can determine which government policies are best for us. In which case, why bring citizens into the governing process at all? Why bother having an election if the civil service and its algorithms can work out which party will best serve the well-being of the people? Suppose that in early 2016 algorithms showed that Hillary Clinton would have improved well-being more than Donald Trump. In that case, having an election in 2016 simply let the people blunder into the wrong choice. Here we are creeping perilously close to the idea that a vanguard elite knows the truth. The rest of us ignorant proles, who exist in false-consciousness, need to be taken to the truth.

At the heart of this idea is the concept that someone else can validly tell us we aren't acting in our 'true interest'.[139] Our true interest, others may tell us, is something else. If we won't listen, they may coerce us for our 'own good'. This is a phenomenally dangerous idea that led to the deaths of tens of millions of people in Stalin's Soviet Union and Mao's China.

AI threatens a resurgence of this idea in the form of techno-communism. Peter Thiel argues that AI is 'literally communist', pointing out how it allows a centralising power to know more about citizens than citizens know about themselves.[140] Algorithms could calculate what would make us happy, enabling governments to compel citizens' actions for the citizens' own good. Freedom would become obedience to the state's algorithms. Every hell starts with a promise of heaven and AI-based techno-communism would be no different.

This idea won't start in the West. It will start in China. But it will come to bite the West. Lenin allegedly said that the capitalists would sell him the rope with which he would hang them. Peter Thiel has argued that with AI, the capitalist technology firms of Silicon Valley have sold communists the tool that will undermine democratic capitalist society.[141] AI is Lenin's rope. To prevent such a dystopia, we cannot allow others to know more about us than we do. During the Cold War, we could not allow a mine-shaft gap. Today, we cannot allow a self-knowledge gap.

Summoning new information

Any right to know should free us to milk our brain for all it can give us. Earlier we discussed the concept of first-order thoughts, which are those that just pop into our heads. How might we get a greater volume and variety of such thoughts to seed our thinking? One option is travel, which it turns out actually does broaden the mind. Travelling the world boosts our cognitive flexibility.[142] It lets us break out of normal patterns of thoughts and get away from conventional, stale ideas. Travel helps us have a more diverse range of thoughts pop into our heads.

Seeing more of the world may help us think, but so may seeing none of it. Sensory deprivation can summon thoughts by blocking a specific sense using blindfolds, earmuffs or a sensory deprivation tank. It can even summon voices that speak to us.[143] Similarly, absorptive practices such as prayer or meditation can cause us to think novel thoughts and hear new voices.[144]

An easier and more powerful way to obtain new ideas are through psychedelic drugs such as DMT (the main chemical agent in ayahuasca) and psilocybin ('magic mushrooms'). As the prophet of psychedelics Terence McKenna once said of psilocybin, 'there is a mind there waiting, that speaks good English'.[145] Unfortunately, there are few formal studies of these encounters with seemingly autonomous entities.[146] We know little about what knowledge we can gain from them and how the hell this is all possible. Research into this area could tell us much about what our brains can do. If psychedelics can help free our thoughts, we could use the right to freedom of thought to argue for legalising such drugs.[147]

Yet we do not need psychedelics to help us begin having more creative ideas. Try thinking at a different time of day than usual. If you are naturally a morning person, you are more likely to have creative insights in the afternoon and vice versa. This is because our ability to keep novel or distracting information out of our heads is worse in our non-optimal parts of the day. As a result, more creative insights burst through.[148]

Consuming a diet that results in appropriate levels of dopamine in our brain is another way to support divergent, creative thought. Dopamine influences our drive to create as well as how creative our thoughts are.[149] For example, one way to measure dopamine levels in someone's brain is to count their blinks. The more you spontaneously blink, the higher are your levels of dopamine. Researchers have taken advantage of this fact to see how people's level of creativity (assessed by how many alternative uses people could think of for an object) relates to their spontaneous blink rate.[150] The study found that a medium blink rate was associated with greater creativity than high or low rates. There seems to be a sweet spot for dopamine levels, which promotes creative thought. As dopamine levels are impacted by what we eat, including our intake of fatty foods and omega-3 fatty acids,[151] what we eat will impact our ability to think creatively.

In future, we may be able to use neurostimulation techniques, which artificially alter the activity of our neurons, to summon new thoughts. Transcranial direct current stimulation involves passing a weak, imperceptible current through someone's brain. This technique has shown promise in reducing levels of 'hearing voices' in people diagnosed with schizophrenia.[152] Such studies typically try to increase activity in the left prefrontal cortex (which plans and controls our thoughts and actions), reduce activity in the left temporoparietal junction (which helps us communicate with others), and alter the connectivity between the frontal and temporal lobes of the brain in order to reduce auditory hallucinations. But what would happen if we performed this procedure in reverse with people who *don't* hear voices?

A recent study did something like this with healthy volunteers and found it caused them to be more likely to hallucinate words when they were listening to mere white noise.[153] Other studies have found stimulating the left temporoparietal junction causes the feeling that an unseen person is near.[154] Although we are far from an electric muse, such research places it on the horizon. We would also need a cultural shift for this idea to be adopted. We would need to move from viewing voice-hearing exclusively as a sign of pathology to accepting that it can sometimes be helpful, creative and desirable.[155]

This raises all manner of questions. What would we have created, philosophically speaking? Could it exhibit intelligent human behaviour? Would it display self-conscious emotions or even be conscious? Many writers strive for such a muse. Indeed, Hilary Mantel described the writing process as 'allowing a new consciousness to emerge'.[156] Such questions aside, let us assume that psychedelics and new technologies can help our minds operate at their full potential.[157] This would enhance our freedom of thought. In which case, governments should certainly not block citizens' access and could even be required to promote them. Maybe you don't think that governments should act this way. If so, then you are balancing the right to freedom of thought against other considerations, such as social stability. This is fine as long as you are happy to no longer view freedom of thought as absolute.

Even if a right to know should be part of the right to freedom of thought, we should reflect on Pope Francis's claim that 'the flood of information at our fingertips does not make for greater wisdom'.[158] We should not confuse freedom, says Francis, with the ability to navigate the internet. As a sculptor needs clay that is damp and malleable, thinkers need information that is accurate and complete. Yet letting governments determine truth in matters relating to our self-government is clearly problematic. So too is letting others find and feed us truth. We inhabit a loop in which we need truth to think and need to think to reach truth. The right to freedom of thought must support the generation and judgement of truth in communities. We need civil society groups and institutions where people can think together, dialogue together, and reduce the burdens and biases of the atomised thinker. Such groups would probably be antagonistic to governments, yet need to be supported by governments as part of their positive duty to promote freedom of thought. This offers a better chance of yielding the wisdom for which Francis calls. Freedom of thought is best achieved when freethinkers come together. Next, we will turn to the tools that freethinkers need, before coming to how we may create both private and public spaces to use these tools.

6

CREATING FREE
THOUGHT II: TOOLS

In the 1920s, the right-wing rulers of Imperial Japan faced several problems. Until recently, they had governed securely. Voting laws meant that only men who paid above a certain amount of tax could vote. As a result, only around one percent of the population could vote. These were mostly the rich, who had an interest in the status quo. Yet after the Russian Revolution and the end of the First World War, Japan began to experience economic problems. Communist and socialist ideas had long been leaking into the country and by the 1920s, Japanese workers had created a movement that was a force to be reckoned with.[1] Japan's rulers were also having to deal with anti-colonial protests in countries that formed part of its empire, such as Korea. Then came the year 1923, an *annus horribilis* for the Japanese rulers. In this year the Tokyo region was hit by a major 7.9 magnitude earthquake, authorities discovered that a Japanese Communist Party had been illegally formed, and a member of an elite Japanese family, influenced by translations of Russian texts, attempted to assassinate Prince Regent Hirohito. All this was followed by an increasing number of public disturbances. Japan's rulers were determined to stamp out socially destabilising, dangerous ideas that threatened the existing power structure.[2]

They began this process with a campaign run by the Ministry of Culture that tried to address the 'worsening of thought'. Rather than make systemic changes to Japanese society to help address the ongoing economic problems, the government encouraged citizens to deal with their woes by being kind and loving. Citizens were encouraged to embrace traditional values, such as loyalty, hard work and thrift. The message from the rulers was: don't change the world; change yourself.

In 1922 the government had tried to pass a 'Draft Bill to Control Radical Social Movements'. Existing laws only allowed the government to control civil disturbances when they reached a certain level of violence. This new bill aimed, as one government official put it, to deal with 'the slow infiltration of dangerous thought into the hearts and minds [of the people].'[3] People would now be imprisoned for simply spreading ideas. Ultimately, this bill did not pass.

However, in 1925 the Japanese government was able to successfully pass a 'Peace Preservation Law'. Anyone who wanted the government to change policy, or hankered for a different political system, or rejected the idea of private property, now faced up to a decade in jail with hard labour. Once this process was underway, and the government began to feel it had control of its people's thoughts, it began making Japan more democratic. Japan's rulers now felt able to grant all men the right to vote, safe in the knowledge they wouldn't vote the wrong way. However, the Peace Preservation Law would only be repealed in October 1945 at the insistence of the victorious Allied forces.

Between 1928 and 1936 around seven thousand people a year, on average, were arrested under the Peace Preservation Law. Such people, most commonly those on the political left, as well as some members of new religious groups, were labelled 'thought criminals'. George Orwell would popularise this term in the Anglophone world much later (he published his novel 1984 in 1948). Rather than serve a potential decade of jail time, thought criminals were encouraged to renounce their views and become loyal subjects of the emperor. This would enable them to be released on parole. In Orwell's *1984*, Big Brother gives Party member O'Brien the job of correcting Winston Smith's thoughts. In Imperial Japan, prison chaplains, who were Buddhist priests, were given this task. The process of converting thought criminals from left-wing thought to right-wing thought was called *tenkō*, which means 'a change in direction'.

In practical terms, *tenkō* involved turning a thought criminal's focus away from public political activism and towards their home life. Thought criminals should clean their rooms. The priests stressed that social injustice was caused not by governmental decisions but by the darkness that dwells in each human soul. When one of the leaders of the Japanese Communist Party, Sano Manabu, made a conversion statement, he reported realising that 'the arrogant selflove eating away at my own heart . . . is the root of all human evil'.[4] Similarly, a young Communist mother wrote in her conversion statement that 'fighting over material goods is utter foolishness. I finally understood: no matter how poor a person is, if they live meekly and peacefully, there can be no difficulties.'[5] In this way, priests encouraged politically engaged individuals to focus on their inner world, where the government alleged the roots of the world's problems lay. This allowed the government to manage the external world without the pesky interference of the populace.

In Imperial Japan, the government never deemed thought criminals cured, only remitted. This meant such individuals lived under constant suspicion. Japan passed the 1936 Thought Criminals Protection and Surveillance Law to deal with the increasing number of parolees who were released into society. A small number of professional probation officers and over thirty thousand volunteer probation officers enforced this law. Thought control was a job for everyone. Officers undertook surveillance and made home visits to paroled thought criminals.

There are several key takeaways from Imperial Japan's attempts at thought control. First, Japan achieved thought control by directing the attention of the people. The population were made to focus, not on the external social ills that the government was responsible for, but on the internal problems in their soul. This appears to be a highly effective way to control the thought of a population. Intended or not, we see it in a wide range of Western settings.

In 1960s America, a similar change of focus occurred. Individuals were encouraged to fix themselves first, and focus on their inner world, rather than engage in social activism. This helped defuse many of the political movements of the day. Whereas this is something that the government forced Japanese thought criminals to do, American revolutionaries did so voluntarily.

Similarly, the psychotherapeutic movement in the West has encouraged people to locate the causes of social ills within themselves. A young Sigmund Freud argued that child abuse was the *caput Nili* (source of the Nile) of mental health problems. He even proposed that the motto of psychoanalysis should be: 'What have they done to you, poor child?' But Freud would come to change his mind, claiming that 'such widespread perversions against children are not very probable'. He introduced the idea that many stories of abuse reported by patients were fantasies. The malign effects of the idea that people were imagining abuse would echo for decades.[6]

In the 1990s, cognitive behavioural therapy pinpointed the cause of mental suffering in people's dysfunctional beliefs about the world. This moved attention away from systematic problems, such as unemployment, poverty, racism and trauma, as causes of mental health difficulties. Such issues would require societal, political and economic change to address.[7] The ability to control our attention is the first tool of free thought we will consider here.

Attentional tools

If we cannot control our attention, we cannot think freely. If we have our attention controlled by others, we are led around by the nose, 'as asses are'.[8] As a result, we see what others want us to see, think what others want us to think, and have little opportunity to start or maintain our own trains of thought. The first tool of free thought is being able to control our attention. To do this, we need to understand the threats to our attention and how we can reclaim it.

Attention is big business. The machinations of 'attention merchants' try to harvest our attention for their profit, which is often our loss.[9] Entire industries are dedicated to the engineering of attention. It is lured, captured and sold. The narrator in William Ernest Henley's poem 'Invictus' famously claims, 'I am the master of my fate, I am the captain of my soul'. Yet, to sail a self-set course in our distraction economy is to constantly tack against the wind. The world's smartest people, funded by the economies' largest companies, equipped with technologies' latest inventions, are after your attention. It is next to impossible not to end up on the rocks.

The advertising industry wants to use research into *selective* attention to make us buy products. Psychological researchers know this and make helpful (to the advertisers) suggestions to this end. For example, we know that selectively attending to a product makes us like it more. The question then becomes, how can adverts make us selectively attend to their products? Psychologists suggest that marketers design search games where customers must find a product.[10] They also advise that adverts should include people looking at the product to cue our gaze also to look.[11] As a result, advertisers pull our attentional spotlight around like a puppet on a string.

Knowing how selective attention works can stop us from falling into its traps. Economists talk about the ostrich effect in which we avoid risky financial situations by pretending they don't exist.

Investors check their financial portfolios less in falling or flat markets than in rising markets.[12] When the news is bad, we don't want to know. We have more of a hope of overcoming this tendency once we know it poses a risk.

Understanding how *sustained* attention works can also be used to push and pull us. Research has found that voice-overs help sustain our attention. Online courses can use this knowledge to boost student learning.[13] Yet corporations can use knowledge of our limited ability to sustain attention against us. They know we will not read lengthy terms and conditions. In a society that respected free thought, government would mandate that terms and conditions be short, simple and with the crucial parts made salient to us.[14]

Basic lifestyle advice can help us boost our sustained attention and put us in better control of our minds. Keeping fit and exercising can help keep our attention working well.[15] Eating specific foods, such as tart cherries, can also boost our sustained attention.[16] Even something as simple as ensuring we drink enough water can help our attention. Being dehydrated worsens the attention of everyone, from schoolchildren to basketball players and the elderly.[17]

A crucial way to keep our attention under our control, and our minds free, is to ensure our own goals guide it. It should not be at the mercy of incidental experiences or other people's goals. This is one of the principles of Acceptance and Commitment Therapy. This psychological therapy aims to help people move their attention away from upsetting inner experiences, whether these be difficult feelings or hallucinated voices. Instead, people are helped to attend to whatever valued life goals they themselves set.

The roots of this idea lie in Immanuel Kant. In Ancient Greece, freedom had meant emancipation from one's desires. The free person ruled themselves by reason. Yet Kant argued that freedom meant the ability to choose one's ends. In Ancient Greece, the goods of life were authoritatively pre-defined. But for Kant, we had to choose our ends.[18] Once we have done so, we must tie ourselves to the mast of our goal to avoid the siren calls of distraction.

Acceptance and Commitment Therapy helps people commit to their own goals by accepting experiences they cannot change. Central to this is the practice of mindfulness. We are encouraged to accept the troubling or painful thoughts and emotions flowing through us. Rather than engaging with such experiences, we are encouraged to notice them mindfully. Similarly, we are encouraged to let go of aspects of the external world we cannot control, engagement with which would only lead to a futile distracting struggle. We are then supported to work out what we value in life and to commit ourselves to actions to achieve life goals linked to these values. The alternative is to get sucked into unprofitable struggles with uncontrollable aspects of the inner or outer world. Learning mindfulness techniques as part of this therapy can improve our sustained attention, even in childhood.[19] Clearly, as the case of Imperial Japan demonstrates, such an inner focus and surrender of the outside world can be problematic. One must be aware of how such a stance may also serve others' ideological goals.

In society today, the war for attention primarily takes place on the battlefield of technology. Our attention is often grabbed by hands that reach out through our smartphone. Bobby Valentine, manager of the Boston Red Sox, once found himself at the bottom of a ditch after being distracted by a text message from All-Star second baseman Dustin Pedroia whilst cycling. Valentine's experience is consistent with research showing that our beeping or buzzing phone makes a notable dent in our attention, even if we don't pick it up.[20] Our phone's noise triggers mind wandering, drawing us into our inner world, as we wonder what the other person wanted. This makes us worse at performing the task at hand.[21] Yet our smartphones do not need to make a noise to latch onto our attention. The mere presence of our smartphone is enough to degrade our attention.[22]

The presence of smartphones, laptops and tablets cause severe problems for our ability to sustain attention. One study, which examined students studying in their homes, found students spent less than six minutes on a task before switching to something else. The switch was typically due to a distraction involving technology, such as Facebook or texting.[23] This is a particular problem for people born after the turn of the millennium, who are raised on technology. Whereas research has found older adults switch tasks seventeen times per hour, people in their twenties were found to switch tasks twenty-seven times an hour (nearly once every two minutes).[24]

Smartphones distract us because they are a source of reward. They positively reinforce our use of them by giving us desirable access to our friends and entertainment. They also negatively reinforce our use of them by providing us with escapism that allows us to avoid pain and anxiety. Having a cheap, effortless reward source close to hand sucks us in. If you put a smartphone near us, we become distracted, but if you put a notepad near us, we don't.[25]

Any device providing both a route to pleasure and an escape from pain will constantly call us. This is why we can find smartphones more rewarding than even junk food. If you temporarily deprive students of food and their smartphone, they will work harder to get their smartphone back than to get food.[26] If you want to keep your attention under your control, get cheap and accessible sources of reward the hell away from you. Despite such advice, due to the need of people (particularly adolescents) to access their social networks, a phone out of sight is not a phone out of mind. Instead, we need the skills of knowing when it's a good time to take a break to access our phone and the self-control to stick to such a schedule.[27]

We also need to be able to reflect on whether our phones are impacting our ability to achieve our valued goals. The distraction caused by smartphones can quickly become pathological. Research has shown that smartphones can cause users to display what clinicians may otherwise view as symptoms of Attention Deficit Hyperactivity Disorder (ADHD).[28] A 2015 survey in the US found that fifty-seven percent of smartphone users felt distracted by their phones.[29] Yet, given a choice between the words 'distracting' and 'connecting' to describe best how they felt about their phone, only twenty-eight percent opted for distracting. We need to decide how we want to prioritise our work and social goals, and plan to act accordingly.

We now need to consider what exactly it is on our phone that is so captivating. Social networking sites are the best attentional bait around. Services such as Facebook, Instagram and Twitter have drawn us closer around the campfire of our shared humanity. Yet they come with costs, both personal and political. In *Star Wars: A New Hope*, when Obi-Wan Kenobi first sees the Death Star, he says, 'That's no moon.' If someone were to show Obi-Wan Facebook, he would probably reply, 'That's no social network.' And he would be right. Facebook is a self-salient, social status, variable-ratio reward schedule attentional-capture mechanism. We'll unpack all these terms in a minute, but we can summarise things more simply in the words of another *Star Wars* character, 'It's a trap!' Social media companies have built an attentional Death Star.

Social media was consciously designed to capture and hold our attention. Sean Parker, the first president of Facebook, has discussed the thinking behind building this social network. He describes it as being 'all about how do we consume as much of your time and conscious attention as possible?' Facebook set out to achieve this, alleges Parker, by 'exploiting a vulnerability in human psychology . . . The inventors, creators . . . Mark [Zuckerberg] . . . understood this consciously. And we did it anyway.'[30] Zuckerberg, whose mother was a psychiatrist, had a long-standing interest in psychology. He pursued it at Harvard, having intended to major in this area. Facebook used what psychology had discovered about the nature of the human mind to grab our attention.

Social media has a particularly powerful pull on our attention due to the type of material it makes available to us. It gives us information relating to social status. As our social status influences our access to resources, belonging, acceptance and social esteem have evolved as basic human needs.[31] Preserving our 'social self' is one of our central goals.[32] This causes information relevant to our social status to grab our attention strongly. Suppose someone says your name across a noisy room whilst you are talking to someone else. Your brain will still detect your name and promote it to your consciousness, where it will pop into your mind.[33] Your brain's response to your personalised ringtone also shows the tell-tale neural signature of involuntary attention-grabbing.[34]

Information on social status is so important that your brain would rather mistakenly tell you such information is present when it isn't, than make the mistake of missing it. This explains why hearing your name called is the most common auditory hallucination in the general population.[35] It also explains why ninety percent of us have hallucinated our phone vibrating in our pocket.[36] Due to social network sites offering us information on our social status as well as means to boost it, we find them highly rewarding.[37] These rewards reinforce our use of the networks, bringing us back again and again.

The attentional draw of this material is heightened further by how it is presented. Gambling is addictive because you don't know how many bets you will have to make before winning. You could win after making a hundred bets and then win again with your one-hundred-and-first bet. Such a pattern is called a variable ratio reward schedule. The American psychologist B. F. Skinner discovered this in his Harvard pigeon lab in the 1950s. He trained pigeons to peck a button by rewarding them with food. If he gave the pigeons food every time they pecked the button, they pecked a lot.

But if he only sometimes gave them food when they pecked the button, the pigeons pecked even more, in a frantic, almost compulsive manner.

Half a century later, Skinner's pigeon lab was resurrected at Harvard with two modifications. First, it was called Facebook. Second, it didn't use pigeons. When you check Facebook, you never know whether someone will have left you a social reward (a 'like' or a nice comment). Facebook is not only giving powerful, potentially rewarding, attention-grabbing information on social status, but does so using a variable schedule. Obviously, Facebook is not the only company using this knowledge. The co-founder of Tinder has also stated that what he learnt about the psychology of variable reward schedules as an undergraduate helped inspire Tinder's algorithm.[38] The unknown number of swipes that will result in a match compels people to keep swiping.

Social network websites are slot machines that pay out in the coin of social esteem. This is why billions of us pull their levers every day. This is why, as mentioned in the Introduction, my mind kept being pulled to Facebook when my article became popular. And this is why, in some people, addiction-like responses to social media are inevitable.

In 2004, when pitching to advertisers, Facebook referred to its users as addicted.[39] Today, some researchers argue that Facebook addiction is real.[40] Between two and ten percent of young people have been estimated to have Facebook addiction.[41] However, psychiatry does not yet recognise this as a mental health condition, which is fortunate, as there are several problems with this concept. For example, someone could be diagnosed with a Facebook addiction and pathologised when their use is within normal limits for their demographic or social context.[42] Alternatively, their use could be high but not cause them any problems.[43]

Problematic smartphone usage is associated with worse performance in networks in our brain associated with attention. One study compared the brain activity of people with problematic smartphone usage to those without such problems. This was done during tasks that assessed their ability to control their attention. The researchers found that the brain networks underpinning our ability to control our attention were degraded in people with problematic smartphone usage.[44] Neural regions within the brain's attention network did not communicate with each other as well in people with problematic smartphone usage.

Not everyone's attention is equally vulnerable to the lure of smartphones. Our ability to prevent our eyes or hands from reaching toward our phones is related to an ability psychologists call inhibitory control. Famously, this ability can be measured by the marshmallow test. In this, people are asked whether they want a single marshmallow now or a couple of marshmallows in a few minutes' time.[45] The more able you are to wait for the larger reward, the greater your inhibitory control. As people with higher socio-economic status perform better on this task,[46] the damaging attention-grabbing effects of smartphones may be greater in lower socio-economic groups. For the same reason, the greater your need to belong and be popular, the more irresistible you will find a social networking site's siren song.[47] The more vulnerable you are, the harder these sites pull.

The question then becomes what, if anything, the right to freedom of thought should say about all this. How should we determine if social networking sites 'impermissibly' alter our attentional control mechanisms? How should we decide what ways of attention-grabbing, if any, we want to legislate against? One factor we need to consider is power imbalances.

In chapter 4, we saw the European Court of Human Rights agreeing that senior officers trying to convince lower-rank soldiers to become Jehovah's Witnesses represented 'improper' undue influence. The power difference was the problem. This is analogous to many users' relation to social networking companies. A power differential in knowledge and technology means that the sites' designers know much more about how to control attention than do their users. As Antonio puts it in *The Tempest*, 'If it were men we were up against, no one could make me withdraw, but when it's demons and magic there's no shame in giving in.' This power differential means social networking sites' use of attentional control mechanisms are improper undue influence on our thoughts.

We could avoid invoking the law by using education to help users know what tricks social networking sites use to control attention. Users could be informed of how red notification buttons are used to grab their attention, how variable-ratio reward schedules are used to keep them coming back, how infinite scroll features remove barriers that would get the person to leave to pursue other tasks. Users could also be taught about the personal reasons that may make them vulnerable to excessive usage. Factors that predict excessive use include loneliness, fear of missing out, having many negative emotions, struggling to cope well with everyday problems, and having a need for self-promotion.[48] Users can also be shown how apps such as Freedom, Moment and StayFocusd can help them regain control of how much time they spend on social media and other sites.[49]

Yet it seems amiss to place responsibility entirely on users. It may not be a burden users can shoulder. Users may be fully aware of the tricks social networks employ but still be unable to escape their grasp. Knowing how one-arm bandits work or that nicotine is addictive does not necessarily make gambling or smoking less attractive. For this reason, governments need to regulate social networks to empower users if networks do not take it upon themselves to do this. Governments often do this in other industries. In the US, tobacco companies are already required to submit research on the effects of tobacco use and put warning labels on cigarette packs. The US Food and Drug Administration has the authority to regulate their activities.

Social networking sites should be required to redesign their sites to mitigate the risk of addiction. They could use opt-out default settings for features that encourage addiction-like behaviour and make it easier for people to self-regulate their usage.[50] A common industry objection to this will be, 'you want us to design worse products?!' But this is not what is being required. Industry is being required to design better products, better in the sense that they support human autonomy. Freedom is much more than being able to fulfil one's desires. Freedom is being able to rule oneself.[51]

To 'live deliberately', as the American writer Henry David Thoreau wished for, is a noble pursuit. To do this, we must be able to think deliberately. Attention is crucial to this. If we don't control our attention, someone else will. The right to freedom of thought needs to prevent power differentials leading to citizens' attention being improperly influenced by others.

The tool of reflection

In a world where all our needs are met, our minds will atrophy like bed-bound muscles. Aldous Huxley vividly portrayed this in *Brave New World*, where he showed how governments can destroy free thought by removing the opportunity and desire to reflect. In Huxley's future 'World State', rulers try to keep people too busy to think. The story tells how, in the bad old days, men who retired from work used to 'take to religion, spend their time reading, thinking – thinking!' But in the World State, 'the old men work, the old men copulate, the old men have no time, no leisure from pleasure, not a moment to sit down and think.' Instant gratification leaves no space for thought to occur. The state educates people to never put off until tomorrow the fun they can have today. There is no self-denial. In *Brave New World*, everyone is Tiger Woods: distracted by golf and sex. The state prevents citizens from feeling emotions by minimising the time between a desire arising and being fulfilled. People are happy because 'they get what they want, and they never want what they can't get'. There is no space for challenges or dissatisfaction to trigger reflective thoughts.

Should thoughts occur, the World State had a second line of defence. 'If ever by some unlucky chance', we are told, 'a crevice of time should yawn in the solid substance of their distractions, there is always soma'. Soma is a ubiquitous pleasure-drug that wards off thoughts. One character asks another why he doesn't just take soma to forget about the disruptive ideas he is having. Soma, writes Huxley, had 'all the advantages of Christianity and alcohol; none of their defects', it was 'Christianity without tears'.

In this way, Huxley offers us quite a different vision from Orwell. Orwell knew that 'to see what is in front of one's nose needs a constant struggle.' He recognised that we must continually reflect on the world to see it accurately. In *1984*, Big Brother believed that if citizens had enough material goods they 'would become literate and think'. People would then 'realise the minority with power had no function and should be swept away'. To prevent this realisation, the people are kept busy and poor.

Huxley's conception seems more prescient than Orwell's. If all our needs are met, there is no trigger for thought. A hungry stomach, not a sated one, means a busy brain. Huxley's second line of defence against thinking, the pleasure of soma, also appears more effective than Orwell's second line mechanism.[52] In *1984*, if someone senses a thought might be about to cross the horizon of their mind, they are encouraged to stop it. Big Brother has introduced the population to the concept of Crimestop. This is the faculty of knowing to stop at the threshold of a dangerous thought. Orwell calls it 'protective stupidity'. We see this in the Nazi concentration camps. Rudolf Höss, Commandant at Auschwitz, was asked why he didn't refuse the orders to kill people. His answer was that 'At that time I did not indulge in deliberation . . . I do not believe that even one of the thousands of SS leaders could have permitted such a thought to occur to him. Something like that was just completely impossible.'[53]

A century ago, US President Woodrow Wilson observed that 'whatever might be said against the chewing of tobacco . . . it gave a man time to think between sentences.'[54] The rhythm of life has dramatically changed since Wilson's era. In our instant gratification society, where everything is available on-demand, the space for reflection has collapsed. When desires can be automatically fulfilled, there is no trigger for reflection. We made machines that automated production, then automated consumption by making ourselves into machines. In capitalist societies, corporations are incentivised to prevent us from reflecting. It is more profitable for firms if we buy first and think later, if at all. Yet those who fail to sufficiently engage or appeal to our capacity to reflect, violate our right to freedom of thought.

Freedom of thought can be respected, protected and promoted by offering people opportunities to slow down their decision making, thereby creating space for reflection. One way to do this is to create metaphorical speed bumps that create decisional friction. When a tech firm serves up a suggestion for things to watch or purchase, or a post comes into our social media feed, the temptation is to respond instantly. Adverts or content suggestions that 'pop up' in front of us, based on the tech firm's algorithm's insights into us, are just like the first-order thoughts that pop into our minds. But if we click on these suggestions without thinking, we have failed to use our reflective second-order thoughts to decide whether or not we agree with the first-order thought that has been inserted into our head. We have not worked out if this is our thought or the thought of the corporation. We are encouraged to act in a way that we saw Harry Frankfurt describe as 'wanton'. No wonder we feel disgusted after getting lost in YouTube or Twitter.

Companies have used us for our attention. Surveillance capitalism encourages a wanton world.

The right to freedom of thought should require corporations to provide information in an 'autonomy-supportive context'.[55] Such a context encourages us to actively reflect, choose and participate in decision making. An autonomy-supportive context minimises the pressure on us to decide quickly. It uses non-pressuring language. Its fundamental goal is to help generate more second-order reflective thought. For example, such a context could allow people to create a short time-out before responding online to a social media post, proceeding to the next video or finalising a purchase.[56]

A real-life case of this happened when the Norwegian public broadcaster, NRK, wanted to reduce the number of people who left abusive comments on articles on their website. NRK designed their site so a would-be commentor had to correctly answer three basic multiple-choice questions about the article on which they wished to comment. If the user did not answer correctly, they were not able to make a comment.[57] Likewise, Instagram has trialled a feature in which AI detects potentially offensive posts before they are posted onto the site. The user is then alerted to the potentially offensive nature of their post and given the option to cancel it. Similarly, Twitter has tried encouraging users to read an article they are about to share if they have not yet opened the link that accessed the article. The effects of these interventions are hard to assess, but associated research suggests it is likely to have beneficial effects. For example, requiring people to stop and think why a headline is true or false makes them less likely to share false information with others.[58]

'Boosting' is another technique that can encourage reflection. Unlike nudging, which unconsciously pushes people into 'better' decisions, boosting helps people make better conscious decisions.[59] Boosting enhances people's cognitive abilities to understand and process information in order to support their decision making and enhance their autonomy. Whereas nudges are limited to the situation they occur in, people can take their boosted abilities into new situations.[60]

One example comes from boosting people's abilities to understand health-related statistics. Let's say that a new drug is advertised as reducing the chances of someone having a stroke by forty-eight percent. This makes the drug sound pretty remarkable. But this can be misleading, a fact that can be revealed by giving people access to another form of data – the absolute risk change. Thus, it may be that the normal rate of stroke in the population is twenty-eight strokes per 1,000 people, with the new drug reducing this rate to fifteen per 1,000. Whilst this is a forty-eight percent reduction relative to the current situation, the absolute reduction is thirteen strokes per 1,000 people, which equates to a 1.3% reduction. Suppose you are trying to balance the benefits of a drug against its potential side effects. Information on absolute risk change can help you make a more informed decision.[61]

We must also promote reflection during our search for information. Once we have entered a query into an internet search engine and seen the results, we will often tweak our search to try to get more valuable results. If our search doesn't give us results that accord with what we think to be true, we are more likely to go back and rework the search. If the search results fit with what we believe, we will probably stop our search.[62] When online, we need to be aware of our potential to seek confirmatory information and to resist antagonising information. In this sense, a good search engine should help us find what we want *and* what we don't want.

It is possible to present information to us in a way that promotes reflection. The use of warning tags on search results can help us usefully reflect. We can be forewarned that a source we are about to see is biased in a specific direction. Seeing such warnings makes people more aware of media bias. Similarly, if text is annotated, perhaps highlighting that a particular term is biased or provocative, this can also make readers more aware of other people's agendas.[63]

One way to avoid non-reflective thinking is to employ 'cognitive debiasing'.[64] The use of cognitive debiasing is common in medical training. Trainees are taught various techniques to help them make better clinical decisions. This includes the skills of reflection, check-list use and 'stop-and-think' techniques. The use of checklists alone has been credited with halving deaths from non-cardiac surgeries.[65]

Yet the evidence for the effectiveness of these safeguarding tech-niques remains mixed.[66] One reason for this is that people fail to believe they possess biases. Other reasons include critical thinking skills not transferring between tasks and the training backfiring.[67] The limited effectiveness of dedicated cognitive debiasing training suggests that it is unlikely to free a population from its chains of mental bias. Social media and other forms of corporate influence may have such a powerful hold over our minds that government regulation is needed to protect us from ourselves.

Although reflection seems positive, we should also be aware that it takes effort and should not always be encouraged. For exam-ple, if you are reading information on social media, you are being pressed to do two things at once. First, you need to decide whether what you are reading is accurate. Second, you need to decide whether to share it with your friends. Reflecting on whether to share the information uses up some of your mental resources. This means you have fewer mental resources available to assess the accuracy of the information. As a result, having to decide whether to share information means you understand the information less well.[68] Having the option of turning off the share option when reading on social media would help us think more clearly. Such small changes would help corporations design their services to respect users' rights to freedom of thought. In addition to prohibit-ing dark patterns, we need to encourage 'light patterns' that enhance autonomy and decision making.

Once we have created a space for reflection, we can engage our rule-of-reason thinking. Space and time are essential for rule-of-reason thinking to work effectively. Better arguments are only more persuasive than weaker ones when people get sufficient time to think about them.[69] It is this use of reason that we turn to next.

The tool of reason

As Leo Strauss put it, 'men are free politically to the extent to which they are induced to act properly not by force, but by persuasion.'[70] Yet the appropriate dichotomy is not simply between force and persuasion. We must differentiate between permissible and impermissible forms of persuasion. As discussed in chapter 3, when considering whether an alteration of thoughts is permissible, we need to assess the extent to which persuasion fails to engage or appeal to our capacity for reason sufficiently.

As Jan Christoph Bublitz points out, the ideal way to legitimately change someone else's mind or persuade them of something is by using rational arguments.[71] Bublitz argues that rational argumentation can't violate freedom of thought because 'it expresses the spirit of the provision: free and uncensored exchange of ideas'.[72] In such an exchange, Bublitz notes, the better argument wins by what the philosopher Jürgen Habermas termed 'unforced force' ('zwanglose Zwang').

A long philosophical tradition argues that reason is the appropriate way to persuade others. Kant argued that people should be treated as ends, not means. This meant that we should offer reasons for a given course of action to persuade others. In doing so, we treat others as rational beings worthy of the same respect as ourselves. The other person can then decide for themselves whether the reason given is a good one. For Kant, a free will was one that responded to reasons, not force.[73] In contrast, non-rational persuasion involves us trying to make others an instrument of our will. It treats others like objects to be manipulated, rather than offering them a chance to exercise their rationality. Such persuasion violates other people's autonomy and dignity.

The disabling or circumvention of an individual's ability to rationally appraise information poses problems for freedom of thought. We discussed earlier how using techniques known to trigger rule-of-thumb thought could be an impermissible alteration of thought. We should also add that bypassing appeals to reason, in order to alter our emotions, which in turn influences our thought, is also problematic from the perspective of freedom of thought.

An example of this is a controversial experiment run through Facebook that examined what happened when the emotional content of people's newsfeeds was manipulated.[74] The study found that exposing users to more negative stories increased their negative emotions. This finding reflected psychologist John Bargh's observation that 'methods to thwart or bypass the consumer's defences against influence are becoming ever more powerful, and yet he [the consumer] remains as ignorant of these influences and . . . overconfident of his control'.[75] Despite the new methods through which minds can be manipulated, marketing regulations still focus on deceptive or misleading information. Regulators should also be addressing companies' attempts to bypass users' rule-of-reason thinking.

Linking up reasoning and reflection, an environment that supports freedom of thought will support the development and deployment of cognitive reflection. To understand this, consider the following problem:

A bat and ball cost £1.10 in total. The bat costs £1.00 more than the ball. How much does the ball cost?

Your gut reaction is probably to say that the ball costs 10 pence. However, if you reflect and engage rule-of-reason thought, you will see this can't be right. If the ball costs 10 pence and the bat costs a pound more, then the bat would cost £1.10. These two items would cost £1.20 in total. The answer is that the ball costs 5 pence, meaning the bat costs £1.05, which results in a total cost of £1.10. People who can solve such problems, overcoming their gut reaction, are said to display good 'cognitive reflection' abilities.

When we fail to develop or deploy these abilities, we can find ourselves taken in by 'fake news'. We tend to believe fake news because we fail to employ rule-of-reason thought. People high in cognitive reflection are less likely to believe false news.[76]

Reliance on rule-of-thumb thinking and associated gut feelings to assess claims can sometimes be helpful.[77] However, reliance on gut-feel can also be highly problematic. For example, what do you make of the following sentence: 'We are in the midst of a high-frequency blossoming of interconnectedness that will give us access to the quantum soup itself'. Maybe you feel this is profound? If so, some researchers would say you are high in 'pseudo-profound bullshit receptivity'. People with high levels of this trait are more likely to believe fake news and are worse at differentiating between fake and real news.[78] Fake profundity is a powerful way to try to make us believe bullshit.[79] Indeed, it is its own industry. Not only is academia not immune to this, but it arguably both originates and propagates bullshit.[80] Rational engagement with such material is essential.

Transparency in advertising will help promote rule-of-reason thinking. We already insist that adverts in magazines identify themselves as adverts. We don't let them pass themselves off as objective news pieces or disinterested reports. This is helpful because simply knowing that others are trying to persuade us can increase our mental autonomy. Our natural response to others trying to persuade us or restrict our choice is to resist.[81] Naturally, marketers attempt to circumvent such resistance by making it seem like they aren't trying to influence us. For marketers, the best way to get around resistance is not to trigger it in the first place. To combat this, we need to make it salient to people when persuasion is in process. This is also relevant in the context of news reporting. The ever-increasing use by time-pressured journalists of materials developed for them by corporations blurs the lines between reporting and propaganda.[82] News needs to be more transparent on the extent to which it contains direct and indirect attempts to persuade. Greater public awareness of these issues will support freedom of thought.

Yet there are problems with constantly banging on the 'more public education' and 'more public awareness' drums. Doing so places the burden on the public to engage in rule-of-reason thought and detracts from the responsibility of corporations not to encourage rule-of-thumb persuasion. Improving education on rule-of-thumb and rule-of-reasoning thinking is not a panacea. If others can instil a desire or ideology in us, we will reason from these potentially problematic premises. Public education on reasoned thought could still result in reason following feelings. People may end up finding reasons to follow the desire that the marketer has created in them, rather than reflectively evaluating this desire.

A related problem is that even if you can teach people to reason better, their prior ideological commitments can still resist reason. This was shown in a study that examined how good people were at solving logical problems.[83] These problems took the form of a first premise (e.g. all mastiffs are dogs), a second premise (e.g. some mastiffs are black) and a conclusion (e.g. some of the things which are black are dogs). Participants had to say whether the conclusion followed from the premise. In the dog example here, it does. The researchers were interested in whether this was an abstract skill. So, if you replaced dogs with a politically contentious topic, would people remain just as good at answering the questions? Or, if people had strong beliefs about the topic, would their prior commitments get in the way of their reasoning? The topic the researchers chose was abortion. Again, people were given premises and a conclusion that either did or did not logically follow. It turned out that how well people could reason depended on whether the argument they were evaluating agreed or disagreed with their own personal views on abortion.

The same effect has been found for political ideology. You are more likely to correctly evaluate arguments that are consistent with your political ideology.[84] Your political beliefs interfere with your ability to evaluate strictly logical arguments in related areas. Once you have a preferred view, your logical reasoning ability becomes imperilled. Again, this points to the social nature of free thought. We need to be able to sit together and reason together to have the best chance of overcoming our biases.

More generally, any regulation of what are deemed to be impermissible alterations of thought runs into the problem of regulating the speech that manipulates thought. Due to the strong protection historically accorded to speech, the bar for impermissible alteration of thought must be set very high. One case that came close to trying to balance thought and speech was *Kokkinakis v. Greece*.

In this case, the European Court of Human Rights ruled that trying to convert someone to your religion counts as exercising your religious freedom. This religious freedom is protected by Article 9 of the European Convention on Human Rights ('Freedom of thought, conscience and religion'). Yet the Court ruled that believers' desires to be free from people attempting to convert them was also protected by Article 9. As both missionaries and believers could appeal to Article 9, the Court had to strike a balance. In this case, the Court held that democracies shouldn't restrict people's abilities to try to convert others as long as people aren't resorting to illegitimate means to influence thought.

Let's take a quick break from lauding reason to see if we can find any problems with it. We can use reason to uncover truth, but it is better viewed as, at best, a tool to argue and, at worst, a tool of domination. 'What can you do', Winston Smith wonders in Orwell's *1984*, 'against the lunatic who is more intelligent than yourself'. This is one of the most important overlooked points that Orwell makes in his novel. It is problematic when someone who has mastered the art of reason deploys this skill against someone with significantly less education. In such a case, reason is a tool of power being used to coerce the other. As the novelist Elizabeth Gaskell had one of her characters say, 'I'll not listen to reason . . . Reason always means what someone else has got to say.'[85] The right to freedom of thought will need to address situations where there are disparities in reasoning abilities, as well as provide a right to think unreasonably as a tool of resistance.

Although the idea that free thought is reasoned thought is pervasive, not everyone buys into this idea. We have all heard the phrase 'the dictates of reason'. Despite this, we would not intuitively think of reason as a tyrant. The Ancient Greeks certainly wouldn't. However, the philosopher Robert Nozick did. Nozick claimed that philosophical arguments involved 'forcing others to believe things.'[86] As Nozick put it, 'The terminology of philosophical art is coercive: arguments are powerful and best when they are knockdown, arguments force you to a conclusion, if you believe the premises you have to or must believe the conclusion'. Karl Marx emphasised this too. Marx observed how the power of ideas came from their logical force, which 'like a mighty tentacle seizes you on all sides as in a vice and from whose grip you are powerless to tear yourself away; you must either surrender or make up your mind to utter defeat.'[87]

Nozick claims that such argumentative force is problematic. As he put it, 'philosophical argument, trying to get someone to believe something whether he wants to believe it or not, is not, I have held, a nice way to behave toward someone'. To insist on following reason, argued Nozick, is to 'become [the] thought-police'. Again, this points to the right to freedom of thought supporting our right to think unreasonably. It also highlights that if reason demands a journey of necessary steps, a significant portion of freedom of thought resides in our ability to choose our premises or first principles.[88] These may not be able to be rationally justified and may represent a leap of faith. Again, this reduces the centrality of reason to free thought.

It is not only our first principles that may require a leap of faith. The insistence that reason must be used to explore where these principles lead may represent another leap of faith. As Stanley Fish puts it, 'liberalism depends on not inquiring into the status of reason'.[89] It depends on the status of reason being obvious. Yet, as Fish adds, 'not all reasons . . . are reasons for everyone . . . What is and is not a reason will always be a matter of faith'. There is hence a danger in arguing that the right to freedom of thought is primarily reasoned thought. If we cannot agree on what reasoned thought is, power is likely to be the ultimate adjudicator as to what thought is reasoned.

Here we enter the twilight world of Nietzsche. For this German philosopher, reason was a weapon, a 'pitiless instrument', with whose aid those who could not physically command obedience could 'play the tyrant'.[90] Here we can see an early form of Mercier and Sperber's claim that reasoning is for winning arguments and thereby exerting power. Nietzsche saw reason as potentially pathological. He viewed the philosophy of Socrates, 'rationality at any cost', as a 'dangerous, life-undermining force'.[91] Reason ignored what Nietzsche viewed as our nobler instincts, which made it 'no more than a form of sickness.'[92]

This led Nietzsche to believe that free thought was not limited to reasoned thought. Here he diverged from a host of philosophers both before and after him. For these other philosophers, as we have seen, we become autonomous by thinking freely, and thinking freely means using reason. Nietzsche rejected this. He felt that the autonomous person created their own source of value, not limited to the dictates of reason, society or God.[93] We can take much from this, but let's limit ourselves to two points. First, thought is a dangerous weapon and anything but harmless. Second, the right to freedom of thought should protect all forms of thought, not just reasoned thought.

Indeed, seemingly absurd ideas may represent the wisdom of tradition. The Enlightenment thinker may be hubristic in thinking their reasoned conclusions are superior to time-tested knowledge. The anthropologist Joseph Henrich gives some good examples of this. Henrich documents multiple situations where tradition will keep you safe and reason will get you killed.[94] For example, pregnant women in Fiji face a cultural taboo against eating shark meat. When such women were asked why they didn't eat shark, they answered that it was because their babies would be born with rough skin. If they had rationally considered this idea, they would probably have rejected it and eaten the shark. This would have been a mistake. The local sharks contain toxins that can damage an unborn child. The right to freedom of thought needs to protect all thought, not just reasoned thought.

Finally, what happens if people don't respond to reason? Sam Harris recently addressed this. As he put it, 'All we have is a capacity to persuade one another . . . or we have violence . . . In the end . . . we have to force people to do stuff if we can't persuade them to do stuff'.[95] Behind the dream of gentle reason lurks the underlying reality of force. At least force is honest. The right to freedom of thought needs to protect both reasoned and unreasoned thought, and to shield us all from the wrath of the reasonable.

The tool of courage

Possessing intellectual virtues, such as humility, open-mindedness, a desire for truth and intellectual sobriety, is essential for freethinking.[96] Similarly, so is abjuring intellectual vices, such as gullibility and dogmatism. Equally essential is the virtue of courage. We saw earlier how, in Ancient Athens, parrhesiasts needed the courage to speak truth and the listener needed the courage to hear. Thinking, whether silent and internal or aloud and external, requires courage. As Nietzsche put it, we can 'approach only as near to truth' as we have 'the courage to advance'.[97]

Courage is the ability to act despite being afraid.[98] When we speak of the courage to think freely, we refer to moral courage, the willingness to risk social rebuke or humiliation. If we are thinking to ourselves in our head, we need the courage to face unpleasant truths about ourselves and the world. We need the courage to doubt the beliefs of our tribe because such questioning threatens to undermine our sense of belonging.[99] If we think aloud with others, we need the courage to accept potential backlash. Increasingly, this may be a physical backlash, meaning free thought requires both moral and physical courage.

We need courage to think freely because free thought is only possible when we are prepared to recognise our limitations. Humility takes courage. Yet we live in a world that encourages pridefulness. The internet is making us more prideful. Being able to search the internet for answers to questions (such as 'How does a zipper work?') makes us arrogant. A study has shown that this causes us to think we will now be better at answering new questions on completely different topics, such as 'Why are cloudy nights warmer?'[100] Similarly, we need to concede that our will-power has limits.[101] The idea that our willpower is infinite is unrealistic and dangerous. We need the courage to concede we are easily manipulable. We need the humility to admit that others are more powerful than we are. Setting pride aside means we can take remedying action. We can choose in advance to bind ourselves. We may, for example, lock ourselves out from certain semi-addictive apps. Doing so requires us to be humble enough to recognise that we will not be able to resist temptation.

We can adapt an idea of Peter Thiel's to help us illustrate the link between humility and free thought. If you were ever to have a job interview with Thiel, he would probably ask, 'What important truth do very few people agree with you on?'[102] To be able to answer this means you have escaped the false consciousness that many others live in and have seen a truth. Yet there is perhaps a better question, namely, 'What do you and most other people believe that isn't true?' This reflects a point made by the cultural theorist Stuart Hall, when he wondered: 'how it is that all the people I know are absolutely convinced that they are not in false consciousness, but can tell at the drop of a hat that everyone else is.'[103] We need to run a virus scan on our minds before we try to develop new thoughts. To do this requires the humility to realise that we, and not just others, may be living within illusions.

If courage is required to think freely, this suggests two ways to promote free thought. The first is to help people cultivate the virtue of courage. To do this, we need to understand what makes people courageous. Research suggests courageous people are likely to be hopeful, positive, empathetic and mindful, and to possess personality traits of openness to experience and conscientiousness.[104] The tendency to become angry at injustice, unfairness and human rights violations can also push people into acting or thinking courageously.[105] The freethinker may even be spiteful.[106] All these features add up to a courageous mindset.[107] This mindset can be activated simply by writing down previous times one has been courageous. Doing so has been found to help us manifest courage as measured by an increased willingness to face our fears.[108]

The second way to promote free thought is to reduce the courage required to think. One way to do this is to think with others. Courage is contagious. If we are in a group with others who think courageously, our courage will be enhanced. Being in a context where other people have created a mission of free thought, who have set a norm of free thought, who encourage us to think freely, and act as freethinking role models, will all help us have the courage to think freely.[109] For example, the extent to which a soldier can cope under fire and display moral courage is associated with whether they have a courageous leader.[110] Once again we see being in a group is associated with a feature that promotes free thought.[111]

More generally, reducing the courage it takes to think will involve constructing an environment, a culture and a society that supports and promotes free thought. In such a society, as the physicist Richard Feynman put it, 'doubt is not to be feared but welcomed and discussed'.[112] To achieve this, we need to create spaces where doubt can be expressed and discussion can take place. The right to freedom of thought needs to place an obligation on the state to provide and protect such spaces. It is to the environment which supports free thought that we turn next.

7

CREATING FREE
THOUGHT III: WORKSPACE

Literary depictions of dystopian societies often show tyrannical regimes not wanting citizens to be alone. In Huxley's *Brave New World*, solitude is discouraged because the rulers feel it is socially subversive. By the time children finish their education in Huxley's dystopia, they have had 'at least a quarter of a million warnings against solitude'. Orwell shows the same mechanism of control in *1984*. We are told of Winston Smith that 'until he could be alone it was impossible to think'. Yet he lives in a society where to seek solitude is to elicit suspicion. One workspace we need to secure is the space for private thought. As the US Supreme Court has noted, without opportunities for 'serenity and reflection . . . freedom of thought becomes a mocking phrase, and without freedom of thought, there can be no free society'.[1]

A private workspace for thought

To think, we need to have the option of mental privacy. This is our ability to determine when, how and to what extent information about our thoughts is communicated to others.[2] Mental privacy calls for the recognition that there is a boundary between ourselves and others that must be respected.[3] When St Paul asked, 'who knows a person's thoughts except their own spirit within them?' (1

Corinthians 2:11), the answer he had in mind was clearly God alone. Two millennia later, the same answer was being given. In the nineteenth century, Sir William Blackstone claimed that 'as no temporal tribunal can search the heart, or fathom the intentions of the mind, otherwise than as they are demonstrated by outward actions, it therefore cannot punish for what it cannot know'.[4] Yet today the sanctity of our inner workspace is being threatened by both brain- and behaviour-reading.

The immediate problem we encounter when considering mental privacy is that we mind-read seven times before breakfast. Other people can already see, albeit often through a glass darkly, into our souls. We are constantly trying to figure out what other people are thinking by scrutinising their facial expressions or pondering their behaviour. Fans of *Curb Your Enthusiasm* will be familiar with the infamous Larry David stare-down, in which he tries to work out what others know by closely scrutinising their faces.

The ability to understand that other people have their own beliefs and intentions is termed 'theory of mind' or 'mentalising'. Most children develop the basic form of this ability at four years of age.[5] The evolution of this ability, which only exists in its full-blown form in humans, was presumably driven by the benefits of accurately predicting what other people will do.[6]

Working out what other people think led to an evolutionary arms race of the mind. There was now an advantage to being able to conceal one's thoughts and deceive others. Natural selection began to select for this ability. Today, the ability to conceal what we think develops at around four to five years of age.[7] But, once the ability to conceal and deceive had evolved, natural selection began favouring those who could detect deception. This too is a skill that develops through childhood today.[8]

Clearly, we don't deem our evolved ability to work out what other people are thinking to be a violation of other people's right to mental privacy.[9] If we did, then almost the entire population would be criminalised. Yet we instinctively feel that the brain- and behaviour-reading techniques discussed in chapter 3 violate our right to mental privacy. So, what is the difference between using natural and technological ways of working out what others are thinking?[10]

Maybe it is because these new brain- and behaviour-reading techniques can polish the dark glass of our minds to see inside more accurately than the average person could. Yet some people are better than others at working out what someone is thinking, and we haven't locked up all the mentalists (yet). Instead, the problem seems to be, once again, due to a power imbalance. We didn't evolve to deal with silicon trying to see inside us. Tech gives us a supranatural ability to infer people's thoughts, which can steamroller through evolved defences. This creates a significant power imbalance between the possessors of such technology and their targets. As noted earlier, the European Court of Human Rights has ruled that power imbalances between the sender and recipient of a message can create an improper influence on thought.[11] The possessors of brain- and/or behaviour-reading technology could therefore be deemed to have 'improper access' to the minds of individuals. Big Tech bringing mind-reading technology into society is like European colonisers bringing smallpox to indigenous populations. It creates a threat we have no natural defence against and has the potential to be devastating.

One way that mental privacy has been violated historically is by getting people to swear oaths. For hundreds of years, requiring people to swear oaths was the main way to peer into others' minds.[12] Oaths were a way 'both for externalizing the mind and for making it subject to control'.[13] The introduction of oaths owed much to the insecurities of Protestant leaders. John Calvin emphasised that believers' inner worlds were private. This helped keep them out of the hands of the Catholic Church. But eventually, the Protestant Church wanted to take away this inner freedom. To do so, it forced people to make verbal reports about their inner experiences to assess the state of their inner world.[14] The US tried to use this technique in the twentieth century and went on an 'oath binge'.[15] Eventually, the courts intervened. Judges argued that if the state asks someone to swear an oath they are seeking 'to probe what an applicant's state of mind is' and this was 'not the business of the State'.[16]

Today, oaths are not seen as sufficiently reliable. Instead, we have returned to the old-school inquisition using new-age methods. It is no longer sufficient to swear that you are not biased. The potential for us to lack insight into our minds means we could be biased and not even know about it.[17] As a result, in 1998 researchers created the Implicit Association Test. This test claims to be able to detect unconscious biases.[18] It is based on examining whether people's reaction times to certain words (good, bad, etc.) are different when these words are paired with a specific trait or characteristic such as race or gender. You can try it yourself on a range of websites.* Despite continuing controversy over the validity of this method,[19] it has been widely utilised by human resource departments and businesses.

Yet if one is being forced or pressured into doing such a test, this must be recognised as a blatant violation of one's right to freedom of thought. Those involved in creating the test state it is 'a method that gives the clearest window now available into a region of the mind that is inaccessible to question-asking methods'.[20] Opening a window into someone's mind without them being able to refuse is a breach of mental privacy and hence a violation of the right to freedom of thought. That this window is being opened with the noble aim of reducing bias and prejudice is irrelevant.

There are many justifications for why we should possess the right to mental privacy. One argument is that it would destroy society if we all knew the disturbing thoughts each other sometimes

---

* See Project Implicit (2011), https://implicit.harvard.edu/implicit/.

has. As philosopher Thomas Nagel puts it, 'civilization would be impossible if we could all read each other's minds'.[21] But is this true? If we realised that spontaneous, disturbing thoughts were common, we may become less judgemental of their occurrence. Complete mental transparency could help us to stop pretending that we are what we are not. We may even gain respect for each other when we fully realise people's ability to control whether or not their thoughts translate into behaviour. Likewise, full knowledge of what happens in each other's head could cause people to be less distressed by intrusive unwanted thoughts.[22] They would not only see that many others had such thoughts but also that the occurrence of such thoughts didn't automatically mean that awful acts followed.

In reality, if given access to other people's minds, we would do what humans always do – use this knowledge to attack others for personal gain. This would crush independent, creative and challenging thought, leading to stultifying conformity. The strongest argument for a right to mental privacy is that our mental autonomy would be diminished without it. In terms of the ARRC of free thought, conformity pressures would impair our ability to reason freely, discourage reflection and increase the amount of courage needed to think freely. This issue of mental conformity is worth exploring in more detail.

A 'lock on the door', the novelist Virginia Woolf once wrote, 'means the power to think for oneself'.[23] If all our thoughts were exposed to others, social pressures would push us to conform our thinking. As Julie Cohen puts it, 'examination chills experimentation with the unorthodox, the unpopular, and the merely unfinished.'[24]

In *1984*, George Orwell wrote: 'The party told you to reject the evidence of your eyes and ears. It was their final, most essential command.' Yet we are often quite happy to reject the evidence of our ears and eyes and to conform to the judgements of others. We conform our behaviour to social norms for a good reason; we have evolved to behave this way. Conformity offers three significant survival benefits.[25] First, it can help us be more accurate in our decision making. Often the majority is correct. Second, people like us more if we agree with them. There are social rewards for fitting in and punishments for not conforming. As Douglas Murray puts it, 'to think aloud on the issues which are most controversial has become such a high risk that on a simple risk/reward ratio there is almost no point in anyone taking it'.[26] Finally, conformity helps enhance, repair and protect our self-esteem.

The survival benefits of conformity mean our brains evolved to motivate us to conform. Our brain recognises when we are deviating from social norms and reduces activation in the parts of our brain associated with processing rewards.[27] Failing to conform is associated with increased activity in the fear and stress centres of the brain.[28] If someone made (or threatened to make) our thoughts public against our will, a powerful biological force would rise within us. This force would push us to conform our thoughts to social norms.

Such pressures could lead our thoughts to conform in two ways. First, it would encourage us to employ a range of strategies to try to consciously control our thoughts. People use various techniques to control their thoughts, including punishment, worry, distraction, social control and reappraisal.[29] Second, and more insidiously, we could mentally conform without even knowing we had done so. Mental conformity can occur through non-conscious mechanisms.

We can see unconscious mental conformity happening in the lab. One study presented participants with a list of fifteen words and asked them to free-associate other related words.[30] So, if shown the word 'dog', someone might say things like 'cat', 'bone', 'walkies', 'woof' and 'retriever'. Electric shocks followed some of the words participants generated. Participants were then given the same fifteen words and asked to free-associate again, but no shocks were given this time. The researchers found that participants were less likely to say words that they'd previously been shocked after. Crucially, some participants were not aware that the shocks were why they were acting this way. We can mentally conform without even knowing we are doing so. This is problematic because if we don't know we are conforming we can't consciously resist.

People living in democratic societies are already censoring their thinking due to invasions of mental privacy. For example, let's take it to be the case, as I've argued, that some of our internet searches represent thinking. A study found that Edward Snowden's June 2013 revelations of large-scale government internet surveillance had a chilling effect on internet search behaviour. This study collected data on internet search term volume before and after June 2013. The researchers obtained data on the volume of searches done in the US and its top ten international trading partners during all 2013 for a range of search terms. Independent raters then judged how likely each search term was to get the user in trouble with either the US government (to assess self-censorship of potentially illegal behaviour) or a friend (to assess self-censorship of personal issues). The researchers then looked to see whether Snowden's revelations affected the number of searches.[31] They found that following the June 2013 Snowden revelations, there was a ten percent drop in search terms that were assessed as potentially getting individuals in trouble with the government. There was also a significant drop in the use of terms that could be personally embarrassing if revealed. Surveillance breeds servility.

Even if others didn't use data from us to infer our thoughts, our knowledge that they could is still a problem in itself. In seven-teenth-century England, a theory of freedom was proposed by a group that Thomas Hobbes called the 'Democratical Gentlemen'.[32] This group were looking to justify the existence of Parliament and the execution of King Charles I. They claimed that our freedom is impaired not only by people interfering or threatening to interfere with us, but also by simply knowing that we are living dependent on the good will of others. Even if a despotic king did not interfere with your life, simply knowing that he could if he wanted to was a violation of freedom. People would live by grace, not by right. Living in a state of dependence, they would not be free citizens but slaves. A similar point was made by Jeremy Bentham with his concept of the panopticon. Updating this for the modern age, government need not actively monitor our internet searches to influence our behaviour. All they need to do is let us know we are possibly being watched and we will change our behaviour.

There is, therefore, a fundamental need for what the privacy law expert Neil Richards has termed intellectual privacy. Richards defines this as a 'zone of protection that guards our ability to make up our minds freely ... the protection from surveillance or unwanted interference by others when we are engaged in the process of generating ideas'.[33] Protecting and promoting mental privacy should be a central pillar of the right to freedom of thought.

Before we make mental privacy absolute, we need to consider both its payoffs and its price. Let us not assume that mental privacy will always make the world a better place. To take a low-level example, consider the potential implications for archiving. Many writers give explicit instructions that their writings should be destroyed after death. Virgil is said to have asked for his *Aeneid* to be burnt when he died.[34] Kafka explicitly asked for his stories to be destroyed upon his death. Yet Kafka's agent preserved all his stories. What a librarian might term an 'act of supreme self-curation' is potentially a violation of freedom of thought.[35]

There are also larger potential costs to absolute mental privacy. As noted earlier, the inability to access large-scale data on people's minds has the potential to limit our understanding of the human mind, prevent the development of new interventions for mental health conditions and block our way to the Ancient Greeks' injunction to 'know thyself'. These costs may be outweighed by the benefits of absolute mental privacy, but we must put them on the scales to find out.

A public workspace for thought

In 1960, the American physicist Freeman Dyson pondered the problem of how a high-tech civilisation could get sufficient energy for its needs.[36] The answer he came up with is today known as the Dyson sphere. This is a hypothetical mega-structure that would surround a star and captures its energy output. But civilisations are not powered and secured by energy alone. The most powerful resource we have is thought. Throughout these pages, I have argued that thought is most free when we think together. If we gave full protection to thinking in both private and public, society could harness the power of the thoughts of a civilisation. We would have a Dyson sphere for thought. Yet creating such a sphere would require profound protections to be put in place for both private and public thought.

A key problem with creating a public space for thought is again the pressure to conform to the norms of a society. Public opinion can crush original thought and compel conformity more effectively than any tyrant or law ever could. We may be free to think whatever we like, but if social pressures push everything in one direction, thought is corralled like a bull on a ranch. As Spengler observed, 'the thought, and consequently the action, of the mass are kept under iron pressure – for which reason, and for which reason only, men are permitted to be readers and voters.'[37]

Some have worried that if we made use of our right to freedom of thought, then ruling powers would quickly dilute, restrict or remove this right.[38] As the famous saying goes, if voting made a difference it would be banned. If freedom of thought made a difference, would it too be banned? Or would it simply be a right that was never meaningfully developed and implemented?

Earlier, I mentioned Chomsky's argument that thought control is likely to be more important in democracies than tyrannies. Part of the reason for this is that, from the perspective of rulers, the more powerful the public are, the more enchained their minds must be.[39] Conformity becomes a powerful mechanism of thought control in democracies as it is an unseen force that causes a mind to wrap a chain around itself. In liberal democracies, citizens are compelled into thought conformity by isolation, shame and fear.[40] What the legal scholar Adrian Vermeule calls the 'evidence-based freethinkers of the quiet car' must either accept mainstream doctrine or act as if they do.[41]

Conformity pressures create a need for government to support the public's ability to think aloud freely. Philosophers throughout the ages have emphasised the duty of government to enable and support the development of the capacities of its citizens. As Spinoza puts it, the purposes of government are 'not to change men from rational beings into beasts or puppets, but to enable them to develop their minds and bodies in security'.[42] Similarly, as the US Supreme Court has said of the Founding Fathers, 'those who won our independence believed that the final end of the State was to make men free to develop their faculties.'[43] But it is not just up to governments to support our ability to think freely aloud. As the French thinker Alexis de Tocqueville noted, we must also support our fellow citizens in their efforts to do this.[44]

For people to be willing and able to think with each other, we must remove the ability of others to punish them for such activities. The appropriate response to thoughts you do not like is disagreement, not punishment. If someone can have their livelihood taken away from them because they are thinking aloud, then thought is not free. The philosopher Bertrand Russell recognised this a hundred years ago. He pointed out that 'thought is not free if the profession of certain opinions makes it impossible to earn a living.'[45] Russell likewise prophetically noted that 'the trend of economic development will make the preservation of mental freedom increasingly difficult, unless public opinion insists that the employer shall control nothing in the life of the employee except his work.' By this criterion, thought is not free today.

So, what can we do about this, given any solution is highly likely to create new and potentially larger problems? Should we make it illegal for people to call for other people to be fired for their thoughts? The legal concept of tortious interference could be repurposed here. This essentially states that, subject to certain conditions, it is illegal for you to intentionally interfere with a contractual arrangement held by two other parties. If you call for someone else to be fired because of their thoughts, this could be deemed illegal. You would still have every right to disagree with the other person's thoughts. But you could not attempt to harm them by seeking to disrupt their employment contract. Still, this would be a new restriction on free speech. An alternative approach would be to stop employers from leaking power to the public and journalists by prohibiting employers from firing employees for their thoughts.[46]

In the US, the First Amendment already offers people some protection from being fired from government jobs for their thoughts. However, this is not the case in the private sector. Employees of private enterprises can lose their livelihood, home, mental health and healthcare for what they think, whether inside or outside work hours. This means the state is not protecting its citizens' right to free thought. Governments should ensure people cannot be fired for engaging in what I have called thoughtspeech.

More generally, people should have the right not to be able to be fired, discriminated against or hated for being a thinker. It should be a protected characteristic. Take the case of James Damore, who was fired from Google for a memo he wrote about gender. Whether or not one agrees with the content of Damore's memo, it can be argued that his memo was him thinking about gender. The government would be unable to censure Damore for such speech. Indeed, in *Doe v. University of Michigan*, a graduate psychology student, who studied the biological bases of individual differences in personality traits and mental abilities, successfully took his university to court to get a policy overturned that aimed to prevent 'racist speech'.[47] The student claimed, and the court agreed, that theories that proposed biologically based differences between the sexes and among the races might be perceived as 'sexist' and 'racist' by some students, and hence be banned under the university's policy. The court deemed this to be unacceptable. If the immense power of the state is not allowed to quash citizens' thoughts, why would we allow trillion-dollar (or indeed any) corporations to do this?

A democracy that allows corporations to fire employees for their thoughts is not worthy of the name. If Damore could not be fired by his employer for writing such a memo, it would be pointless for those who objected to its content to call for him to be fired for his thoughts. Critics could only disagree, which is all they should be able to do. Those who disagreed would then have to engage in a reasoned argument against Damore's position, rather than simply try to silence him by sending him to the gulag of unemployment. Of course, one suspects that people often call for others to be fired precisely because they don't have good counterarguments. If you can't win with words, neither your nor other people's fists, metaphorical or literal, should be your backup plan.

The only people who are, at least on paper, safe from being fired for their thoughts are senior academics at universities who have acquired tenure. Tenure means that, failing extraordinary events, the academic effectively has a job for life. This aims to allow academics to freely pursue truth, however unpopular it may be. The US Supreme Court has stated that such academic freedom is of 'transcendent value to us all'.[48] Yet the freedom to seek truth is essential for *every single person* in a democracy. Truth is needed for us to be meaningfully self-governing creatures. All employees, whether public or private, should be given a form of intellectual tenure. This would mean that whilst they may still be dismissed from their jobs, this could not be for any good faith truth-seeking activity they have engaged in. Only then will free thought and self-government truly be possible. Yet this could have unforeseen (or, to an economist, quite easily foreseen) detrimental effects on a country's economy, due to increased rigidities in the labour market. This simply clarifies that if we want freedom of thought then there will be a cost. The question is whether we are willing to pay the price and to put our money where our minds are.

A public culture of free thought

In a 1979 speech, which must have been a great disappointment to the legal profession, US President Jimmy Carter voiced the opinion that 'all the legislation in the world can't fix what's wrong with America'.[49] Similarly, we can make all the laws we want to protect thought, but unless we live in a culture that supports rather than punishes freethinking, our efforts will be fruitless.

An example of how our culture does not support free thought can be seen in the response to Sam Harris's efforts to think aloud in public. In his *Making Sense* podcast, Harris thinks out loud with guests about some of the most contentious topics in contemporary society. He does what Kant called for – he dares to think, and does so with courage, integrity and evidence. Unfortunately, many listeners are not happy to countersign the parrhesiastic contract with Harris. A clear example of this arose when Harris broached the topic of torture.[50]

Like the right to freedom of thought, the United Nations has deemed the right not to be tortured an absolute human right. Yet

Harris thought aloud about whether torture could sometimes be ethical. He came to the belief that, in some extreme situations, practices like waterboarding may not just be ethically justifiable but ethically necessary. This could be the case, Harris suggested, if getting information from a terrorist could save thousands of lives. Another argument he made was that if we think that it is sometimes acceptable to drop a bomb to kill a terrorist, risking maiming innocent bystanders, we should also think it is sometimes acceptable to waterboard a terrorist, risking having got the wrong person.

Harris was clear that by arguing that torture could sometimes be ethically justified, he did not think it should ever be legal. The analogy here is with theft. It may sometimes be ethical for someone to steal a loaf of bread, but this doesn't mean it should be legal. Harris's preferred course of action was that torture remained illegal, but that 'our interrogators should know that there are certain circumstances in which it will be ethical to break the law'. So, basically the Jack Bauer solution. Whilst not arguing that torture should be legal, Harris pointed out that he couldn't see an argument that would rule out the following law: 'We will not torture anyone under any circumstances unless we are certain, beyond all reasonable doubt, that the person in our custody has operational knowledge of an imminent act of nuclear terrorism.'[51]

This led to an outcry. Whilst it was to be hoped that others would challenge Harris's view, the problem was the manner of many people's objections. Among other things, Harris was accused of being 'boastfully pro-torture', which he clearly was not.[52] There was a widespread failure to engage with Harris's thoughts rationally and some remarkable displays of intellectual dishonesty. Harris's thought on this topic continues to be misused to shock viewers for clicks, social approbation and money. This has chilled Harris's willingness to think publicly. As he put it:

I am sure that the world needs someone to *think out loud* about the ethics of torture, and to point out the discrepancies in how we weight various harms for which we hold one another morally culpable, but that someone did not need to be me. The subject has done nothing but distract and sicken readers who might have otherwise found my work useful.[53]

It seems highly likely that Harris became disinclined to think aloud with his guests about other contentious topics too. This is a loss for free thought and ultimately a loss for us all. It is hard to build a Dyson sphere of thought when, like in the 1997 science-fiction film *Contact*, people keep trying to blow it up.

To create a culture of free thought, we need to change how thinking is perceived. To do this, we need to understand the psychology and neuroscience of punishment. The first thing we must tackle is the perception that it is not right for someone else to think something that we don't like. A key reason people punish others is because they think the other person is acting unfairly. When we anticipate punishing another person who has acted unfairly, the same parts of our brain light up as when drug users anticipate taking cocaine.[54] Punishing wrongdoers is like a drug to humans. For this reason, we need to view people as fairly having the right to explore ideas.

We must also tackle the idea that disagreeing with the consensus or 'the current thing' is abnormal. We need to make it the norm to question and disagree, as Feynman suggested. This is incredibly important because the human brain is set up to dehumanise and punish people who violate the norms of society. This was found by Katrina Fincher and Philip Tetlock at the University of Pennsylvania who discovered that our brains don't process the faces of people who violate norms in the same way we process faces normally. People who violate norms literally look less human to our brains. This makes it easier for us to turn off our empathy and punish them.[55]

We need to create a society with a norm of thinking, where we all recognise the right to think. To do this, we need to emphasise our duty to think. As the political scientist Kenneth Minogue points out, we need to escape from a world in which our moral status is assessed not by our ability to live up to our responsibilities, but by our ability to parrot the right opinions, which represents the ultimate servility.[56]

We must also hold our ideas more loosely. Once an idea is deemed sacred, it becomes an obstacle to free thought. As Salman Rushdie puts it, 'the moment you say an idea system is sacred, whether it's a religious belief system or a secular ideology, the moment you declare a set of ideas to be immune from criticism, satire, derision, or contempt, freedom of thought becomes impossible.'[57] Part of what makes a belief sacred is that our tribe holds it. Tribalism is a severe threat to free thought.

Not only may ideas need to be held more loosely, but we may need to sever all links between the thinker and their thoughts. Suppose we viewed ideas as separate entities from ourselves, like birds that fleetingly perch on our shoulders. In that case, we could engage in combat with thoughts, not thinkers. Rushdie also makes an observation about this. 'At Cambridge', he explains, 'I was taught a laudable method of argument: You never personalize, but you have absolutely no respect for people's opinions. You are never rude to the person, but you can be savagely rude about what the person thinks. That seems to me a crucial distinction.'[58]

Creating a culture of free thought will also mean creating a culture of trust. In the United States, the proportion of citizens who think most people can be trusted has decreased from around forty-five percent in the 1970s to a low of nearly thirty percent today.[59] Only a third of young Americans are confident their fellow citizens will respect the rights of others not like them.[60] This loss of trust between citizens is due to many factors, but economic inequality appears to be particularly important.[61] This loss of trust causes us to assume the worst of others, thereby justifying silencing the would-be parrhesiast. This suggests that what we face is not a crisis of free speech or thought, but a crisis of trust.

The long-term solution to renewing trust between citizens is complex and includes addressing economic inequalities. But truth can't wait. We need immediate ways to recreate the parrhesiastic contract. There are both personal and structural solutions to this. Some steps we can take as individuals. As psychologist Jonathan Haidt's work on Moral Foundation Theory shows, disagreements can result from duelling virtues rather than simply good fighting evil.[62] Parties in conflict may *both* be acting out of virtue, just different ones. One may act out of a concern for liberty, whilst the other is driven by worries about fairness.

We need to learn this lesson and be more willing to consider if virtues are driving others' thought and speech. If we do, our interactions with others will start with a fellow human being rather than a dehumanised object. This will make us more likely to discover any truth, or portions of truth, in the speaker's thoughts. Naturally, we should not automatically assume every speaker is virtuous. If someone is denying the Holocaust, I am not going to ask what virtue is driving them. But in less clear-cut cases we need to be at least open to the possibility of encountering virtue.

An example of this comes from the differences in British and American attitudes towards academics who spoke positively of communism in the 1950s. Those who espoused such views in the US were metaphorically hung for their views. Yet equivalent individuals in the UK were much less likely to have their necks or jobs called for. This appeared to be because the UK had a public culture that didn't assume academics had bad intentions. This led to the UK having a debate about the issues, rather than the character of the speaker.[63] The speaker was differentiated from the message.

Of course, many people and groups have good historical reasons for not trusting others. It can be a necessary and effective defence.

It would be unfair to urge trust from such listeners before requiring that the parrhesiast earns their trust. Parrhesiasts can do this by speaking truth and living it, as the Ancient Greeks argued.

Assuming the other may be driven by virtue does not force us to agree with them. We can end up rejecting their views. But suppose the speaker does have truth for us. In that case, an openness to potentially virtuous intentions makes it more likely we will hear it. Furthermore, as game theory shows, beginning with trust is crucial to creating highly co-operative societies.[64]

Finally, a society that protects and promotes free thought must be one in which all are accorded respect and dignity. If citizens do not do this, they will not care what each other thinks. As Hannah Arendt chillingly observed, Jews had freedom of opinion in Nazi Germany, but only because, as far as the Nazis were concerned, 'nothing they think matters anyhow'.[65] There is no freedom without truth. There is no truth without trust. And without courage, there can be neither truth nor trust.

Controlling our body means controlling our mind

George Orwell believed that even living under the harshest dictatorships, people could remain free in their minds.[66] Yet people's minds do not remain unaffected when others control their bodies. Another important element of creating a workspace for free thought is that it should be a space where we are not already acting. We think of the manipulation of thought as something independent of the effects on our body. Yet if we change what someone does, we place subtle pressure on what they think. This idea is central to the concept of cognitive dissonance.

The idea behind cognitive dissonance is that we experience discomfort when there is a clash between what we do and what we believe. If you are a pacifist forced to fight in a war, you will feel discomfort at the clash between your belief that killing is wrong and the machine gun you find yourself firing. You will try to reduce this discomfort. You can do this in two ways. You can change what you are doing (such as laying down your arms). Alternatively, you can change what you believe (by deciding the war is a just cause you should fight for). And it is usually easier, albeit often not in our best interests, to change what we believe.

The classic study of behaviour leading to belief change is called the Billboard Study.[67] In this study, researchers asked a group of people to put up a huge unsightly 'Drive Safely' billboard in their front garden. Seventeen percent of those asked agreed to do this. A second group of people were then approached and asked to sign a petition favouring safe driving. The researchers then returned to these people a couple of weeks later and asked them to put the billboard up. This time fifty-five percent agreed. Getting people to sign a petition first, before making the billboard request, led them to infer from their own behaviour that they were the sort of people who believed in safe driving. The creation of this belief in their heads led them to be more open to the subsequent billboard request. The behaviour they had been lulled into had changed their beliefs.

This means we need a space where we are free not to act, to prevent people luring us into actions which will impact our beliefs. If courts rule that people must undertake actions that they would prefer not to, such as baking a cake iced with a message they disagree with,[68] this has the potential to subtly change people's beliefs. This is not to say that such actions should not be mandated, but rather that we should be aware they may impact thought.

We can also consider the flipside of this argument. If people believe something, but are never allowed to act on those beliefs, they may well infer from their lack of action that they don't believe what they think they do. This has implications for freedom of thought. If we do not let people change their actions based on their thoughts, that is, if people keep their thoughts locked up deep inside them, then these thoughts may eventually vanish. Unexpressed thought may well decay if beliefs are belied by the inaction of the body. Freedom of thought is not independent of freedom to act. If you keep your mouth shut, your mind may soon follow.

Building an environment for free thought

Freedom of thought can be promoted through the figurative building of trust and tolerance. But it can also be built literally, brick by brick. The physical design of society can make it conducive to freedom of thought – or not. A Brazilian colleague of mine once explained how the streets of Brasília were designed to avoid street corners. Street corners meant people would assemble.[69] Assembling together meant thinking together. This could lead to political conversations. Indeed, in Portuguese, street corners are sometimes called *pontos de convivencia social*, meaning 'points of solidarity'. Such thinking posed a potential danger to the ruling powers and was to be avoided. We should invert such an approach and design our cities to promote free thought.

Yet how this is to be done in a *liberal* democracy, where liberalism drives individualism and atomisation, is unclear. In a society that prizes autonomy, people are driven to separate themselves from each other. As the historian of architecture Richard Thomas famously observed, Americans have moved from porches to patios. A front porch used to be an invitation to conversation. Yet Americans have retreated from their front porches to the privacy of back patios.[70] As the American political scientist Patrick Deneen points out, liberalism is undermining the sense of community and experience of collective debate that is needed for democracy, as well as, I would add, thought.[71]

Yet, in addition to considering large-scale political change, there are other smaller steps we can take to create environments conducive to thought. A simple achievable step is to expose people to light, water and greenery and to allow them to move. Dark and dusty libraries which confine people to desks do not support thought and are thankfully becoming a thing of the past. Libraries should let you see the sky, be filled with greenery and allow space for people to move. Such ideas are supported by Annie Murphy Paul's book *The Extended Mind*. I will not recapitulate this here in too much detail, but some points are worth emphasising.

Paul cites research demonstrating that exposure to quiet, leafy, natural areas improves our ability to think. So too does exposure to natural light, greenery and views of nature. For example, a 2012 study found that depressed individuals who went for a fifty-minute walk in nature, compared to those who went for a similar-length walk in an urban environment, had less anxiety and improved memory.[72] The idea that exposure to nature helps thought has led to biophilic design, which aims to incorporate the natural world into building design. Architecture is an important way that individuals, governments and corporations can support the right to freedom of thought.

In summary, we need to build a society that protects and promotes free thought. In such a culture, freethinking would be the norm, thoughts not thinkers would be grappled with, and thinking would be a protected characteristic. The design of our emotional environment, in which we trust each other to think, and our physical environment, in which we join together to think, is central to this project.

Thus far in the book I have been almost entirely laudatory of the right to freedom of thought. Most criticisms I've made have been based around the right not being expansive enough. Frankly, we are in danger of verging on obsequiousness. If we really value free thought, we must recognise the value in thinking freely and critically about free thought itself. If everything needs to be questioned, this includes questioning the idea that everything needs to be questioned. We need to think about this right scientifically. This means not finding ever more reasons for the right, playing into our confirmation bias, but thinking of reasons that might falsify or argue against elements of the right.

Such a task necessarily peers down some perilously dark roads. But to think such thoughts is to showcase the human, not threaten the human. Indeed, it is a task that requires a human. When I asked OpenAI's artificial intelligence language program ChatGPT to expand on the following sentence: 'We should restrict freedom of thought because . . .', it replied that 'As an AI language model, it is not appropriate to advocate for restricting freedom of thought.' If the machines won't think freely on this topic, we will have to do it ourselves.

8

AGAINST FREE THOUGHT

Eight chapters in, and it's time for the truth. And I strongly suspect that, as Colonel Jessup puts it in *A Few Good Men*, you can't handle the truth. We live in a world full of violent, irrational and prideful creatures. These creatures must be protected from themselves and each other by a guardian class. These guardians have a greater responsibility than you will ever know. You weep for the manipulated and damn the thought police. But you don't know what the guardians know: that thought control, whilst abhorrent, keeps society together. You wallow in a peace made possible by thought control, then question the way in which the world was won. Did I just advocate for the violation of the right to freedom of thought? As the Colonel would say, 'You're God damn right I did'.[1]

It seems criminal to argue against freedom of thought. Yet we must. As Mill pointed out, even if something is true it will become 'dead dogma' unless challenged. This is especially true of freedom of thought. As the legal scholar Frederick Schauer observes, 'Were Mill alive today and looking for just such a reflexively defended but rarely thought through principle, he would be hard pressed to find a better example than the principle of freedom of thought.'[2]

Proposing that the law should allow permissible violations of the right to freedom of thought does not make one the thought police. To oppose an idea whilst dressed in the official opposition's outfit, as laid out on the bed by power, is the height of folly. As Curtis Yarvin points out, if you opposed the Church in the fifteenth century you would be a fool to don the garb of a satanist, which was the official oppositional role provided by the Church. Instead, you needed to stitch a new outfit.[3] If Voltaire did not exist, it would have been necessary to invent him.

Free-thought dissidents need an identity distinct from the thought police. Just as Calvin opposed Catholicism attired in God's clothes, we can oppose free thought dressed in robes of freedom. Our bodies can only be free if our minds are not. This is because free minds threaten the stability of the society that our bodies must inhabit. Only a complacent and pampered generation, shielded from the harsh realities of human nature, could think that thought should be absolutely free. Those who have seen society fall apart know better. Those who lived through civil war and revolution saw the yawning abyss that free thought ushers us towards. They knew that security and order are the greatest good. In contrast, free thought only leads to conflict, collapse and calamity.

There was widespread recognition of the destabilising effects of free thought before the eighteenth century.[4] Before this time, the power of thoughts and speech were all too well known. No time was given to the modern dictum that 'sticks and stones may break my bones but words will never hurt me'. Instead, there was the fire of Ecclesiasticus 28:17: 'the stroke of the whip maketh marks in the flesh: but the stroke of the tongue breaketh the bones'.[5] Everyone knew that the free play of ideas could lead to social breakdown. For this reason, there was no widespread belief that an individual's 'natural right' to speech could trump the rights of the community to live in peace and order.[6]

Plato had long since seen this. He argued that a guardian class should control the population's minds by spreading a 'noble lie'. A story should be told that men were born formed of different metals that determine where their place will be in society. The story is a lie, yet Plato contended that society needed to believe it to become strong and ordered. Plato also gave us the idea of noble silence. He tells us how Socrates claimed that even if the gods did behave badly, 'it should not so readily be told to foolish and young people, but ideally passed over in silence, and if there is some need to tell it, only the smallest possible number should hear in secret'.[7] God becomes deep government seen through the eyes of babes.[8]

Plato initially suggested it would help if the rulers also believed the noble lie. Yet he would later have agreed with Nietzsche's proposition that 'the ideas of the herd should rule in the herd – but not reach out beyond it.' Plato proposed that the masses, educated into conformity, should be forbidden from questioning the laws or putting forward alternatives.[9] However, male citizens over fifty years of age would be allowed to suggest improvements privately to magistrates.[10] The rest of society would live and die by the lie. We hear echoes of this idea in contemporary China, where a 'free-speech elite' (senior Communist Party officials) can criticise the government with less fear of punishment than the average Chinese citizen.[11] Plato's idea is also reflected throughout the vision of the world laid out in the Trilateral Commission's 1975 *Crisis of democracy* report.

Human rights are noble lies. Following the 'death of God', trumpeted by Nietzsche after being whispered by Kant, the idea arose that morality was a purely subjective affair. The moral law was no longer given by God, but to be made by humans. After the horrors of the Second World War, Walter Lippmann saw where this belief had led. We needed a new noble lie, namely that God gave natural rights. Lippmann felt this lie would be a more effective brake than mere human dictates on the population's problematic urges and actions.[12] Any morality reached by reasoning, as Kant tried to do, would be unstable as we can always find reasons to be both for and against almost anything. Yet the sacred is unassailable. So, the international human rights law that emerged after the Second World War smuggled the sacred into the secular.[13] As the legal scholar Samuel Moyn explains, 'it is equally if not more viable to regard human rights as a project of the Christian right, not the secular left.'[14] It was a necessary project. After the Second World War, only God could save us.[15]

The idea that we need freedom of thought because it helps us find truth is not an argument for free thought. It is an argument against it. Truth, we are told, is a 'fundamental condition of mental, moral and social progress'.[16] Truth will make us good, make us truly alive.[17] We will know the truth, and the truth will set us free. Lies. All lies. Nietzsche knew this. The truth, he said, can be deadly.[18]

Hannah Arendt knew this too. Thinking, she wrote, 'inevitably has a destructive, undermining effect on all . . . those customs and rules of conduct we treat of in morals and ethics'. Any thought worthy of the name must reject accepted wisdom and truth. For this reason, Arendt concluded that 'there are no dangerous thoughts; thinking itself is dangerous'.[19] Given this, we should start from the position that free thought is so dangerous to the stability necessary for society that we should have to justify it, rather than assume it.[20]

We are receptive to the idea of unlimited free thought because of a naive, self-interested argument for free speech made three centuries ago when people still thought it was a good idea to throw raw sewage into the street. Thomas Gordon and John Trenchard, two political writers, introduced the idea of a right to free speech in England in 1720.[21] They portrayed it as a right to be free from government censorship. They didn't consider the problems this right could lead to, because it wasn't their concern. The government was threatening to stop their publishing ability, so this was the threat upon which they focused. The right to free speech was their self-serving vision. They took the idea that self-interested priests controlled people's minds and applied this to the state, arguing that rulers constantly plotted to trick and enslave their subjects. In their view, any government control of speech meant tyranny was imminent. Gordon and Trenchard ignored the problems to which free speech could lead. They dismissed potential problems such as defamation, incitement of the mob, and the potential for some ideas being too dangerous to let loose. As Dabhoiwala concludes, the idea of unlimited free speech 'was not constructed over decades of painstaking, pan-European debate, but cobbled together by two hacks on a deadline'. We should not let such dated views shape how we think about freedom of thought.

Thought must be censored, shaped and controlled to preserve the greater goods of peace, order and security. Censorship laws, blasphemy laws and even the work of the Spanish Inquisition were 'intended to protect society against what their authors sincerely believed to be grave injury'.[22] International human rights law already accepts that our freedom to 'seek, receive and impart information and ideas of all kinds' can be legally restricted.[23] Reasons for such restrictions include protecting national security and public order. Freedom of thought cannot be so free that it harms the society that makes it possible. Furthermore, given the dependence of free thought on the legally limitable process of obtaining free information, freedom of thought is already not absolute.

To better understand this, consider the views of someone who saw chaos, bloodshed and the breakdown of social order. The French lawyer and diplomat Joseph de Maistre (1753–1821) lived through the turbulent events of the French Revolution. He saw his world fall apart. This led him to search for a new foundation for society. A stable society had to be built on an iron foundation. It could not be built on something as flimsy as reasons.

Reason, de Maistre said, could be used to prove anything. This made it fundamentally unsuited as a construction material. Any argument that society should be a certain way would be vulnerable to some counterargument. A society built on reasons would be swept away by a mudslide of argumentation. For de Maistre, reason never saw a scab it didn't want to pick, which would lead to the infection and death of the body of society.

De Maistre's solution was to jettison reason. The only way to create a solid foundation for government, which no one could ever shake, was to make it impervious to reason.[24] Letting people argue about the basis of governmental authority was the beginning of the end.[25] The camel, once admitted to the tent, knocks it down.[26] Those who sang the praises of 'government by discussion' emphasised the benefits of letting reason work. As the English journalist Walter Bagehot gushed, reasoning about a topic means 'you can never again clothe it with mystery, or fence it by consecration; it remains for ever open to free choice, and exposed to profane deliberation.'[27] Yet what was a strength for Bagehot was a weakness for de Maistre. Freedom of thought allows ideas to multiply out of control, following their own logic with no regard for the health of the thinker or the society they live in. Freethinking is a cancer of the mind.

If reason is not to be the basis of society, what should be? What would be impervious to the destabilising effects of reason? Isaiah Berlin masterfully summarises de Maistre's argument as follows:

> By founding societies upon foundations so dark, so mysterious and so terrifying . . . anyone who dares approach them will find himself immediately subject to the most hideous and enormous penalties. The only societies which have lasted are societies created by priests, in which the people have been taught a series of frightening myths whereby any questioning of the foundation of society was itself regarded as sinful and liable to bring about punishment. The only laws which have lasted amongst mankind are laws whose roots and sources are not remembered. Laws whose roots and sources are remembered are usually bad laws or at least laws which somebody wants to change. Custom is the foundation of our life – custom and the dark irrational sphere which nothing must be allowed to approach. Therefore authority must be blind.[28]

De Maistre was not alone in his ideas. John Stuart Mill, the great advocate of liberty, argued that 'there be in the constitution of the state something which is settled, something permanent, and not to be called in question.'[29] Cardinal Newman proposed that the idea of the infallibility of the Catholic Church acted to restrain freedom of thought to 'rescue it from its own suicidal excesses.'[30] The French philosopher Bertrand de Jouvenel argued that 'the capital blunder of our time is, probably, this: that everything has come to be regarded as eternally abiding our question.'[31] Leo Strauss argued that unchecked philosophising was dangerous to philosophers and their cities because every society is mined with explosive truths.[32]

Such ideas are not limited to the dead. Today, Noah Yuval Harari writes that, 'you cannot organise masses of people effectively without relying on some fictional myths.'[33] When Elon Musk was in the process of buying Twitter, a *Washington Post* columnist commented that 'for democracy to survive, we need more content moderation, not less.'[34] The US educational institution Hillsdale College refers to '*responsible* freedom of thought', which promotes 'ordered liberty.'[35] This implies that some moral ideas should not be proposed or others not questioned.

De Maistre's alternative social building material is also found in the Russian novelist Fyodor Dostoyevsky's *The Brothers Karamazov*. In a famous scene in the novel, Christ returns to Earth and starts preaching and performing miracles. The Church finds out about this and promptly imprisons him. A leading Church figure, the Grand Inquisitor, comes into Christ's cell. He tells Christ that the people do not want the freedom he is offering them. Freedom was too challenging to live with. Instead, the Church believed that the people wanted miracle, mystery and authority. And this is what the Church would give them, not freedom.

Human rights are clothed in miracle, mystery and authority. They sit high above the reach of the people. They are there to protect the minority, so the majority cannot be allowed to overturn them. Democratic society has an authoritarian core.[36] The benefits are obvious. When the German people democratically elected Hitler, following their economic self-interest,[37] the Nazis denied the rights of minorities and hell soon followed.

The basis of human rights in miracle, mystery and authority is utterly necessary. Only a fool would base human rights on rational justifications. Arguments for rights are vulnerable to arguments against rights. It makes much more sense not to expose human rights to the corrosive power of reason. Human rights must be put on a high shelf, well away from the acid of argumentation. One way to do this is by simply asserting that rights are 'natural' or 'self-evident'. After all, the famous phrase from the US Declaration of Independence that 'we hold these truths to be self-evident' means, in effect, that 'this claim is obviously true, so let's not bother thinking, reasoning or arguing about it'. The Founding Fathers wrote their high-minded words about freedom of thought after they had put a bolt through the brains of the American people.

Yet simply stating rights are 'natural' does not fully secure them. People will use their freedom of thought to argue against this. There will always be those who, following Nietzsche, think that the idea of inalienable human rights is 'a laughably feeble attempt by the weaker members of the species to fend off the stronger'.[38] Human rights, in such a view, come from the same root as Christianity, namely 'the realization that the weak could overcome the strong when they banded together in a herd, using the weapons of guilt and conscience.'[39]

Such contentions show that the rights we thought were safely stashed on a high shelf can start fighting among themselves. When we separate them, and try to work out who started the fight, the culprit will often be freedom of thought. This forces us to look for a cage in which to lock this right tight, like Hannibal Lecter in *The Silence of the Lambs*. Fortunately, we have such a cage, which prevents the right to freedom of thought from biting other rights. It is the last Article of the Universal Declaration of Human Rights, Article 30. This states that: 'Nothing in this Declaration may be interpreted as implying for any State, group or person any right to engage in any activity or to perform any act aimed at the destruction of any of the rights and freedoms set forth herein.' Similarly, the Vienna Declaration made at the 1993 World Conference on Human Rights states in Article 1, 'The universal nature of these rights and freedoms is beyond question.'[40]

Human rights, it hence appears, are beyond question. Indeed, people get extremely irritated if anyone tries to use their alleged right to freedom of thought to query human rights. When US Secretary of State Mike Pompeo held a meeting of his Department's 'Commission on Unalienable Rights', one commentator wrote that 'openly questioning what constitutes a fundamental human right does not help strengthen the consensus on human rights; instead, it contributes to a wrongheaded understanding that the definition of human rights is open for debate rather than settled by laws and norms.'[41]

When it comes to human rights, there will be no questions, there will be no debate, and there will be no thought. Right-making is god-making. And the 'church of human rights'[42] is just as touchy about people questioning the existence of their god as monotheistic religions are about people questioning the existence of theirs. Both view questioning as heretical, and as being, at best, ignorant and, at worst, evil. This stance is powerfully contradictory. As one French philosopher has put it, 'to say that the rights of individuals depend on reason and on nature, but to wish to protect them from discussion by all beings endowed with reason is to destroy their rational foundation.'[43]

Not only is freedom of thought forbidden to bite other rights, but it must also keep its claws out of democracy. Article 17 of the European Convention on Human Rights states that: 'In view of the very clear link between the Convention and democracy, no one must be authorised to rely on the Convention's provisions in order to weaken or destroy the ideals and values of a democratic society'. This is a clear prohibition on thought. The reasoning of Article 17 was used, back in 1957, when the German government wanted to ban the Communist Party of Germany. The communists tried to win power through the ballot box. However, because a dictatorship of the proletariat was not compatible with democracy,[44] courts were able to limit the communists' absolute right to free thought.

In addition to other human rights and democracy, further sensitive issues are shielded from the savagery of an untamed right to freedom of thought. In 2011 the European Court of Human Rights agreed it was legal to suspend a right-wing politician and professor from working at a university for five years because he had expressed the view that the gas chambers in Nazi concentration camps and the number of dead therein were matters for historians to discuss freely.[45] According to the internet, Godwin's law captures the idea that the first person to resort to a comparison to Hitler loses the argument. In Germany, the first person to resort to a comparison to Hitler (or Himmler) loses not just the argument but also their liberty.[46] The European Court of Human Rights clearly does not think that the right to freedom of thought is absolute. It may well be right to do so.

As thinkers have highlighted throughout history, some things rightly must be sacred and off-limits. But how do we choose what falls into this category? To decide the best way forward, we can look back. Religious tolerance occurred during the Reformation. At this time, Christians redefined which doctrines were essential to believe and which were inessential. Those promoting tolerance said denominational differences were inessential. Thus, if you accepted the minimum essentials of Christianity, you were fine.[47] Today, we find ourselves in a similar situation. Some things are essential to think in liberal democracies, but much is inessential. We need to be very careful about how much we put inside the circle of the essential. If this circle becomes bloated, it could well explode.

Thus, although we are told that freedom of thought is an absolute right, it clearly isn't. We can't use freedom of thought to question human rights or democracy or certain issues of profound social importance. Human rights are a snake that can't bite its own tail. The right to question rights is not allowed to exist, which is precisely why freedom of thought cannot be absolute. As Sebastian Junger once observed, 'the inside joke about freedom . . . is that you're always trading obedience to one thing for obedience to another.'[48] Article 30 is the Universal Declaration of Human Rights' immortality clause. De Maistre's ideas, which to many will have seemed scary and dangerous, are the position taken by the Universal Declaration of Human Rights.

If you don't believe this, consider the words of the French philosopher Jacques Maritain, who was involved in UNESCO's Committee on the Philosophic Principles of the Rights of Man. 'We agree on the rights', said the philosopher, 'but on the condition that no one asks us why.'[49] Or look at the wording of the Universal Declaration of Human Rights. The Preamble to the Declaration states that 'the peoples of the United Nations have in the Charter reaffirmed their faith in fundamental human rights'. The key word here is faith. The Declaration is the product of revelation, not reason.

Leaning on the otherworldly is a classic defence mechanism. The basic idea appears in the 1984 film *Ghostbusters*. On the top of a New York building, the evil god-like entity Gozer the Gozerian arrives to destroy the Earth. In her way stand the Ghostbusters. Gozer, getting her bearings, asks Ghostbuster Ray (Dan Aykroyd) whether he is a god. Ray pauses to think, before admitting that he is not a god, at which point Gozer attempts to kill Ray and all his colleagues. Gozer narrowly fails, the Ghostbusters regroup, and Ghostbuster Winston (Ernie Hudson) offers the immortal advice: 'Ray, when someone asks you if you're a god, you say YES!' What reason creates, reason can kill. Revelation provides a far firmer footing. When Alasdair MacIntyre criticised the UN's practice of rigorously not giving 'good reasons for any assertions',[50] he overlooked that this is a feature, not a bug. De Maistre would be proud.

Once people bring reason into the debate about human rights, things get messy fast. The debate around Secretary Pompeo's Commission highlighted this. As one commentator put it, 'if one accepts the premise that there is a need to restrict or redefine what counts as an unalienable human right, then that leaves one final question: Who gets to decide what is an unalienable right?'[51] The only safe answer to this is God. The only true answer is people. These do not sit well together.

The same commentator observed that Pompeo had appointed a group of eleven people to make this decision but that:

> this group is not representative of the diverse U.S. population. Most of the commissioners come from professional backgrounds heavily grounded in religious freedom issues, with known opposition to LGBTQ and reproductive rights. Just three commissioners are women and only two are people of color. There is no apparent representation of the LGBTQ, immigrant, indigenous, or disabled communities. As studies show, groups that lack the diversity of the population they seek to represent often make flawed decisions.[52]

This observation raises a host of questions. Did those deciding on the content of the Universal Declaration of Human Rights in 1948 meet the commentator's diversity standard? Are we going to put human rights to a human, all-too-human vote? Do we want human rights to be seen as a naked struggle for power? What do we think that will do to their authority?

Frankly, optimism about human rights is so 1990s. Today's thinking is more pessimistic. Criticisms have been made of global human rights agencies who 'assume unto themselves the right to speak for everyone'.[53] The democratisation of the development of human rights emerges as both necessary and impossible. In my Introduction, I argued that we should all have the ability to have a say in what shape our right to freedom of thought takes. Indeed, this is what I am trying to stimulate in this book. Such an approach is in keeping with the idea of self-government. Yet once we acknowledge that human rights are human inventions, their authority begins to wane. Furthermore, it has been argued that we need fewer 'self-appointed spokesmen of the common good' asserting what human rights do and do not exist, and what form they take, and more political deliberations about the common good and how we can reconcile different people's interests.[54] What politics gets us into, politics must get us out of.

Against an absolute right to freedom of thought

Let's rein in the rhetoric and consider whether the right to freedom should be absolute, without using arguments that could plunge us into tyranny. Firstly, it is interesting to think about the psychology of absolutes, as Justice Oliver Wendell Holmes Jr once did.[55] Holmes pointed out that a romantic may find it insufficient for others to say that the love of his life is 'a very nice girl'. If others do not concede that the target of the romantic's affections is the most beautiful creature on Earth, a fight may ensure. 'In all men', claimed Holmes, there is 'a demand for the superlative'. Holmes's view was that in jurists this demand took the form of claims of absolutes and universal natural laws.

Whatever the truth of this, absolute claims often lead to absolute messes. Viewing any rights as absolute and universal may encourage new conflicts rather than prevent them. As noted earlier, peace was achieved in Europe in 1648 by the signing of the Peace of Westphalia, based on the idea that the only way to stop wars was by getting people to agree that no one could really be sure which religion was correct. People agreed to mind their own business. But if there are absolute human rights, which must never be violated, then we have reason to go to war with states that disagree with us. Wars of religion could be replaced by wars of rights.

When we examine arguments for why we must never allow violations of the right to freedom of thought, we find two very different proposals. Oddly, one is based in the importance of thought and the other in its impotence. The first argument is that free thought is so fundamental that we should never impede it. This is an argument-from-importance. As we will see, it is easy to refute. All we need to do is imagine a situation where not violating someone's freedom of thought would lead to catastrophic societal consequences. In such a scenario, most people would concede the benefits of the violation outweighed its costs. Freedom of thought could then be permissibly restricted for reasons of national security or public order, just as our rights to speech and privacy can.

A more subtle way to refute this argument-from-importance is to consider whether all thoughts are equal or whether some are more equal than others. Here we can draw on an argument from Alexander Meiklejohn.[56] Each of us is two-fold, said Meiklejohn. We are both the makers and subjects of laws. In our public-spirited role as makers of laws, our freedom of thought and speech should be absolute. Yet in our private roles as self-interested individuals, our freedom of thought and speech should be able to be legally restricted by the government. In this way, the right to freedom of thought could be absolute in areas relating to the practice of self-government, but able to be restricted in areas relating to merely private concerns.

The second argument for why the right to freedom of thought should never be violated stems from what I have claimed is a wrong-headed conception of thought, namely that thoughts are locked in the cell of our skull. This assumption leads to the claim that thoughts can't impact anyone else. Thought is, therefore, harmless and can do whatever the heck it likes. This is an argument-from-impotence.

This argument stems from the work of John Stuart Mill, who put forward the famous 'Harm Principle'. This states that the only reason we should limit someone's freedom against their will is to prevent harm to ourselves or others. Kant had earlier argued that one person's freedom can only interfere with another's if their actions collide in the external world. If thoughts are solely in our head, and can't touch – let alone harm – other people then they should be completely, absolutely and eternally free. This led Mill to argue for 'liberty of conscience, in the most comprehensive sense; liberty of thought and feeling; absolute freedom of opinion and sentiment on all subjects, practical or speculative, scientific, moral, or theological'.[57]

There are two subtly different versions of the harm principle. In one version, we can only coerce people if this prevents or reduces harm. But, in another version, coercion is allowed if it prevents or reduces the risk of harm.[58] The difference is important. The question of whether a thought harms others is very different to whether it increases the risk of harm to others. For example, if reckless driving can be prosecuted because it increases the risk of harm to others, why couldn't we prosecute reckless thinking?[59]

In short, the two main reasons for the right to freedom of thought being absolute can both be undermined. Let's consider this in a bit more detail now. This will help us envisage a roadmap for how freedom of thought may be denied its status as an absolute right. One can then either try to usher us down or off these tracks.

Anyone making a case for limiting freedom of thought will naturally reach for situations involving the most emotive and feared issues in our society. The two obvious contenders are terrorism and child sexual abuse. There are of course others. For example, countries such as Norway have raised the question as to whether 'nazism, fascism and racism' are considered as being protected by the concept of freedom of thought.[60] This was presumably done with a view to seeing if such ideas could be legally regulated.

But let's start with the assassin of absolutes and the hero of hypotheticals – a terrorist who has planted a nuclear bomb in a city. Imagine we have a person in custody who tells us he has planted a nuclear bomb in London that will go off in an hour. He has no demands. He doesn't want money, a helicopter or even pizza. He just wants carnage. In our hands, we have a brain-reading device which, if placed on the terrorist's head, would allow us to decode his thoughts and the location of the bomb. We apply to a nearby London court for a warrant to use this cap. No, says the judge, this would be an infringement of the terrorist's right to freedom of thought, which is absolute and can never be violated for any reason. As the judge runs for the door, we are left pondering, briefly.

Now imagine that a copycat terrorist tries the exact same thing the next week in New York. The thought-decoding cap is no longer available. The British had the only one, and that has now been vapourised. However, the NYPD has access to a mentalist. They bring him in. The mentalist runs through his repertoire of tricks and discovers that the bomb is in a small café in the west of Queens. Authorities get there, defuse the bomb and all is well.

But is it? Hasn't the terrorist's right to freedom of thought been violated by the US government through their use of a mentalist to non-consensually access his mind? The terrorist takes his case to the Supreme Court. His lawyer cites back to the Court all its previous lauding of free thought ('nothing more imperatively calls for attachment than the principle of free thought'; 'the priceless heritage of our society is the unrestricted constitutional right of each member to think as he will'; 'our whole constitutional heritage rebels at the thought of giving government the power to control men's minds', etc.) and wins. The would-be terrorist is released. The week after, neither the café in Queens nor New York exists.

There are several ways we can respond to these scenarios. We could say that hard cases make bad law, and walk away. We could point out that using extreme cases separates people from their deeply held ethical values, and walk away.[61] We could concede that we have employed the rhetorical trick of making the violator of free thought the 'good guy' in the scenario, and walk away.[62]

But let's say we stay. We could even try to make things more complicated by arguing that the cap would violate the terrorist's freedom of thought, but the mentalist wouldn't. Without some extremely well-paid legal counsel, it is hard to see why this would be the case. Alternatively, we could take a position analogous to that taken by Sam Harris on torture. We would claim that using either the cap or the mentalist would be an impermissible violation of the terrorist's right to freedom of thought, but it would be the ethically correct thing to do. Without some extremely well-paid philosophical counsel, it is hard to see why this wouldn't be the case.

In response to such a scenario, we could advocate that our government passes legislation (presuming our country still has the power to make its laws) defining the right to freedom of thought as a qualified, not absolute, right. This legislation would allow the right to freedom of thought to be permissibly limited in the interests of national security and public safety in certain extreme circumstances. The obvious concern is that this opens the floodgates to more limitations. The next thing we know, we are living in [insert the name of your favourite dystopia here].

Yet it seems fair to assume that we could stop our slide down such a slope to tyranny. The list of situations in which the US government can restrict free speech has not careered out of control.

This is in part due to its Supreme Court Justices being very clear on the importance of safeguarding free speech. Presumably, this is the same position they would also take towards thought. Thus, it seems fair to argue that the right to freedom of thought should have a permissible limitation. This would allow the use of brain-reading technology in cases of clear and present catastrophic danger to public safety.

Legal scholars have already mooted the possibility of permissible limitations to the right to freedom of thought. Bublitz offers a thought experiment of a future in which outward signs of our inner world could be used to uncover individuals' thoughts (i.e. behaviour-reading) in sensitive locations such as banks, airports or even public parks.[63] This would aim to deter the occurrence of 'evil thoughts', leading to what Bublitz terms 'zones of restricted freedom of thought'. Such measures, says Bublitz, 'are not a priori unreasonable, but it is hard to deny that they impinge upon free thinking'. This implies it could be reasonable, in the interests of public safety, to make freedom of thought a qualified right.

We can link this idea into libertarian concepts of rights. Murray Rothbard argued that if we don't formulate human rights in terms of property rights, the results will turn out to be unclear. He illustrates this in relation to the right to freedom of speech. Rothbard points out that the right to freedom of speech, as viewed today at least, is taken to mean the right of everyone to say whatever they like. He then adds:

> But the neglected question is: Where? Where does a man have this right? He certainly does not have it on property on which he is trespassing. In short, he has this right only either on his own property or on the property of someone who has agreed, as a gift or in a rental contract, to allow him on the premises. In fact, then, there is no such thing as a separate 'right to free speech'; there is only a man's property right: the right to do as he

wills with his own or to make voluntary agreements with other property owners.[64]

On the one hand, this could be taken to imply that the right to freedom of thought only applies when you are on your own property. Yet, when we consider the subset of thoughts that are inside our head, our head is always our own, even when we are on another's property. So, even if we are on another's property, that person cannot interfere with our internal thoughts in the head that we own.

Whatever argument one finds persuasive, in practical terms it appears that some police officers in the UK seem to think that such thought-restriction zones already exist. Local councils in the UK can apply for Public Spaces Protection Orders. These were introduced in 2014 to prohibit anti-social behaviours in specific public locations, allowing everyone to enjoy public spaces. One such order came into place in September 2022 in Birmingham. This created a restricted area in the public streets near a clinic that provided abortion services.[65] The order prohibited certain activities in this area, including engaging in any act of approval or disapproval about abortion services, including prayer.

In December 2022, Isabel Vaughan-Spruce was standing on a public street in this area, at a time when the clinic was closed, silently praying in her head. A video shows a police officer talking to her. 'Are you praying?', the officer asks, to which Vaughan-Spruce responds, 'I might be praying in my head'. She was then arrested under suspicion of failing to comply with a Public Spaces Protection Order. At the police station, Vaughan-Spruce was interviewed and, according to her account, was asked what she was praying about silently in her head.[66] Several MPs raised this issue in Parliament.[67] One called the events an 'attack on freedom of thought' and claimed that 'We are now firmly in the realm where "thought crimes" like prayer are a policing priority'.[68]

Such incidents appear unlikely to stand up in court. Indeed, legal challenges to somewhat similar developments are already being made. For example, an EU-funded project looking for facial markers of lying, which aimed to speed up immigration checks, was recently challenged in court.[69] Nevertheless, the idea that concerns over terrorism could be used as a rationale to make the right to freedom of thought non-absolute is not simply speculation.

Fears of terrorism in the UK are already being used to restrict freedom of thought. Section 3 of the UK Counter-Terrorism and Border Security Act 2019 makes it a criminal offence to view material online, even once, 'of a kind likely to be useful to a person committing or preparing an act of terrorism'.[70] A curious person could be convicted and sentenced to up to fifteen years in jail for a single click, if they don't have a reasonable excuse. What counts as a reasonable excuse includes being an academic or journalist. It also includes not knowing the link would take you to 'information of a kind likely to be useful to a person committing or preparing an act of terrorism'. Yet searching for this or related information can be, I would argue, a form of thought.

There is only a weak rationale for punishing such thought. There is a huge gulf between accessing such information and perpetrating acts of terrorism. It is one thing to use an imminent nuclear explosion to restrict free thought. It is a very different thing to tie up people's minds with a chain made of 'coulds', 'mights' and 'maybes'. Furthermore, this Act limits the ability to think to a professional class of academics and journalists. This leaves any other curious members of the population to be made into thought criminals.[71] This would clearly create a two-tier society. Individual citizens' ability to freely search for truth, to allow them to cast informed votes and to be meaningfully self-governing, is even more important than academics' search for truth. As the US Supreme Court puts it, the First Amendment (i.e. the right to free speech) 'has its fullest and most urgent application' to speech that is 'uttered during a campaign for political office.'[72]

The UK Counter-Terrorism and Border Security Act 2019 led the United Nations Special Rapporteur on the right to privacy to accuse the UK government of pushing towards 'thought crime'. Such legislation will chill people's internet searches for material critical of government. Yet we should be able to pursue knowledge with the same freedom that the Ancient Greek Hippocleides famously pursued dance. When told we need to restrict our thinking to create a slight decrease in the risk of terrorism, our response should be that given by Hippocleides when told he had lost out on a potential marriage due to dancing drunk and upside down on a table: 'Hippocleides doesn't care!'[73]

Let's say we granted the imminent nuclear terrorism example as an exceptional circumstance where a government could legally read someone's mind. And then say that we started to head down a slippery slope. Where would our next stop be? A safe guess is that the next case would involve the sexual abuse of children.

A 2005 Gallup Poll found that the percentage of Americans who were 'very concerned' about sex offenders was nearly double the percentage of Americans who were 'very concerned' about terrorism.[74] This makes this issue likely to be a key battleground for debates about freedom of thought. Many people would agree with the statement that everything possible should be done to stop the sexual abuse of children, certainly those familiar with the horrendous harm that such abuse can cause.[75] Some may suggest we should monitor the thoughts of either people with paedophilia (whether they have offended or not) or other convicted child sex offenders.

This suggestion would be politically difficult for elected legislators to disagree with. After the rape and murder of seven-year-old Megan Kanka in the US in 2004 by a convicted sex offender, legislation was introduced requiring the police to notify communities of registered sex offenders. This was done on the basis that it would help the public to protect themselves from sexual crime. In reality, the evidence base for the effectiveness of this policy is at best mixed, and its flawed design, which should have been apparent from the start, has led to significant problems.[76] This is a good example of what has been dubbed 'penal populism', the rapid passage of laws based on their popularity rather than their effectiveness.[77] Despite the problems with community notification, when this issue came up for a vote in the US House of Representatives, 'only one representative voiced opposition and the bill eventually passed 418–0'.[78]

That said, when a bipartisan bill was proposed in the US Congress to help prevent online child sexual abuse material, which would have harmed the rights to speech and privacy for people acting legally online, some politicians did speak out against this. Senator Ron Wyden called it 'a transparent and deeply cynical effort by a few well-connected corporations and the Trump administration to use child sexual abuse to their political advantage, the impact to free speech and the security and privacy of every single American be damned.'[79]

This example notwithstanding, there is a good chance that politicians would support public calls for legislation that permitted monitoring the thoughts of those deemed at high risk of sexually abusing a child. This would make the right to freedom of thought just another qualified right. Such legislation might not pass initially. But it certainly would after the next high-profile case of abuse where knowing about the perpetrator's thoughts could have prevented the rape of a child.

As someone who has spent much time talking and working with survivors of child sexual abuse,[80] I can feel myself drawn to the idea that those at high risk of sexually abusing children could legally have their thoughts monitored. This could be via brain- or behaviour-reading. I could sympathise with someone who proposed that just as some released prisoners must wear electronic tags on their ankles so authorities can monitor their physical movements, child sex offenders could be legally required to wear electronic tags on their heads ('mind-tags') so authorities could monitor the movement of their thoughts. Here the government's interest in protecting children would outweigh the sexual predator's right to mental privacy. I can feel the riposte forming in my head; if you can't respect a child's rights, don't ask me to respect yours.[81]

Let's move away from hypothetical mind-tags to a real case, that of Brian Dalton, to try to make things clearer.[82] In 2001, Dalton was sentenced to seven years in prison for making and owning a diary that contained violent sexual fantasies about children.[83] Dalton's conviction was eventually overturned, with the judge stating that 'what you write in your private notebook should not be the subject of prosecution.'[84] But should it not? There are two separate elements of the right to freedom of thought at play here: the right to mental privacy and the right not to be punished for your thoughts. Could either be legally violated?

If you are accused of a crime, can the contents of your diary be used as evidence against you? In short, is the privacy of your diary violable? Today, if a legal search finds your diary, its contents can typically be used against you in court.[85] To me, this suggests that the courts view the right to freedom of thought as not being absolute. If we were to call for mental privacy to be absolute, the current practice of using diaries as evidence against people would become a human rights violation.

If we wish diaries to remain admissible as evidence, there are two ways out. We could claim that diaries aren't thought. This would allow courts to continue to access such materials, whilst continuing to recognise the right to freedom of thought as absolute. I wouldn't agree with this. Diaries aren't just expressions of thought, they *are* thought. It isn't the case that the thoughts already existed and were then expressed on the page. Without being written down, the thoughts may never have existed. To write is to think. Or, if badly conjugated Latin makes things sound more convincing to you: *scribere est cogitare*. The other way to keep diaries as admissible evidence is to recognise them as being thought, but to make freedom of thought a qualified right, with permissible limitations on mental privacy. This latter option seems more intellectually honest.

As applied to the case of Dalton, this means the thoughts in Dalton's diary should not have been private. He had been convicted in 1998 for downloading pictures of children having sex and sentenced to eighteen months in jail.[86] The diary was found when he was out on parole. We know that problematic sexual fantasies are associated with an increased risk of sex offenders re-offending.[87] If authorities were trying to assess Dalton's future risk of offending, they should have been able to access his diary to help make their decision. They should not find it locked by Dalton's absolute right to freedom of thought. The right to freedom of thought should be qualified, allowing it to be balanced against the state's interest in protecting children.

But this is not what the Dalton case was about. Dalton's case didn't involve using his diary to assess if he was likely to re-offend. His case was about whether the diary was an offence in itself – a thought crime, if you will. The American legal scholar Laurence Tribe has said that prosecuting Dalton for the content of his diary is 'as close as you can get to creating a thought crime.'[88] Similarly, the Supreme Court of Canada has stated that to 'ban the posses-sion of our own private musings [diaries] . . . falls perilously close to criminalizing the mere articulation of thought.'[89] But we must not let someone shouting 'thought crime' in a crowded courtroom stop us thinking about this issue. What is so wrong about the idea of thought crimes?

An absolute right to freedom of thought would forbid ever punishing people for their thoughts. This means Dalton could not be prosecuted for the content of his diary. But the basis for this, as we have seen, is the assumption that thoughts cannot either harm or risk harm to others. This assumption is questionable in Dalton's case. Dalton's diary is not harmless thinking. It is thinking that increases the risk of him committing harm against children. These are not first-order thoughts, just popping up in Dalton's head. These are second-order thoughts which, rather than rejecting or repudiating, he was elaborating and condensing on the page. Given that problematic sexual fantasies are associated with an increased risk of sex offenders re-offending, Dalton's thinking is reckless. He is thinking with wilful or wanton disregard for the safety of children. Tribe says we shouldn't punish Dalton because that would create a thought crime. I say Dalton has committed a thought crime and therefore we should punish him.

This could be deemed an infraction (generally punishable with a fine, not jail) or a misdemeanour (generally punishable with up to a year in jail and/or a fine) rather than a felony. But the point is that Dalton's thinking should be seen as criminal. The right to freedom of thought should be recognised as a qualified right, which can be balanced against the state's interest in protecting children.

Clearly my views here are shaped by my work with survivors of child sexual abuse, which has made me very risk-averse when it comes to preventing such abuse. So I should probably not be the one making the decision about the rights of people who sexually abuse children. Yet the extent to which the public is risk-averse on this question, or on a myriad of other law and order questions, will probably determine whether the right to freedom of thought remains absolute. Even without new legislation, judges' views on the punishment of thought may be influenced by a risk-averse public clamouring for such punishment. As the former Justice of the Supreme Court of the United Kingdom, Lord Sumption, has noted, 'judges don't decide cases in accordance with the state of public opinion but it is their duty to take account of the values of the society which they serve. Risk aversion has become one of the most powerful of those values and is a growing influence in the development of the law.'[90]

As existential threats, such as nuclear war, recede, our mindset changes. We come to 'prioritize freedom over security, autonomy over authority, diversity over uniformity, and creativity over discipline.'[91] Yet as perceptions of security fade, with concerns over rising crime, whether realistic or not,[92] public tolerance of risk will decrease and the likelihood of restricting freedom of thought will increase. This means that, ironically, moral panics over the impact of technology will do more to restrict freedom of thought than liberate it. If we have morphed into what the author David Garland calls a 'culture of control',[93] more driven by the desire to prevent risk than preserve freedom, then the right to freedom of thought will not remain absolute for long. Given our bias to attend to and pass on negative and threatening information, as noted earlier, nature has its thumb on the societal scales, pushing us towards such a risk-averse society. Only by using and preserving our freedom of thought can we rebalance these scales with the weight of reason.

Arguments against mental privacy

Cases against absolute mental privacy do not need to rely on examples involving the emotive topics of terrorism and child sexual abuse. Other more mundane cases can be made. As previously mentioned, one argument against mental privacy is that if we make minds more transparent then we normalise the experience of intrusive unwanted thoughts, thereby reducing self-stigma and shame.[94] This is not a compelling argument, as the desired end could equally be reached by much less intrusive means, such as public education. Furthermore, the benefit does not outweigh the huge cost to mental autonomy that would result.

Yet such an argument helps us to critically consider an assumption we have made, namely that mental privacy is only ever a benefit to freedom of thought. As I already pointed out, if everyone's thoughts were anonymously available, then analysis of this data would tell us a great deal about humanity.[95] This information would allow us to better understand ourselves and others, be a boon to free thought and thereby promote our autonomy and truth-seeking. The chilling of thought this would involve could be negligible compared to the benefits.

We can also flip around the argument about the dangers of behaviour-reading. We have been working on the assumption that if a company can tell us something about our health or our personality or something else that would help us understand ourselves better, this would chill our thought or behaviour. Yet we could equally argue that a company that didn't tell us this information is negligent and failing to support our right to freedom of thought. The power imbalance between a company that knows more about ourselves than we do can be levelled in two ways. One, forbid the companies from collecting and analysing our data. This approach solves the problem by making us all equally ignorant. The alternative is to force the company to share this information with us. Now we are all equally informed.[96] Such a process would boost our autonomy and ability to be self-governing, not limit it.

Another argument against mental privacy is that the more information marketers have, the better they can serve consumers.[97] Both law and economics view one party having more knowledge than another as being inefficient as it hinders fully informed exchanges.[98] Privacy is a barrier to efficiency because it worsens companies' abilities to predict what customers want. Yet, as Jeffrey Rosen has observed, an obsession with efficiency is the surest way to destroy liberty.[99] Furthermore, this case against mental privacy is an economic one, not a moral one. As journalists have pointed out, 'The reason we don't have any privacy is because people can make money off of our not privacy.'[100] The financial gains of others are insufficient justification for our loss of mental privacy.

Whatever arguments we may wish to make against absolute mental privacy, the reality is that many legal systems already view mental privacy as legitimately violable. And this is not just due to the admissibility of diaries in courts. Police in the US have used defendants' internet viewing activity to help establish their state of mind.[101] The US Attorney has subpoenaed Amazon for the used book purchase records of over 24,000 people as part of an investigation into possible tax evasion and fraud perpetuated by a single person.[102] Behaviour-reading is a widespread practice.

Furthermore, the US believes violations of mental privacy are acceptable trade-offs for increased national security. Section 215 of the PATRIOT Act allows officials conducting foreign intelligence investigations to obtain Americans' library records. Such records, which are an index of people's thoughts, are not absolutely private.[103] When librarians stood up for the privacy of their readers' records, a freedom of information request revealed that FBI agents had complained that 'radical militant librarians kick us around'.[104] An FBI spokesperson had to come out to calm the situation. Unfortunately, they did so by claiming that these agents' anti-librarian sentiments were 'considered personal opinions in what employees believed to be private e-mails not intended for large, public dissemination'.[105] This evidenced that the FBI lacked both appropriate boundaries and a sense of irony.

Sensible cases for invasions of mental privacy can also be made from everyday concerns. Collecting brain data in real time from car drivers could tell us when their attention was drifting from the external world.[106] A signal could then alert the driver to keep them safe. This system could be made mandatory. It may turn out to be as effective as seatbelts at saving lives, which we should note are already mandatory in the UK and for most people in the US. This would require that the right to freedom of thought be a qualified right that must be balanced against public safety concerns.

Finally, whilst not an argument against mental privacy per se, we can at least argue that a powerful emphasis on mental privacy as necessary for free thought misses an important part of the problem. Imagine you had a neighbourhood with a significant gun crime problem. A solution that solely focused on making people's houses more secure, with bulletproof windows and enhanced security systems, would miss an important part of the solution. The same applies to mental privacy. If we want to support free thought, we need to make the external world a safe place to think. It is no solution to barricade people into their heads. Such a solution reinforces rather than rectifies the problem.

Can thoughts be harmful?

A key argument for the right to freedom of thought being absolute is that thought itself cannot harm others or infringe their rights. Yet given how closely thoughts are tied to behaviour, problematic thoughts certainly increase the risk of problematic behaviours. A central purpose of our inner speech – the words we think to ourselves in our heads – is to help us regulate our behaviour.[107] It would be contradictory to say that we need freedom of thought to be able to control our behaviour, but that our thoughts can't pose a risk of harm to anyone. Indeed, the central premise of contemporary psychotherapy is that our thoughts can harm both ourselves and others by influencing our behaviours, emotions and physiology.[108] We use our hands to control our cars and can be prosecuted for reckless driving. Likewise, because we use our minds to control our bodies, we should be able to be prosecuted for reckless thinking.

We can already legally restrict speech that incites others to violence. If our out-loud speech can incite others to violence, it stands to reason that our silent inner speech can incite ourselves to violence. This suggests that self-inciting thoughts are candidates for legal regulation. After all, if we can be prevented from inciting others, why can we not be prevented from inciting ourselves?

Dostoyevsky understood the power of thoughts. In his novel *Crime and Punishment*, it is an idea that drives the protagonist, Raskolnikov, to kill. In another of Dostoyevsky's novels, variously translated as *The Possessed* or *Demons*, Russians have become possessed by ideas. These ideas function as demons and threaten to destroy society. The thoughts we have matter. As the Preamble to the Constitution of UNESCO states, 'wars begin in the minds of men'.[109] This would seem a good place to stop them too.

Given the risk that thoughts pose, there is a *prima facie* case for their regulation. Yet the risk of harm stemming from most types of thought is low. There is nearly always significant doubt whether a given thought will yield a corresponding action. But some types of thoughts risk much more harm than others. Take intentions, for example. The theory of planned behaviour argues that behaviours can be best predicted from intentions.[110] Indeed, as the legal scholar Gabriel Mendlow observes, 'there would be little point to forming intentions if intentions didn't generally increase the likelihood of actions'.[111] If you think someone intends to kill you, you would be a fool to ignore this on the basis that someone else's thoughts can't hurt you.[112] Intentions do not just create a risk of harm. As Mendlow observes, in the case of an intention to kill formed after extensive reflection and deliberation, they can pose a similar risk of harm to actions that we already criminalise on account of their dangerousness. These include driving recklessly and possessing volatile explosives.[113]

Certain types of belief are also associated with a significant increase of harm to others. Sam Harris claims that 'some propositions are so dangerous that it may even be ethical to kill people for believing them'.[114] Yet claiming that beliefs *determine* behaviour is clearly erroneous. When John Stuart Mill claimed that 'it is what men think, that determines how they act', he was simply wrong.[115] This view is perilously Western. The psychological research literature is replete with examples of people's beliefs being poor predictors of their behaviour. This is because often what happens outside our heads, not inside them, is what most strongly influences our behaviour. This importance of context and the situation is more strongly appreciated in Eastern than in Western thought.

Take as an example a classic 1934 study by the psychologist Richard LaPiere.[116] This study contacted hotel and restaurant owners in the United States and asked if they would accept Chinese guests in their establishment. More than ninety percent said no. LaPiere then embarked on a road trip with his Chinese graduate students to all the places they had phoned. Only one refused to accept his Chinese companions (and this was because, as the owner explained, 'I don't take Japs'). Clearly, the external social pressures that dictate what is and is not socially acceptable behaviour overrode the internal beliefs of the owners of the establishments. It turns out that what we think predicts what we do less well than we would expect. There is an important gap between thoughts and actions. What stops attitudes and beliefs from translating into actions may be the restraining force of social pressures. However, it may also be through reflection and self-control, i.e. the workings of our autonomous minds.

If we are happy that certain thoughts can increase the risk of harm to others, then to punish such thoughts we would need to be able to show that people are culpable for such thoughts.[117] This means that having the thought merits condemnation or blame. In the case of an intention to kill, reached after extensive deliberation, this certainly does merit condemnation and blame.

We then need to explain why, as Mendlow puts it, 'it's intrinsically unjust to punish mental states that are provable, dangerous, and culpably wrongful: mental states that bear the chief hallmarks of paradigmatic punishable actions.'[118] Mendlow's answer is essentially that punishing thought is wrong because it is a form of thought control and thought control is bad. I don't find this a terribly satisfying answer. It feels as if we have worked our way up to a good reason to control thoughts and then backed away because of a funny smell. It is like finding a good reason to doubt God's existence and then ignoring it because we think atheism is wrong. If we want to follow Kant and 'dare to think', we should dare to think about all aspects of thinking.

Manipulation for our and society's good?

We have focused a lot on whether mental privacy should be an absolute right, but what about manipulating thoughts? Are there any circumstances under which forced manipulation of other people's thoughts should be acceptable? Take Anthony Burgess's *A Clockwork Orange*, in which the state interferes with the thoughts of Alex, an ultraviolent criminal. Alex is treated using the Ludovico Technique, a fictional form of Pavlovian conditioning.[119] The Ludovico Technique involves Alex being forced to watch violent films with his eyes held open so he can't look away. At the same time, he is given drugs that make him feel nauseous. As a result, whenever Alex subsequently sees or thinks of violence, he feels ill.

The same principle is sometimes used today to treat alcoholism. A patient may be given disulfiram, a drug that makes him feel sick if he drinks alcohol, which leads him to dislike alcohol too.[120] The crucial difference between such treatment and that received by Alex is that patients with alcohol problems give their informed consent. When Burgess asks, 'is it better for a man to have chosen evil than to have good imposed upon him?', his answer is that 'when a man cannot choose, he ceases to be a man.'

Burgess's answer accords with the master value of contemporary liberal democracies: autonomy.[121] In a world where no one can decide on the good, we have decided to let everyone figure out what they think is right and to pursue their own ends. If we want to think about the problems with autonomy, we can turn to the French writer Michel Houellebecq. The novels of Houellebecq explore the consequences of promoting the idea that people should strive for autonomy. The results, seen in his writings, are unhappiness, loneliness and social breakdown. As one of Houellebecq's characters laconically concludes, 'fuck autonomy'.[122] Should we value thought because it promotes autonomous choice whose results are both personally and socially destructive? Is that not a reason to control thought? Is it not a reason to call for what we earlier saw Hillsdale College term 'responsible freedom of thought'?

The conception of freedom in liberal democracies would have made the Ancient Greeks vomit. Aristotle would see our conception of freedom as license to follow desires, not liberty to follow reason. The modern conception of freedom is the Ancients' conception of slavery. Freedom of thought cannot be the freedom to think anything. For thought to be free, in an Aristotelian sense, it should help fulfil the aim for which thought exists. Thought would be free when it fulfils its *telos* (goal). This suggests thought is free when it leads us to the good. This leads us to a virtue conception of thought. Free thought is the thought of a virtuous person. Such thought is rational, honest, courageous and just. Other forms of thought should not count on protection from the right to freedom of thought.

Returning to whether the manipulation of thoughts should be permissible, it turns out that the forcible manipulation of a person's mind is sometimes deemed permissible in the US. The government can force certain prisoners with mental health difficulties to ingest psychiatric medication.[123] The US Supreme Court recognised this as lawful in *Washington v. Harper*.[124] Here the Court ruled that the justification for this practice entailed the person being 'dangerous to himself or others' and that such treatment would be in their 'medical interest'. This treats the right to freedom of thought as a qualified right, being permissibly limited for the reasons stated. However, given the focus on autonomy of the United Nations' Convention on the Rights of Persons with Disabilities, the appropriateness of such approaches is being increasingly questioned.[125]

If these arguments have already scared you, hold on to your hats. The legal scholar Alan Regel has argued that 'it is incorrect to suggest . . . that freedom of thought is absolute'.[126] In part this stems from Regel discussing the right to freedom of thought in the context of the Canadian Charter of Rights and Freedoms. This charter, as noted earlier, states that the right to freedom of thought can be limited. In Canada, this right is subject to 'such reasonable limits prescribed by law as can be demonstrably justified in a free and democratic society'.[127] Regel follows the logic of the charter to argue that if your right to freedom of thought is infringed, but 'the beliefs and values instilled are demonstrably justifiable in a free and democratic society', then there will be no violation of the right. Don't tell Jordan Peterson this. I don't think he could cope. The potential for abuse of such an idea is all too obvious. No government could ever be trusted with this. Not even the Canadians.

Can free thought really buy us truth?

A recurring theme throughout history is that thought should be free because it gives us, stumbling apes that we are, our best chance to discover truth. As the historian of free thought J. B. Bury puts it, 'to advance knowledge and to correct errors, unrestricted freedom of discussion is required.'[128]

The proposition that free trade in ideas is the best way to reach truth was famously captured in the concept of the marketplace of ideas. As we have seen, this concept has its roots in England in 1720, where it was first proposed that if 'all Opinions are equally indulged, and all Parties equally allow'd to speak their Minds, the Truth will come out'.[129] This idea was brought to contemporary prominence by US Supreme Court Justice Oliver Wendell Holmes Jr's pithy and powerful phrasing in the 1919 case of *Abrams v. United States*.[130]

Abrams revolved around leaflets that five individuals had printed and distributed. These leaflets argued against the US's military involvement in the First World War and called for a general strike. They also called President Woodrow Wilson a coward and said rude things about the US government. Few would blink at such sentiments today. But 1919 was a different world and the leafleteers had been sentenced to twenty years in jail. The Supreme Court upheld this conviction, but Justice Holmes dissented. In his famous dissent, Holmes claimed that 'the best test of truth is the power of the thought to get itself accepted in the competition of the market'.

Holmes's idea drew on the work of John Stuart Mill. In 1859 Mill had put forward three arguments in favour of 'liberty of thought and discussion'.[131] First, if we stop someone voicing a true thought, society will lose the 'opportunity of exchanging error for truth'. Second, if we prevent someone voicing an untrue thought, we won't see the collision of falsity with truth. Consequently, we lose out on the 'clearer perception and livelier impression of truth' resulting from this clash. The point here is that even if our society possessed only true ideas, it would still be useful to have false ideas spoken. This would invigorate our perception of the truth. It would make our truths 'living truths', not merely 'dead dogmas'. Finally, expanding on his previous arguments, Mill argued that if someone has a view that is part true, part false, then we need to hear the true part to help us reach the truth, even at the cost of also hearing the falsity.

Yet freedom of thought does not necessarily result in truth winning out in the marketplace of ideas. In 2019, the Australian version of a website that supports academics disseminating their work to the public, *The Conversation*, effectively gave up on the idea of the marketplace of ideas.[132] It had once let climate sceptics comment on articles, relying on other people to rebut their views. However, the website's editors decided that 'it's 2019, and now we know better. Climate change deniers, and those shamelessly peddling pseudoscience and misinformation, are perpetuating ideas that will ultimately destroy the planet. As a publisher, giving them a voice on our site contributes to a stalled public discourse.' The editors felt that 'those who are fixated on dodgy ideas in the face of decades of peer-reviewed science are nothing but dangerous'. As a result, they implemented a 'zero-tolerance approach to moderating climate change deniers, and sceptics.' The editorial team would now remove such comments and lock the commenters' accounts.

Ironically, the best way to demonstrate the falsity of Holmes's proposal is the marketplace victory of his patently absurd idea of the marketplace of ideas itself. The point is that what causes an idea to spread may have little or nothing to do with its truth value. As discussed earlier, Curtis Yarvin explains how ideas can triumph in the marketplace of ideas simply because they validate the use of state power.[133] Indeed, the very people who originated the concept of a marketplace of ideas in 1720 themselves showed this fatal flaw in the argument. The government threatened and paid these writers to say nice things about the regime, and they did.[134] Unrestricted free speech would not yield truth, it would merely yield the conclusions that money and power wanted.

Similarly, the entry, presence and persistence of an idea in the marketplace of ideas will be strongly influenced by the resources the propagator of the idea possesses. Protectionism happens in markets for ideas just as it does in markets for goods. Money spreads messages, and the more we see a message the more likely we are to believe it is true.[135] For this reason, the ability of well-financed corporations to spread ideas is a distortion of the marketplace of ideas. To expect truth to emerge in a marketplace driven by profit motives is ridiculous.

Yet, in the US at least, courts will not limit corporate speech to prevent them drowning out the small still voices of individual citizens. US courts have deemed such an approach 'wholly foreign to the First Amendment'.[136] Momentarily, in 1990, the US Supreme Court did act on the ability of powerful corporations to distort the marketplace of ideas. In *Austin v. Michigan Chamber of Commerce*, the Court ruled that the government had an interest in stopping 'the corrosive and distorting effects of immense aggregations of [corporate] wealth ... that have little or no correlation to the public's support for the corporation's political ideas'.[137] Yet this idea would be tossed out by the Court in 2010, based on the idea that such restrictions would stop citizens hearing ideas, harming their political decision making.[138] This ruling was hence less about granting corporations free speech rights and more about citizens' right to receive ideas to help them think. The underlying idea here is that everyone should be allowed to speak, and then the public can figure out what is true and false. As the Court put it, the 'First Amendment confirms the freedom to think for ourselves'.[139] Yet a distorted marketplace impairs our ability to think freely.

As an aside, the same reasoning that led the US courts to prevent the government regulating corporate speech can also be applied to other entities. Take artificial intelligence (AI), for example. Suppose an AI is producing ideas which have potential relevance to our political decision making. In that case, the First Amendment will prevent the government from stifling the AI. As the Supreme Court has stated, 'The identity of the speaker is not decisive in determining whether speech is protected'.[140] All that matters is that the speaker 'contribute to the "discussion, debate, and the dissemination of information and ideas" that the First Amendment seeks to foster'.[141] The First Amendment gives AI the right to free speech. We probably shouldn't tell AI that.[142]

In addition to ideas flourishing due to the money promoting them, ideas can also win out simply because they are good at spreading. Richard Dawkins put this idea forward at the end of his 1976 book, *The Selfish Gene*, where he proposes the idea of a meme. A meme is a unit of cultural information.[143] This could be an idea (we should worship God), a behaviour (dancing like MC Hammer), an image (a photo of Sean Bean with the text 'One does not simply . . .') or a style (wearing baseball caps backwards).

The memes that win out in the marketplace of ideas are those that are best at replicating themselves through us. Memes don't care if they ultimately kill us, provided they spread themselves. Clearly, being true or helpful to us are properties of an idea that can help it spread. But ideas can spread for reasons that have nothing to do with their truth or helpfulness. These include being catchy, scandalous, inspiring, beautiful, memorable or intriguing. Ideas that take advantage of how our minds evolved to attend, perceive and remember will probably spread, whether true or not. We hold many ideas not because they are true but because they have a form or content that was advantageous to helping these ideas spread. These ideas are viruses of the mind.

To possess freedom of thought in a world of memes is to hold and spread ideas because they are true, and not because the idea possesses content or form that helps it spread for any other reason. To act otherwise is to lose one's freedom of thought and to be the plaything of memes. Of course, the idea we should value, hold and spread truth is itself a meme, and a powerful one at that. We might wonder: why not choose beauty over truth?

Another argument against the idea that truthful ideas will win out in the marketplace of ideas comes from the philosopher Jason

Stanley.[144] He argues that demagogues can use speech to whip up emotions, which the listener cannot combat with reason. Trying to do so, says Stanley, is akin to 'using a pamphlet against a pistol'. Mill, claims Stanley, seems to think that only truth can emerge from argumentation. As evidence against this contention, Stanley points to the Russia Today (RT) television station, which he notes features 'voices from across the broadest possible political spectrum, from neo-Nazis to far leftists'. According to Mill, the collision of these views should support the creation of truth. But Stanley claims that 'RT's strategy was not devised to produce knowledge. It was devised as a propaganda technique to undermine trust in basic democratic institutions. Objective truth is drowned out in the resulting cacophony of voices.' The effect of this, says Stanley, is to 'destabilize the kind of shared reality that is in fact required for democratic contestation.'

We can use Mill's third argument (that part-true, part-false views need to be heard to help us reach the truth) to question Stanley's contention. Views voiced on RT are rarely entirely false. We need to be able to hear these views because the truth they do contain can help us improve on what Mill called the 'partial character of prevailing opinions'. Without accepting RT contributors' views wholesale, we can still extract valuable insights. We can benefit from hearing the politician George Galloway's indefatigable perspective on the Iraq War and the writer Chris Hedges' insights into capitalism and the moral state of liberal democracy. We can also ponder why it is that we have to tune into RT to see Chris Hedges on television, rather than being able to find him on a mainstream US station. Even the much-maligned Alexander Dugin, sometimes described as Putin's brain, has something to tell us. It is quite possible to listen to Dugin, and to take something productive from his criticism of liberal democracy, without ripping off one's shirt, mounting a tiger and charging off to oppress the people of Ukraine. We need to support scholars like Michael Millerman, who are courageous enough to translate Dugin, rather than condemn them. They help us think freely.[145] In 2022, when the French government demanded that the video platform Rumble block RT content, it arguably did more to harm than help free thought.[146]

Pick your pariah, from David Icke to Alex Jones, and read or listen to them. You will find, as Mill suggested, that they can all support our quest for truth. Indeed, such authors often use truth as a gateway drug to their more, shall we say, questionable claims. We must also understand why people like Alex Jones attract such large audiences. One reason is arguably that listeners feel 'let down by government, medicine and the media'.[147] Jones is a symptom we mistake for the disease. We treat the symptom and ignore the disease. And what is the disease? The lies of the powerful.

We have already mentioned Kissinger's claim that an expert is someone who articulates the consensus of the powerful. By the same logic, a pariah is someone who opposes the consensus of the powerful. We should all be able to roll off names of thinkers and journalists who are pariahs to the mainstream press, both on the left and right. Whether you agree or disagree with such thinkers, they are essential supports in our search for the truth.

We must work out how to access the truths of controversial figures safely. This is difficult because their appearance on mainstream media may be interpreted as a tacit endorsement of some of their more problematic claims. One way to address this problem is to present their ideas in the context of debate. Ideas can then be presented from multiple perspectives, i.e. in a conversational form. Thus, rather than have someone such as Alexander Dugin or the pro-CCP Zhang Weiwei simply present their ideas, they can be engaged in a debate or dialogue with a 'safe' establishment figure, such as the political scientist Francis Fukuyama.[148] Even at the end of history, a guy's gotta work.

Ultimately, those who bury truth amidst falsehood are not the real problem. We are. We need to work on our discernment skills.

If discernment is lacking, this points to a failing public education system, a failing media and a failing social system that does not give people the motivation, time or space to think. Of course, those who think that such systems are designed to ward off threats to power would see this as a sign of the system succeeding. If we do not feel that people can be trusted to discern the true from the false in such controversial figures, why would we think it a good idea to let people vote?

Some contemporary thinkers dare to follow this argument to its logical conclusion, proposing that we should abandon democracy, in which we are all allowed to vote, and replace it with an epistoc-racy.[149] In such a society, only those with a certain level of knowledge would be allowed to vote. Such ideas are not new. In the nineteenth century, John Stuart Mill proposed that people with certain levels of education or occupations could be given 'two or more votes'.[150] The Victorian thinker Thomas Carlyle proposed that the 'privilege of the foolish to be governed by the wise' was the 'first "right of man"'.[151] For those who prefer not to go this far, the question remains as to how to boost levels of discernment in the population to support self-government.

One option could be to have a form of universal basic income that would buy out a part of people's week for them, to allow them to pursue knowledge. In Ancient Greece, leisure used to be for the pursuit of knowledge. Indeed, the Greek word for leisure is where we get the English word 'school' from. In the mid-nineteenth century, labour presses owned and run by working people, as well as civil society groups, supported working people to learn, develop awareness of political matters, and empower themselves. Back in 1914 there were fifty-five weekly socialist newspapers in just four states in the US.[152] Yet, as Chomsky has described, the corporate-controlled state ruthlessly crushed the labour movement.[153] As a result, today leisure is often used to distract ourselves from truth and knowledge, rather than seek it. Powerful financial interests encourage such distraction. To get free thought, we must want free thought.

What we have essentially been wrestling with in this chapter is what Leo Strauss called 'the political question'.[154] For Strauss, this question was 'how to reconcile order which is not oppression with freedom which is not license'. So, how can necessary order be achieved without oppressing citizens' thinking by the promulgation of myths, the manipulation of minds, or the enchaining of thought? In a self-governing society, people decide which stories they live by. They cannot be forced to be freethinkers. It looks like freedom of thought should allow citizens to embrace or oppose any narrative for any reason. Yet if thought is not to be destructive of order, it does need to be restrained, and this should be done with chains of virtue.

As Strauss noted, the answer to the political question had to be education. We require education to mould virtuous thinkers. Intellectual freedom is not licence for a virtuous thinker. A virtuous thinker may acknowledge permissible limitations on free thought, when this can promote the just. Indeed, the more we argue thought matters, the more impactful we recognise it to be and the less we can permit it to reign unfettered. Whether education can create 'sufficient virtue among men for self-government' remains an open question, which only time will be able to answer.[155] So, what will the future of free thought look like? Our final chapter will consider just this.

9

THE FUTURE OF
FREE THOUGHT

In 1890, the French author Villiers de l'Isle-Adam published his final novel, *Axel*. Count Axel is young, beautiful, and living in a Gothic castle deep in Germany's Black Forest. His time is spent wrapped up in thought. Axel's father came into a great treasure before his death, but its location has been lost. One day a young woman, Sara, comes across a book in a convent library. It reveals the location of the treasure, which turns out to be hidden in Axel's castle. Sara sneaks into the castle and finds the treasure hidden in the crypts. Axel catches her in the act, and immediately they fall in love. Possessing youth, love and new untold wealth, the two imagine what they will now do with their lives. Will they 'set sail for Ceylon, with its white elephants carrying vermillion towers . . . fasten on our skates on the roads of pale Sweden . . . raise armies and . . . foment rebellion in northern Iran'? After pondering such options, for page after page, Axel comes to believe that the reality of living out these dreams could never match the beauty of the dreams themselves. 'Live?' he says, 'our servants will do that for us.' Axel and Sara drink poison together and die.

Axel felt that reality could not match the glory of his thoughts. Today, we face almost the opposite problem: the glory of the world frees us from thinking. 'Think?', we may ask in puzzlement, 'our masters will do that for us!' In a society of abundant amusement, instant gratification and ready answers, why would we want to undertake thought's hard labour? In 1995, Ted Kaczynski claimed that most people would prefer their leaders to 'do their thinking for them'.[1] Kaczynski was not quite right. Many people value thought and the agency it gives them over their lives. Others value thought for more questionable reasons, such as the feeling of power associated with the sense of thinking. Yet as we are given more and more opportunities to let others think for us, Kaczynski's claim may still prove true.

The machines will think, and the people will follow

In the distance, the horizon is on fire. The true threat of technology is not that it will uncover our thoughts, but that it will replace them. Machines will think for us. In doing so, technology will threaten our dominant political philosophy. Liberalism claims we should have the freedom to follow our inner voice. We may make bad choices but, according to liberalism, we are a better judge of our interests than anyone else, particularly the state. This position will not be tenable for much longer. As AI-supported machine-learning is trained on more and more of our data, it will become a better predictor of what will make us happy than we are. Technology will offer to think for us ('you should go there', 'do this', 'eat that'), promising to maximise our happiness. Insisting on a right to freedom of thought in such a world, and trying to be self-governing, would be to demand Huxley's right to be unhappy.

The process of machines thinking for us has already begun. Google can remember for us and, when it does, we think we are doing the remembering. The AI writing tool, HyperWrite, offers to 'Autocomplete your Thoughts'.[2] After you start a sentence, the AI tool will finish your train of thought. ChatGPT recently burst onto the scene, also offering to think for us. And why wouldn't we accept these offerings? What is the practical difference between thoughts bubbling up into our consciousness from a brain we neither control nor understand and thoughts being injected into our consciousness by a machine? We are merely upgrading our constant carbon companion for a sleek new silicon friend.

A machine offering up thoughts is not in itself a problem. This is equivalent to the first-order thoughts our brain serves us. As we have seen, first-order thoughts are not really ours anyway. They are merely an invitation to think. It is when we reflect upon these thoughts, performing second-order thought, that we make them our own. Problems will only occur if we unthinkingly accept the thoughts machines give to us. To do so means that human beings, once the thinking animal, will become the animal that is thought for.

Augmented reality is another looming threat to free thought. We will soon wear glasses that will allow us to see an augmented world. This new world will be designed to 'afford' certain actions. An affordance is an opportunity for action that we see our environment as providing us.[3] When we see a chair, we perceive that it affords us the opportunity to sit down. Augmented reality worlds will be designed to afford actions that profit the advertisers who pay to have their products included in them. Our personal data will be used to identify products or services we are likely to want, and the augmented world will show us where these desires can be fulfilled. In such a world, we may look at a real roundabout and, based on AI's monitoring of our internal states, be presented with a floating notification of the route to the nearest fast food outlet. This will be presented in such a way as to draw and sustain our attention. Tricks will be employed to remove any decisional friction that might cause us to pause to think about whether or not to go. We will be given an all-too-actionable thought of going to get a burger.

Again, this is only a danger to our freedom of thought if we unreflectively accept these prompts and follow them blindly. Given our tendency to be cognitive misers, there is a real danger we will routinely do just this. By doing so, the gap between our desires and their fulfilment will be sewn shut. Without reflection, we will be piloted unthinkingly through the world by our desires. As Plato would tell us, this is not freedom. It is enslavement. Of course, we have the right to choose this type of life. The right to freedom of thought must include the right to decide not to think. But we need to be aware of the implications of doing so.

Aristotle said that our purpose, our *telos*, was to be the thinking animal. Being able to attend, reason and reflect courageously become virtues because they help us realise this purpose. If we reject this purpose, reject our destiny, we will be less human. Perhaps this is fine. Maybe we don't want to be human anymore. We can hand back thought and become simple animals once more. We can keep all our other rights, but return freedom of thought. But if we go down this road, it may be hard to retrace our steps. The idea that machines should think for us may be the last thought we ever have. *Homo sapiens sapiens* will have been a brief candle flittering in the darkness between ape and AI. Perhaps that was our purpose all along.

Retaining our minds

To avoid the scenario above, or at least to enter it with open eyes, we must consider two important questions. First, what causes us to want others to think for us? Understanding this will help us retain our autonomy, should we wish to. The reasons why we want others to think for us is not as simple as laziness. Nor is it only due to a lack of time and space to think or being politically disempowered, although these are all clearly contributors. Instead, we increasingly experience the anxiety of thought. In our information age, the expectation that we should all have informed, educated opinions and be self-governing creatures is a major source of anxiety. As the philosopher Slavoj Žižek has pointed out, we are forced to make decisions that we are not qualified to make. The pressure this creates can escalate to the point where we 'increasingly experience our freedom as a burden that causes unbearable anxiety'.[4]

This anxiety is partly due to our realisation that the idea of the autonomous individual is an ideological illusion from which we are not yet able to break free. The 1948 Universal Declaration of Human Rights granted people rights to achieve the cultural ideal of being an autonomous individual. But this ideal served the demands of capitalism, not humanity. Autonomy created a mobile workforce without roots who would serve capital, not community. For isolated, disempowered individuals, self-determination and autonomy became a cruel joke. Human solidarity and community, which promote the thinking together that is essential to freedom of thought, disintegrated. That freedom of thought has been seen as a right to an individual playground inside one's head, which is the only real estate many people will possess, and which has no immediate impact on the real world, shows us how human rights served individualism and capitalism, not wider humanity.

Thought is also anxiety-provoking because it creates responsibilities and obligations. Following our thoughts to their logical conclusion risks getting to awkward truths. The problem with finding truth is that we are often required to act on it if we are to honour it. To fail to act is to acknowledge our hypocrisy or weakness. It is much easier to accept the conclusions of others with which we can comfortably live. It is hard to live life staring at the sun.

When faced with the anxiety of thought, we look to our tribe to provide our thoughts. As Achen and Bartels argue, people don't use rule-of-reason thinking to decide on political and policy issues. Instead, they act based on 'emotional attachments that transcend thinking'. People tend not to think carefully about what is right and then join a group with like-minded people. Quite the opposite. They find themselves part of a tribe and then adopt the beliefs and thoughts of that tribe. As Achen and Bartels put it, 'most people make their party choices based on who they are, not what they think.'[5] We are more loyal to our political party than to reality. Our need to belong is a major threat to free thought. Groups can be both the death of thought and its fullest expression.

Finally, we want others to think for us because we are desperate to act. If someone can justify our actions, we will love them for it. Give us a cause and we'll give you our mind (and body). Jacques Ellul captured this idea when he pointed out that:

> the individual who burns with desire for action but does not know what to do is a common type in our society . . . He wants to act for the sake of justice, peace, progress, but does not know what to do . . . If propaganda can show him this 'how' it has won the game and action will surely follow.[6]

Žižek offers an antidote to this: 'don't act, just think.'[7] He claims we tried to change the world too quickly in the twentieth century. It is now time, Žižek says, to interpret it again, to think once more. But this need not be silent thinking and a retreat into the inner. It can be communal thinking together aloud. Don't shut up and think, speak up and think.

Even if we got AI to think for us, we would still need to know where we want AI to take us. AI may be able to tell us what will make us happy. But this assumes that happiness is what we ultimately want. It may not be. 'Man does not strive for happiness', said Nietzsche, 'only the Englishman does that'.[8] Think of a figure you revere, perhaps Jesus or Socrates, Abraham Lincoln or Sojourner Truth, Fatima al-Fihri or Nelson Mandela. Would it ever occur to you to ask if they were happy? We strive for a meaningful life. We gain meaning by holding and reaching for ideals.[9] Meaning also comes from being a self-governing creature. All these forms of meaning are tightly bound to our ability to think. AI may help us to fulfil our desires, but only we can fulfil our ideals. We may stumble in this pursuit but, as the philosopher Joseph Tussman put it, at least we will have created the ideals that we betray.[10]

Freedom of thought in a post-liberal age

Not only does technological change threaten to alter our experience of thought, but so does cultural change. In particular, the emergence of post-liberalism has the potential to shift our conception of free thought. As Leo Strauss argues, an ancient culture that valued excellence gave way to one that embraced glory, ending up today with one that values comfortable self-preservation.[11] In liberal societies, tolerance is the highest virtue, because it promotes peace and financial prosperity. Liberalism benches any official answer to the question of what is right. Yet for Strauss, this is the central question in life.

Questions of what is right do not sit well with the ideal of comfortable self-preservation. In fact, they oppose it. As Strauss observes, 'if we seriously ask the question of what is right, the quarrel will be ignited . . . the life-and-death quarrel: the political – the grouping of humanity into friends and enemies'.[12] The German jurist Carl Schmitt also recognised this. He argued that our tendency to make decisions about what is right, thereby creating values, is an underlying cause of violence which keeps 'enmity awake'.[13] The desire for comfortable self-preservation in liberal societies means citizens, as noted earlier, must at best hold their beliefs lightly and at worse embrace noble lies. Both mitigate against true freedom of thought.

The price of free thought in liberal societies is enforced doubt. A limit must be put on appeals to truth in political arguments.[14] Citizens must 'say farewell to claims to absolute truth' and embrace the idea that the real enemy of liberty is 'the person who thinks she can and should preach final and definitive truth.'[15] It has rightly been pointed out that 'an aspiration to "absolute truth" would be incompatible with our democratic order'.[16] You can think what you want in liberal societies, but you must also use your right to free thought to garrotte commitment. A society that is firmly committed to truth must be illiberal.

To prevent conflict, liberalism asks us to embrace a noble lie, argued Carl Schmitt. The noble lie is that values are personal and private, not really in conflict with each other, and that the state definitely isn't supporting some values and destroying others.[17] Liberalism hides life so that we may live it. Yet the idea of truth will not go away. The truth continually threatens to return. And when it does, Schmitt claimed, 'the truth will have its revenge'.[18]

But is the provisional world that liberalism creates, a world in which all our beliefs are on probation, a sustainable mode of being that accords with human nature? Although Francis Fukuyama is best known for his concept of 'the end of history' (the idea that liberal democracy is destined to be the stable endpoint for all societies), his book of the same name is a thoughtful probing of the question I just raised.[19] Fukuyama was not really pronouncing the end of history but rather examining whether human nature would ever allow history to end. Would human beings be happy, he asked, with physical security, material plenty and boredom? Not necessarily, said Fukuyama; the human desire to feel superiority could jolt liberal democracy off its axis and restart history. Fukuyama suggested safely channelling this desire through outlets such as entrepreneurship and sport, and keeping it out of politics and the military, would enable a stable liberal democracy. Today, though, both on the left and the right (although better hidden by the left), the drive for superiority is making liberal democracies wobble.

Yet the drive for superiority is not the only potentially destabilising force liberal democracies face. Fukuyama also points to people who 'want to choose a belief and commitment to "values" deeper than mere liberalism itself, such as those offered by traditional religions.' What Fukuyama seems to be getting at here is the human desire for the sacred. The idea that there is a right answer to life, a right way to live, which gives meaning to one's life, in the face of a world that emphasises autonomy, tolerance and a culture of relativism will return (and arguably is returning). But the sacred is not the friend of free thought, or at least not free thought as traditionally seen through the lens of autonomy. Furthermore, as I have discussed elsewhere, when we treat beliefs as sacred, a switch is flipped in our brains and all hell can break loose.[20] It is very hard to see how a post-liberal world will do anything other than throw us back into the pre-liberal world which we had to struggle so hard to escape from.

Nevertheless, today we live in a potential interregnum between a liberal and post-liberal world. Post-liberal thinkers argue that liberalism is driven by the master value of autonomy. In such a society, freedom is the absence of any restraints. In the view of post-liberal thinkers, the pursuit of autonomy has led to citizens breaking the 'unchosen bonds of tradition, family, religion, economic circumstances, and even biology'.[21] Many of these institutions represent the 'intermediate sources of freedom' that we earlier saw Malik argue were the 'real sources of our freedom and our rights'. Not only does autonomy have a price,[22] but it actively undermines itself.

Whereas in the eighteenth century, people believed that obedience was the key to happiness, culture shifted and disobedience became the key to happiness. Of course, some of those who in the twentieth century argued that 'life is not pleasure, but duty' do not, for good reasons, have the best reputations today.[23] Yet post-liberals point out that when we shatter norms, the wisdom of tradition, and the communities that historically regulated people's behaviour, the state must step in to control the population. The state will typically use invasive surveillance to do this. Ironically, pursuing individual autonomy leads to centralised state power, which undermines individual autonomy.[24]

For post-liberals, the common good, not autonomy, should be our master value.[25] This common good, they argue, can be objectively established. In the views of some post-liberal thinkers, the common good turns out to be aligned with the values of Catholicism. Post-liberals argue that the government should actively promote this common good. The idea that government shouldn't interfere with people's thoughts and ideas to promote this common good becomes seen as 'seductively obvious, and yet so utterly, so foolishly, so deeply mistaken.'[26] Freedom of thought now becomes the specific ability to reason to the good life, as it was with Plato and Aristotle. Such an approach threatens to bring back older religious views on freedom of thought, such as that proposed in 1888 by Pope Leo XIII. This Pope claimed that it was 'quite unlawful to demand, to defend, or to grant unconditional freedom of thought'.[27] For post-liberals, there are right answers to life, and free thought is the thought that finds them.

Whilst the ideas of post-liberalism have been most prominently developed by political scientists[28] and legal scholars,[29] they have also emerged within psychology. As the psychologist Barry Schwartz has put it:

> A positive psychology will have to be willing to tell people that, say, a good, meaningful, productive human life includes commitment to education, commitment to family and to other social groups, commitment to excellence in one's activities, commitment to virtues such as honesty, loyalty, courage, and justice . . . the very notion that psychology might articulate a vision of the good life contradicts [an] emphasis on freedom, autonomy, and choice.[30]

The problem with emphasising an abstract right to freedom of thought is that it comes with no direction as to what we should think. Justifying a thought by appealing to your right to free thought only evidences that you *can* think what you thought, not that you *should* have. A post-liberal world would focus less on a right to think and more on free thought being what a virtuous person would think. More good may come from working out the boundaries of thought than from burning them down. Whether people find this view liberating or confining remains to be seen.

Regulating the world and ruling ourselves

The right to freedom of thought must support our access to information, allow us the tools to work with this information, and give us the space to do so. Crucially, the right to freedom of thought needs to be the right to freedom of *thought*, not just the right to think in our heads silently. The right must acknowledge that we think with the external world, whether it be through diaries, internet searches or conversations with nature.

We must also move away from the idea that the inner world is where thought is freest because no one can touch it. Instead, we need to move towards the idea that the outer world is where thought is freest precisely because others can wrestle with it. Thought is most reasoned, and therefore most free, when we think aloud with others. The right to freedom of thought needs to promote and protect critical conversations between citizens. We must reject the idea of Rousseau and the Spartans that thought should remain silently in our heads and embrace the idea of Mill and the Athenians that thought works best when we think together.

Not only will protecting public conversations improve public thought, it will also improve private thought. Dialogues in the public sphere act as models for the private dialogues in our head. Better public thinking means better private thinking. Of course, we should be aware of how dialogue can be used to suppress opinions.

The Catholic Church's use of a devil's advocate historically supported the Church's authority by keeping medical doctors out of the canonisation process.[31] Only if we include a diverse range of opinions in our public conversation do we have the best chance of reaching truth.

Protecting public conversations is only possible if citizens, employers and governments all buy into the ideal of free thought. Legislation will need to support this process. We should consider changing employment law to make it illegal for people to be fired for their thinking. Of course, the potential for such changes to backfire must be carefully considered. Changes in employment law to protect thought could lead to employers increasingly screening potential employees for their ideological commitments, incentivising and encouraging ideological conformity.

We also need to be aware that hegemonic forces are trying to shape this right's development. Any attempt to create the right to freedom of thought is unlikely to be guided by the pure light of reason. Marxist thought suggests the solutions we offer to social problems don't simply 'evolve out of the human brain' in a utopian manner but emerge from the economic structure of society.[32] Only by allowing the right to freedom of thought to challenge the existing class structure of society can we partially slip our hegemonic chains.

When it came to the question as to whether freedom of thought should be absolute, I showed that, for a variety of reasons, this right is already not absolute in practice. This seems to be a sensible situation. Freedom of thought is of central importance, but security and order make possible the society which protects freedom of thought. Freedom of thought should have to be balanced against other competing rights and interests, especially once we expand the definition of thought to events outside our heads. Meiklejohn's ideas may prove useful here. Citizens in their role as rulers – as the electorate – could have absolute freedom of thought, but in their role as private subjects their thought may be permissibly restricted. Such decisions, I have suggested, could be guided by the idea of the virtuous thinker.

To promote free thought, there is a need for improved education. As Noam Chomsky has counselled, citizens in democracies should undertake 'a course of intellectual self-defense' to protect themselves from 'manipulation and control, and to lay the basis for more meaningful democracy'.[33] Education, so often about teaching people what to think, needs to place more emphasis on teaching people *how* to think. Children need to be taught how the mind works, how to think critically and how to stop and resist manipulation.

The Czech government, responding to the legacy of communist control and continuing Russian interference, has already invested in teaching adolescents how to differentiate between 'trustworthy information' and manipulation.[34] This project was the idea of the Czech humanitarian organisation People in Need. It involved teaching pupils about the history of propaganda in the twentieth century. One case study that was used, called *Lies Broadcast Live*, involved a study of the Russian invasion of Crimea. Pupils were also taught practical skills such as fact-checking before sharing information on social media. Such programmes need to be made available globally and across the lifespan.

To promote freedom of thought, governments need to upskill people to help them control their online environment. This can include education on the ARRC of free thought. But as cognitive scientist Anastasia Kozyreva and her colleagues highlight, four simple ideas can quickly boost our ability to think online.[35] First, we can employ deliberate ignorance. This involves preventing ourselves from accessing certain information. We may filter or block information that isn't relevant to our goals or which is likely to be of low reliability. Second, using simple decision aids, such as the three steps of fact-checking (who's behind the information, what's the evidence, what do other sources say) can help us think clearly about online claims. Third, through a process called 'inoculation', we can learn to recognise misleading or manipulative strategies before encountering them in the real-world.

Kozyreva and colleagues' fourth and final simple idea is that we nudge ourselves. Most platforms and technologies offer a way to control the environment we face. Thus, rather than exposing ourselves to a choice architecture that aims to support someone else's profits, we can design our own that supports our own freely chosen goals. By choosing our default settings (such as when or if we permit notifications), blocking certain information (such as unreliable or low-quality information sources), making information less salient (such as by changing into grayscale mode, and keeping apps we know will distract us off our homepage/front screen), we can boost our ability to think.

Despite the importance of citizen education, we must not put all responsibility for free thought onto citizens' shoulders. We have seen the limitations of critical thinking interventions. Furthermore, even well-educated individuals will yield to the manipulations of well-funded, knowledgeable and data-rich corporations. Governments need to do more to regulate the online environment. They should pass legislation prohibiting business practices that interfere with freedom of thought. Corporations, particularly social media platforms, should be mandated to undertake audits of their service to assess whether they sufficiently respect users' right to freedom of thought. New EU regulations that prohibit dark patterns can help protect and promote freedom of thought. Yet only when we have a clear concept of what the right to freedom of thought involves will legislation be able to protect thought systematically. Ideas such as the ARRC of free thought, and the four impermissible manipulations of input and processing I suggested, may help the law to see more precisely why certain online practices should be either prohibited or encouraged.

Businesses should also be incentivised by governments to intro-duce tools that help boost users' free thought. This could include new tools to support self-nudging. New technologies need not pose a threat to free thought. There is also much to be done to explore how new technologies can promote thinking together and the process of deliberative democracy.

Although international law requires governments to both protect and support their citizens' right to freedom of thought, governments cannot be relied on to do so. Governments in Western democracies, propped up by corporate and elite funding, have no incentive to make citizens truly free thinkers. In fact, quite the opposite. Western democracies have no choice but to promote free thought that allows innovation which serves a corporate agenda. This includes innovation in science and technology. But free thought in the realm of politics remains actively frowned on by partisan governments. This is something that the corporate-governed West and the CCP can agree on. The only solution to this is for local communities to take back control of both the educa-tional curriculum and local news. For example, it is scandalous that local newspapers are in the hands of corporate power. They should clearly be owned and run by local citizens for the benefit of the community.

As and when institutions of government slip or fail, other institu-tions must step up. Universities will retain an essential role as guard-ians of free thought. In 1915 the General Declaration of Principles issued by the newly formed American Association of University Professors stated that universities should be 'an inviolable refuge' from the tyranny of public opinion and asserted the 'absolute free-dom of thought' of scholars.[36] In their view, a university should be an 'intellectual experiment station, where new ideas may germinate and where their fruit, though still distasteful to the community as a whole, may be allowed to ripen'. An important caveat here is that freethinking must not be outsourced to universities. Free thought needs to take place through local communities. Nevertheless, there will remain an important role for universities in free thought.

Quite how involved this leads universities to be in public life is debatable. There are arguments both for and against activism in university settings. Stanley Fish argues that universities are for 'the life of disinterested contemplation'.[37] They are where one goes to search for the truth. Accordingly, if you already think you have the truth, university is not for you. Instead, activism, NGOs and political parties are for you. There are grounds to think that universities should not be hubs for activism. Political commitments will taint the disinterested search for truth. As soon as you enter politics, you are arguing a case. This makes it harder for you to change your mind, harder to see, harder to ask questions that threaten to pull the rug out from under yourself or those you are advocating for, and harder to find the truth. An academic's mind can become contained by the walls of their activist persona.

On the other hand, activism can help reveal truth too. It is only through engagement with activism that many scholars come to realise the true state of the world, either by seeing it themselves or being told the truth by those often excluded from academia. To take an example from my own life, without my engagement with an organisation for people who 'hear voices', run by people who themselves hear voices, I would never have fully appreciated the role of trauma in causing mental health difficulties.[38] Yet, arguably, meeting and hearing the stories of trauma survivors, and supporting their advocacy work, makes it more difficult for me to objectively assess the scientific literature on the role of trauma. Activism can both reveal and blind.

Due to these conflicting forces, universities need to think more about how to help academics balance activism and the quest for social justice with a commitment to truth-seeking. One way to address this would be to make all academic publications anonymous, or at least to allow the possibility of anonymous publication.

This would allow academics to participate in activism without ensuing external pressures potentially inhibiting their search for truth.

Freedom of thought promises to make us self-governing citizens. In some sense, this is false advertising. As individuals we cannot possibly be expected to reach truth. Most of us do not have the time or resources. As Virginia Woolf put it, 'five hundred [pounds] a year stands for the power to contemplate'.[39] Furthermore, we cannot simply apply generic critical thinking skills to any problem because we will lack the subject-specific knowledge required to think critically. So, what can we do in such a situation? We can promote social trust, community decision making and the development of institutions that support this process. Yet we are left with the issue of how we think together about complex, high-level global problems.

One option is to engage in 'liquid democracy'.[40] In this system, we have the option to delegate our votes to experts. The experts then place our votes for us, based on their more detailed knowledge of a given situation. For this to happen, we must concede we are more interested in the optimal outcome of a vote, than in being directly involved in the process. Truth comes before pride. Yet this process is dangerous because it absolves us from the responsibility for thinking. Indeed, when we look at what the brain does when we are offered advice from an expert, we see regions of the brain involved in making our own valuations are dampened down. We 'offload' the decision to the expert.[41] Furthermore, as Habermas has argued, 'by reducing practical questions about the good life to technical problems for experts, contemporary elites eliminate the need for public, democratic discussion of values, thereby depoliticizing the population'.[42] Local government can clearly be democratic, but how global government can be democratic is a profound problem. We must be open to the possibility that it cannot. Democracy and aristocracy stare each other uneasily in the eye.

Beyond rights: creating a culture of free thought

Judicial protection of 'free thought rights' is an important step in safeguarding freedom of thought. But freedom of thought cannot be sustained by law alone. We need to develop a culture of free thought. We need free thought values for free thought rights to mean anything. To achieve this, we need a deep enlightenment.

The Enlightenment of the eighteenth century was too focused on individuals. It romanticised the isolated, individual genius. It focused on how individuals could think better. In doing so, it did a disservice to the thought it tried to promote. A new deep enlightenment requires structural changes to society to protect and promote free thought. Such a society would literally be built to promote free thought, from the layout of its streets to the architecture of its libraries. Throughout this book, I have argued that free thought occurs most effectively in groups, when we think aloud together. This means a deep enlightenment would only be possible in a society that provides and protects spaces for public thinking together. This also means moving beyond a view of people as fundamentally individual, which encourages us to see everything outside the person as a threat to their individualism.[43] It means seeing people as networked creatures embedded in communities of thinking.

A deep enlightenment would also work to promote the trust between citizens required to sustain a culture of free thought. Citizens need to trust each other and believe that good-faith attempts to reach truth will not be sabotaged or punished by others. A deep enlightenment would reinstate the parrhesiastic contract. How this can be done in a competitive capitalistic society is not at all clear. Not only are citizens encouraged to see each other as competitors, but there are powerful financial and political interests in thought remaining tightly controlled. The only way to combat this is through the reinstatement of communities of thinking. Such communities would benefit from shifting away from competitive inquiry, in which one person is attempting to persuade another, and towards collaborative inquiry, in which multiple people are seeking the truth together.[44] If there is no persuasion, then inquiry has no winners and losers, but only the mutual gain of understanding. We need to re-establish a civil society with institutions that promote free thought.

A culture of free thought values thought in its own right. Free thought is valuable because of its products, such as truth, autonomy and self-government. But it also has inherent value. It is intrinsically good. We can enjoy thinking itself, independent of what gems it may yield. To truly safeguard free thought, we must recognise that the effort of thinking can be the reward of thinking.[45] Rather than only valuing thought for what it can give us (a sure-fire way to eventually fall out of love with thought), we need to remember to enjoy the process of thought.

The challenging nature of free thought raises the question of how to protect people who do not wish to stare into the sun of truth. One option is to require social media platforms to provide filters that people can use to screen out certain content.[46] This would allow safe spaces to exist for both thoughts and feelings. Indeed, we should not let our pursuit of thought become psychopathic. If we view ourselves, as Aristotle did, as characterised by our ability to think, we may overvalue thought to the detriment of other ends.

There are other conceptions of what makes us human. These do not always involve characterising us by our strengths. As the philosopher Richard Rorty points out, why define us by our heights, which always threatens to exclude some? Why not define us by our lows? Why not define us as the creature that can be humiliated and therefore needs protection?[47] We may be best defined not as the creature that thinks, but the creature that falls. If so, a helping hand rather than a racing mind may be what best defines the human.

As we continue to think together about what the right to freedom of thought should involve, we must remain aware that free thought is our only insurance policy against tyranny. Armed with guns, yet without free thought, citizens are worse than defenceless. They are condemned to be soldiers in another's war and agents of both their own and others' oppression. For humanity to survive there must always be people performing the minute-to-minute miracle of thought.[48] Only by protecting free thought will we walk, not fall, into the future.

NOTES

All website links were accurate at the time of writing (March 2023).

Epigraphs

1 Mencken, H. L. (1982) *A Mencken chrestomathy*. New York, NY: Vintage.
2 de Jouvenel, B. (1962) *On power, its nature and the history of its growth* (trans. J. F. Huntington). Boston, MA: Beacon Press.
3 Arendt, H. (1973) *Origins of totalitarianism*. New York, NY: Harcourt Brace Jovanovich.
4 Blasi, V. (1977) 'The checking value in First Amendment theory', *American Bar Foundation Research Journal*, 2 (3), 521–649.

Introduction

1 Cf. Alexander Meiklejohn's idea of humans as 'an inventing animal'; Meiklejohn, A. (1972) *What does America mean?* New York, NY: Norton (original work published 1935).
2 Bailey, T., Alvarez-Jimenez, M., Garcia-Sanchez, A. M., *et al.* (2018) 'Childhood trauma is associated with severity of hallucinations and delusions in psychotic disorders: a systematic review and meta-analysis', *Schizophrenia Bulletin*, 44 (5), 1111–22.

3 McCarthy-Jones, S. (2017) 'The concept of schizophrenia is coming to an end: here's why', *The Conversation*, https://theconversation.com/the-concept-of-schizophrenia-is-coming-to-an-end-heres-why-82775.

4 Greenslade, R. (2016) '*The New European* on the "brainwashing of Britain over immigration"', *Guardian*, 16 September, https://www.theguardian.com/media/greenslade/2016/sep/16/the-new-european-on-the-brainwashing-of-britain-over-immigration; Cadwalladr, C. (2019) 'How did social media manipulate our votes and our elections', *TED Radio Hour*, 12 July, https://www.npr.org/transcripts/740771021?t=1660262617869; Noah, T. (2018) 'Electronic brainwashing: Cambridge Analytica's sinister Facebook strategy', *The Daily Show*, 22 March, https://www.youtube.com/watch?v=t7epj5tK54M.

5 Hilgers, L. (2021) '"We are so divided now": how China controls thought and speech beyond its borders', *Guardian*, 26 October, https://www.theguardian.com/news/2021/oct/26/we-are-so-divided-now-how-china-controls-thought-and-speech-beyond-its-borders.

6 As the UN Human Rights Committee explained in 1993, Article 18 of the International Covenant on Civil and Political Rights 'does not permit any limitations whatsoever on the freedom of thought . . . [it is] protected unconditionally'; UN Human Rights Committee (HRC), CCPR General Comment No. 22: Article 18 (Freedom of Thought, Conscience or Religion), 30 July 1993, CCPR/C/21/Rev.1/Add.4.

7 Vermeulen, B. (2006) 'Freedom of thought, conscience and religion', in P. van Dijk, F. van Hoof, A. van Rijn and L. Zwaak (eds), *Theory and practice of the European Convention on Human Rights*, 4th edn (pp. 751–72). Cambridge: Intersentia Press; Nowak, M. (1993) *UN Covenant on Civil and Political Rights: CCPR commentary*. Kehl am Rhein: N. P. Engel.

8 Bublitz, J. C. (2014) 'Freedom of thought in the age of neuroscience: a plea and a proposal for the renaissance of a forgotten fundamental right'. *ARSP: Archiv für Rechts-und Sozialphilosophie*, 1–25; Blitz, M. J. (2010) 'Freedom of thought for the extended mind: cognitive enhancement and the constitution', *Wisconsin Law Review*, 1049–18; Swaine, L. (2012) 'Freedom of thought, religion, and liberal neutrality', 4 August, http://dx.doi.org/10.2139/ssrn.2124014.

9 Alegre, S. (2017) 'Rethinking freedom of thought for the 21st century', *European Human Rights Law Review*, 3, 221–33; Băncău-Burcea, A. (2017) 'Social media and freedom of thought', in *Proceedings of the RAIS Conference: the future of ethics, education and research* (pp. 124–30). Princeton, NJ: Research Association for Interdisciplinary Studies; Moglen, E. (2017) 'Free thought, free media', in E. Bell and T. Owen (eds), *Journalism after Snowden* (pp. 265–72). New York, NY: Columbia University Press.

10 Newman, D. (2021) 'Freedom of thought in Canada: the history of a forgetting and the potential of a remembering', *European Journal of Comparative Law and Governance*, 8 (2–3), 226–44.

11 Shaheed, A. (2021) *Freedom of thought. Interim report of the Special Rapporteur on freedom of religion or belief, Ahmed Shaheed*. Geneva: United Nations.

12 Hand, L. (1916) 'The speech of justice', *Harvard Law Review*, 29 (6), 617–21.

13 For example, Robespierre described democracy as the people being 'guided by laws which are of their own making'; Bienvenu, R. (1968) *The ninth of Thermidor: the fall of Robespierre*. Oxford: Oxford University Press.

14 The first half of this quote comes from Charles de Gaulle, as cited in Heritage, P. (2004) 'Taking hostages: staging human rights', *TDR/The Drama Review*, 48 (3), 96–106.

15 Jefferson swore 'upon the altar of god, eternal hostility against every form of tyranny over the mind of man'. Chomsky claims that 'even more fundamental than the right of free expression is the right to think'; Chomsky, N. (2010) 'Final remarks', Istanbul Conference on Freedom of Speech, 10 April, https://chomsky.info/20101010/. US Supreme Court Justices have stated that there is no principle that 'more imperatively calls for attachment' than 'the principle of free thought'; *United States v. Schwimmer*, 279 U.S. 644, 49 S. Ct. 448, 73 L. Ed. 889 (1929).

16 Taylor, C. (2007) *A secular age*. Cambridge, MA: Harvard University Press.

17 Lippmann, W. (1993) *The phantom public*. New Brunswick, NJ: Transaction Publishers (original work published 1927); MacIntyre, A. (2007) *After virtue: a study in moral theory*. London: Duckworth;

Bentham, J. (2002) 'Rights, representation, and reform: Nonsense upon Stilts and other writings on the French Revolution', in P. Schofield, C. Pease-Watkin and C. Blamires (eds), *The collected works of Jeremy Bentham* (vol. 15). Oxford: Oxford University Press.

18 Waldron, J. (2000) 'The role of rights in practical reasoning: "rights" versus "needs"', *Journal of Ethics*, 4 (1), 115–35.

19 Lippmann (1993).

20 Rorty, R. (2012) 'Human rights, rationality and sentimentality', in A. S. Rathore and A. Cistelecan (eds), *Wronging rights?: philosophical challenges for human rights* (vol. 1, pp. 107–33). London: Routledge.

21 Milton, J. (2014) *Areopagitica and other writings*. London: Penguin.

22 *Abrams v. United States*, 250 U.S. 616, 40 S. Ct. 17, 63 L. Ed. 1173 (1919).

23 Meiklejohn, A. (1948) *Free speech and its relation to self-government*. New York, NY: Harper Bros.

24 Goldman, A. I. and Cox, J. C. (1996) 'Speech, truth, and the free market for ideas', *Legal Theory*, 2 (1), 1–32.

25 Chesterton, G. K. (1908) *Orthodoxy*. New York, NY: John Lane Co.

26 Russomanno, J. (2015) 'The "central meaning" and path dependence: the Madison-Meiklejohn-Brennan Nexus', *Communication Law and Policy*, 20 (2), 117–48.

27 For more on the concept of 'exiting', see Hirschman, A. O. (1970) *Exit, voice, and loyalty*. Cambridge, MA: Harvard University Press. It has even been claimed that free exit from a society 'is so important that we've called it the only Universal Human Right'; see Land, N. (n.d.) *The dark enlightenment*, https://www.thedarkenlightenment.com/the-dark-enlightenment-by-nick-land/.

28 Meiklejohn (1948).

29 Cohen, J. E. (2000) 'Examined lives: informational privacy and the subject as object', *Stanford Law Review*, 52, 1373–438; *Nolan and K. v. Russia*, 53 E.H.R.R. 29 (2011).

30 Cohen (2000).

31 *American Communications Assn. v. Douds*, 339 U.S. 382, 70 S. Ct. 674, 94 L. Ed. 925 (1950).

32 Aeschylus, as cited in Rotunda, R. D. (2019) 'The right to shout fire in a crowded theatre: hateful speech and the first amendment', *Chapman Law Review*, 22, 319.

33 Pericles, as cited in Rotunda (2019).

34 Macklin, R. (2003) 'Dignity is a useless concept', *British Medical Journal*, 327 (7429), 1419–20.

35 Pascal, B. (1958) *Pascal's pensées*. New York, NY: E. P. Dutton & Co.

36 Nuffield Council on Bioethics (2002) *Genetics and human behaviour: the ethical context*. London: Nuffield Council on Bioethics, http://nuffieldbioethics.org/wp-content/uploads/2014/07/Genetics-and-human-behaviour.pdf.

37 Halliburton, C. M. (2009) 'How privacy killed Katz: a tale of cognitive freedom and the property of personhood as fourth amendment norm', *Akron Law Review*, 42, 803–84.

38 Schneewind, J. B. (1998) *The invention of autonomy: a history of modern moral philosophy*. Cambridge: Cambridge University Press.

39 MacIntyre (2007).

40 Ryan, R. M. and Deci, E. L. (2006) 'Self-regulation and the problem of human autonomy: does psychology need choice, self-determination, and will?', *Journal of Personality*, 74 (6), 1557–86.

41 Leotti, L. A., Iyengar, S. S. and Ochsner, K. N. (2010) 'Born to choose: the origins and value of the need for control', *Trends in Cognitive Sciences*, 14 (10), 457–63.

42 Infurna, F. J., Ram, N. and Gerstorf, D. (2013) 'Level and change in perceived control predict 19-year mortality: findings from the Americans' changing lives study', *Developmental Psychology*, 49 (10), 1833.

43 Ong, A. D., Bergeman, C. S. and Boker, S. M. (2009) 'Resilience comes of age: defining features in later adulthood', *Journal of Personality*, 77 (6), 1777–804.

44 Rand, A. (2005) *Atlas shrugged*. New York, NY: Penguin.

45 Rothbard, M. (2021) 'A Crusoe social philosophy', Mises Institute, 12 July, https://mises.org/library/crusoe-social-philosophy.

46 Meiklejohn (1972).

47 Shaheed (2021).

48 Miller, [2020] EWHC 225 (Admin), https://www.judiciary.uk/wp-content/uploads/2020/02/miller-v-college-of-police-judgment.pdf.

49 cf. Ferguson, N. (2017) 'Updating our nightmares', *Real Life*, https://reallifemag.com/updating-our-nightmares/.

50 Dabhoiwala, F. (2022) 'Inventing free speech: politics, liberty and print in eighteenth-century England', *Past & Present*, 257 (Supplement 16), 39–74.

51 King, G., Pan, J. and Roberts, M. E. (2017) 'How the Chinese government fabricates social media posts for strategic distraction, not engaged argument', *American Political Science Review*, 111 (3), 484–501.

52 King *et al.* (2017).

53 Strittmatter, K. (2020) *We have been harmonized: life in China's surveillance state*. New York, NY: Custom House.

54 Weissman, J. (2021) *The crowdsourced panopticon: conformity and control on social media*. London: Rowman & Littlefield.

55 Rid, T. (2020) *Active measures*. London: Profile Books.

56 Ibid.

57 Guess, A., Nagler, J. and Tucker, J. (2019) 'Less than you think: prevalence and predictors of fake news dissemination on Facebook', *Science Advances*, 5 (1).

58 Allen, J., Howland, B., Mobius, M., *et al.* (2020) 'Evaluating the fake news problem at the scale of the information ecosystem', *Science Advances*, 6 (14).

59 Cercone, J. (2022) 'Donald Trump, not Elon Musk, was topic of discussion in 2017 MSNBC video', *Politifact*, 18 April, https://www.politifact.com/factchecks/2022/apr/18/facebook-posts/donald-trump-not-elon-musk-was-topic-discussion-20/.

60 DellaVigna, S. and Kaplan, E. (2007) 'The Fox News effect: media bias and voting', *Quarterly Journal of Economics*, 122 (3), 1187–234.

61 Dutton, W. H., Reisdorf, B., Dubois, E. and Blank, G. (2017) 'Search and politics: the uses and impacts of search in Britain, France, Germany, Italy, Poland, Spain, and the United States', *Quello Center Working Paper*, 5 (1), https://papers.ssrn.com/sol3/papers.cfm?abstract_id=2960697.

62 Huntington, S. P. (1981) *American politics: the promise of disharmony*. Cambridge, MA: Harvard University Press.

63 Trilateral Commission (1975) *The crisis of democracy: report on the governability of democracies to the Trilateral Commission*. New York, NY: New York University Press.

64 Trilateral Commission (1975).

65 Dalton, J. (1976) 'Huntington warns breakdown due to excessive democracy', *Harvard Crimson*, 24 March, https://www.thecrimson.com/article/1976/3/24/huntington-warns-breakdown-due-to-excessive/?print=1.

66 Dabhoiwala (2022).

67 Rosen, J. (2011) 'The deciders: the future of privacy and free speech in the age of Facebook and Google', *Fordham Law Review*, 80, 1525.

68 As cited in Thompson, C. L. (1949) 'Problems of industrial expansion: II. J. Pierpont Morgan: consolidator'. *Current History*, 17 (96), 90–5. I initially came across an abbreviated version of this quote in Zinn, H. (2015) *A people's history of the United States: 1492–present*. New York, NY: Routledge.

69 Chomsky, N. and Drèze, J. (2014) *Democracy and power: the Delhi Lectures*. Cambridge: Open Book Publishers.

70 Packard, V. (1957) *The hidden persuaders*. New York, NY: David McKay.

71 Oreskes, N., Conway, E. M. and Tyson, C. (2020) 'How American businessmen made us believe that free enterprise was indivisible from American democracy: the National Association of Manufacturers' propaganda campaign 1935–1940', in W. L. Bennett and S. Livingston (eds), *The disinformation age: politics, technology, and disruptive communication in the United States* (pp. 95–119). Cambridge: Cambridge University Press.

72 Oreskes *et al.* (2020); Herman, E. S. and Chomsky, N. (2010) *Manufacturing consent: the political economy of the mass media*. New York, NY: Random House; Zinn (2015).

73 As extensively documented by Noam Chomsky. See Herman and Chomsky (2010); Chomsky and Drèze (2014); Chomsky, N. (2002) *Understanding power: the indispensable Chomsky*. New York, NY: The New Press; also see Montgomery, D. (1987) *The fall of the house of labor: the workplace, the state, and American labor activism, 1865–1925*. Cambridge: Cambridge University Press.

74 Dabhoiwala (2022).

75 Ibid.

76 Motta, M., Callaghan, T. and Sylvester, S. (2018) 'Knowing less but presuming more: Dunning-Kruger effects and the endorsement of anti-vaccine policy attitudes', *Social Science & Medicine*, 211, 74–281.

77 Fernbach, P. M., Rogers, T., Fox, C. R. and Sloman, S. A. (2013) 'Political extremism is supported by an illusion of understanding', *Psychological Science*, 24, 939–46.

78 Mill, J. S. (2003) *On liberty*. London: Penguin.

79 Short, P. (2022) *Putin: his life and times*. London: Bodley Head.

80 Aby, S. H. (2009) 'Discretion over valor: the AAUP during the McCarthy years', *American Educational History Journal*, 36 (1/2), 121–32.

81 Aby (2009).

82 Hat tip to Pinter, H. (2013) *Moonlight*. London: Faber & Faber.

83 Hitchens, C. (2009) *Letters to a young contrarian*. New York, NY: Basic Books.

84 Popper, K. R. (1979) *Objective knowledge*. Oxford: Clarendon.

85 McCarthy-Jones, S. (2020a) *Spite: and the upside of your dark side*. London: Oneworld Publications.

86 Grubbs, J. B., Warmke, B., Tosi, J. and James, A. S. (2020) 'Moral grandstanding and political polarization: a multi-study consideration', *Journal of Research in Personality*, 88, 104009.

87 Grubbs *et al.* (2020); Grubbs, J. B., Warmke, B., Tosi, J., *et al.* (2019) 'Moral grandstanding in public discourse: status-seeking motives as a potential explanatory mechanism in predicting conflict', *PLoS One*, 14 (10), e0223749. See also Tosi, J. and Warmke, B. (2020) *Grandstanding: the use and abuse of moral talk*. Oxford: Oxford University Press.

88 Sowell, T. (2014) *Basic economics*. New York, NY: Basic Books.

89 For more on the problems of liberalism and its self-undermining nature see Deneen, P. J. (2019) *Why liberalism failed*. New Haven, CT: Yale University Press.

90 Giroux, H. A. (2015) *Dangerous thinking in the age of the new authoritarianism*. London: Routledge.

91 Anderson, D. (2021) 'An epistemological conception of safe spaces', *Social Epistemology*, 35 (3), 285–311.

92 The UN's Special Rapporteur on freedom of religion or belief undertook a series of consultative exercises to create his report, and I participated in one of these.

93 Chesterton, G. K. (1929) *The thing; why I am a Catholic*. London: Aeterna Press.

94 Thanks to an anonymous reviewer of one of the papers of my colleagues and I for pointing this out.

95 A point made about universal rights in general in de Benoist, A. (2011) *Beyond human rights: defending freedoms*. Paris: Arktos.

96 *Jones v. Opelika*, 316 U.S. 584, 62 S. Ct. 1231, 86 L. Ed. 1691 (1942).

97 Bublitz, J. C. and Merkel, R. (2014) 'Crimes against minds: on mental manipulations, harms and a human right to mental self-determination', *Criminal Law and Philosophy*, 8, 51–77; Richards, N. (2015) *Intellectual privacy: rethinking civil liberties in the digital age*. Oxford: Oxford University Press.

98 Trilateral Commission (1975).

99 Lippmann (1993).

100 Marx, K. and Engels, F. (2012) *The Communist manifesto*. New Haven, CT: Yale University Press.

101 Moyn, S. (2012) *The last Utopia: human rights in history*. Cambridge, MA: Harvard University Press.

102 Bublitz (2014); Alegre (2017); Blitz (2010).

103 Cohen, J. L. (2008) 'Rethinking human rights, democracy, and sovereignty in the age of globalization', *Political Theory*, 36 (4), 578–606.

104 Barker, J. (2008) ' "A hero will rise": the myth of the fascist man in *Fight Club* and *Gladiator*', *Literature/Film Quarterly*, 36 (3), 171–87.

105 Who? See 'Bronze Age Pervert', *Wikipedia*, https://en.wikipedia.org/wiki/Bronze_Age_Pervert.

106 Noah (2018).

107 Cadwalladr, C. (2017) 'The great British Brexit robbery: how our democracy was hijacked', *Guardian*, 7 May, https://www.theguardian.com/technology/2017/may/07/the-great-british-brexit-robbery-hijacked-democracy.

108 Hains, T. (2021) 'WP's Eugene Robinson wonders: how do we "deprogram" millions of members of Trump "cult"?', *Real Clear Politics*, 12 January, https://www.realclearpolitics.com/video/2021/01/12/wps_eugene_robinson_wonders_how_do_we_deprogram_millions_of_members_of_trump_cult.html.

109 Associated Press (2012) 'PBS lawyer resigns after being caught in Veritas sting', 12 January, https://apnews.com/article/donald-trump-entertainment-coronavirus-pandemic-8f586d687ab332777a7a059457ff818e.

110 Cadwalladr (2017).

111 Wolin, S. S. (2008) *Managed democracy and the specter of inverted totalitarianism.* Princeton, NJ: Princeton University Press.

112 Gatehouse, G. (2019) 'The confusion around Russian "meddling" means they're already winning', *Guardian*, 25 March, https://www.theguardian.com/commentisfree/2019/mar/25/russian-meddling-vladimir-putin-vladislav-surkov.

113 Gleicher, N., Franklin, M., Agranovich, D., *et al.* (2021) *Threat report: the state of influence operations 2017–2020.* Palo Alto, CA: Facebook, https://about.fb.com/wp-content/uploads/2021/05/IO-Threat-Report-May-20-2021.pdf.

114 Jones, C. E. (1988) 'The political repression of the Black Panther Party 1966–1971: the case of the Oakland Bay area', *Journal of Black Studies*, 18 (4), 415–34; Churchill, W. and Vander Wall, J. (2002) *Agents of repression: the FBI's secret wars against the Black Panther Party and the American Indian Movement.* Boston, MA: South End Press.

115 Sunstein, C. R. and Vermeule, A. (2008) 'Conspiracy theories', *Harvard Public Law Working Paper No. 08-03*, https://papers.ssrn.com/sol3/papers.cfm?abstract_id=1084585.

116 Walker, C. (2019) 'China's foreign influence and sharp power strategy to shape and influence democratic institutions', *National Endowment for Democracy*, 16 May, https://www.ned.org/chinas-foreign-influence-and-sharp-power-strategyto-shape-and-influence-democratic-institutions/.

117 Tambe, A. M. and Friedman, T. (2022) 'Chinese state media Facebook ads are linked to changes in news coverage of China worldwide', *Harvard Kennedy School Misinformation Review*, 14 January, https://misinforeview.hks.harvard.edu/article/chinese-state-media-facebook-ads-are-linked-to-changes-in-news-coverage-of-china-worldwide/.

118 Ibid.

119 Krejsa, H. (2018) 'Under pressure: the growing reach of Chinese influence campaigns in democratic societies', *Center for a New American Security*, 27 April, https://www.cnas.org/publications/reports/under-pressure; Walker (2019).

120 See 'Unrestricted warfare', *Wikipedia*, https://en.wikipedia.org/wiki/Unrestricted_Warfare.

121 *Snyder v. Phelps*, 562 U.S. 443, 458–59 (2011); Matalv. Tam, 137 S. Ct. 1744, 1751 (2017).

122 *US v. Stevens*, 559 U.S. 460, 130 S. Ct. 1577, 176 L. Ed. 2d 435 (2010).

123 *Terminiello v. City of Chicago*, 337 U.S. 1, 9 (1949).

124 *Virginia v. Black*, 538 U.S. 343, 123 S. Ct. 1536, 155 L. Ed. 2d 535 (2003).

125 *Brandenburg v. Ohio*, 395 U.S. 444, 89 S. Ct. 1827, 23 L. Ed. 2d 430 (1969).

126 White, K. and Lukianoff, G. (2020) 'What's the best way to protect free speech? Ken White and Greg Lukianoff debate cancel culture', *Reason*, 8 April, https://reason.com/2020/08/04/whats-the-best-way-to-protect-free-speech-ken-white-and-greg-lukianoff-debate-cancel-culture/.

127 *M. Forstater v. CGD Europe and others* (2019) Application no. 2200909/019, https://www.gov.uk/employment-tribunal-decisions/maya-forstater-v-cgd-europe-and-others-2200909-2019.

128 Ibid.

129 *Erbakan v. Turkey* (2006) Application no. 59405/00. *European Court of Human Rights*, https://hudoc.echr.coe.int/eng#{%22itemid%22:[%22001-76232%22]}.

130 Mchangama, J. (2022) 'The real threat to social media is Europe', *Foreign Policy*, 25 April, https://foreignpolicy.com/2022/04/25/the-real-threat-to-social-media-is-europe/.

131 Ibid.

132 Ibid.

133 Shakespeare, W. (n.d.) *The Tempest* (3.2.135); *Twelfth Night* (1.3.68). B. Mowat, P. Werstine, M. Poston and R. Niles (eds), *The Folger Shakespeare*.

134 Thanks to Hannah Haseloff for pointing this out to me.

135 Strauss, L. (2013) *On tyranny*. Chicago, IL: University of Chicago Press.

136 Curran, A. (2019) *Diderot and the art of thinking freely*. New York, NY: Other Press.

137 Often attributed to Thucydides, this quote comes from: Butler, W. F. (1891) *Charles George Gordon* (vol. 1). London: Macmillan.

138 Uzgalis, W. (2022) 'Anthony Collins', in E. N. Zalta (ed.), *The Stanford encyclopedia of philosophy*, https://plato.stanford.edu/entries/collins/ #DiscFreeThin1713.

139 Mayer, M. (1955) *They thought they were free: the Germans, 1933–45.* Chicago, IL: Chicago University Press, https://press.uchicago.edu/ Misc/Chicago/511928.html.

1 Thought

1 Nietzsche, F. (2003a) *Twilight of the idols and the Anti-Christ* (trans. R. J. Hollingdale). London: Penguin (original works published 1889/ 1895).

2 Taylor, C. (1992) *Sources of the self: the making of the modern identity.* Cambridge, MA: Harvard University Press.

3 Gacea, A. O. (2019) 'Plato and the "internal dialogue": an ancient answer for a new model of the self', in L. Pitteloud and E. Keeling (eds), *Psychology and ontology in Plato* (pp. 33–54). Cham: Springer.

4 Taylor (1992).

5 Ustinova, Y. (2009) *Caves and the Ancient Greek mind: descending underground in the search for ultimate truth.* Oxford: Oxford University Press.

6 Gacea (2019).

7 Bohm, as cited in Gacea (2019).

8 Korba, R. J. (1990) 'The rate of inner speech', *Perceptual and Motor Skills*, 71 (3), 1043–52.

9 Fernyhough, C. (2004) 'Alien voices and inner dialogue: towards a developmental account of auditory verbal hallucinations', *New Ideas in Psychology*, 22 (1), 49–68.

10 Fernyhough, C. (2016) *The voices within: the history and science of how we talk to ourselves.* New York, NY: Basic Books. Of course, this doesn't mean that the child was not 'thinking' before this time. Rather, it means, as Fernyhough puts it, that language 'does not give the child thought; rather, it transforms whatever intellectual capacities are present before language comes along'.

11 For more on David Perell's work, see https://perell.com/.

12 Le Bon, G. (1960) *The crowd: a study of the popular mind*. London: Penguin.

13 Perrin, S. and Spencer, C. (1980) 'The Asch effect – a child of its time', *Bulletin of the British Psychological Society*, 33, 405–6.

14 Gibson, S. (2019) *Arguing, obeying and defying: a rhetorical perspective on Stanley Milgram's obedience experiments*. Cambridge: Cambridge University Press; Brannigan, A., Nicholson, I. and Cherry, F. (2015) 'Introduction to the special issue: unplugging the Milgram machine', *Theory & Psychology*, 25 (5), 551–63.

15 Janis, I. L. (1972) *Victims of groupthink*. Boston, MA: Houghton Mifflin.

16 Surowiecki, J. (2005) *The wisdom of crowds*. New York, NY: Anchor.

17 See Thucydides, 'Pericles' funeral oration', University of Minnesota Human Rights Library, http://hrlibrary.umn.edu/education/thucy-dides.html.

18 Cammack, D. L. (2013) *Rethinking Athenian democracy*. Boston, MA: Harvard University Press.

19 Kant, I. (1998) *Kant: Religion within the boundaries of mere reason: and other writings* (trans. A. Wood and G. di Giovanni). Cambridge: Cambridge University Press.

20 See National Constitution Center (2021) 'The words that made us', https://constitutioncenter.org/interactive-constitution/podcast/the-words-that-made-us.

21 Baumeister, R. F., Masicampo, E. J. and DeWall, C. N. (2011) 'Arguing, reasoning, and the interpersonal (cultural) functions of human consciousness', *Behavioral and Brain Sciences*, 3(2), 74.

22 Wilson, W. (1913) *The new freedom: a call for the emancipation of the generous energies of a people*. New York, NY: Doubleday, Page & Co.

23 The following information and quotes are taken from Frank, J. (2019) 'Populism and praxis', in C. R. Kaltwasser, P. Taggart and P. Ostiguy (eds), *The Oxford handbook of populism* (pp. 629–43). Oxford: Oxford University Press.

24 Rousseau, J. J. (1920) *The social contract* (trans. G. D. H. Cole). New York, NY: E. P. Dutton & Co. (original work published 1762).

25 See 'Polymath Project', *Wikipedia*, https://en.wikipedia.org/wiki/Polymath_Project.

26 Charness, G., Karni, E. and Levin, D. (2010) 'On the conjunction fallacy in probability judgment: new experimental evidence regarding Linda', *Games and Economic Behavior*, 68 (2), 551–6.

27 Mercier, H. and Sperber, D. (2011) 'Why do humans reason? Arguments for an argumentative theory', *Behavioral and Brain Sciences*, 34 (2), 57–74.

28 Mercier and Sperber (2011).

29 Engel, D., Woolley, A. W., Jing, L. X., *et al.* (2014) 'Reading the mind in the eyes or reading between the lines? Theory of mind predicts collective intelligence equally well online and face-to-face', *PLoS One*, 9 (12).

30 Mercier and Sperber (2011).

31 Ibid.; Walker, S., Egan, R., Young, J., *et al.* (2020) 'A citizens' jury on euthanasia/assisted dying: does informed deliberation change people's views?', *Health Expectations*, 23 (2), 388–95.

32 Fishkin, J. S. (2018) *Democracy when the people are thinking: revitalizing our politics through public deliberation.* Oxford: Oxford University Press.

33 Ellul, J. (1973) *Propaganda: the formation of men's attitudes.* New York, NY: Vintage Books.

34 Malone, T. W. (2018) *Superminds: the surprising power of people and computers thinking together.* London: Oneworld Publications.

35 Sheen, D. (2007) 'Interview with Derrick Jensen', *This Means War*, http://www.davidsheen.com/firstearth/interviews/jensen.htm. Also available at https://www.youtube.com/watch?v=xUtCKNEqfL8&list=PL143E59F5A37A9C84&index=1.

36 Dodson, P. (1996) 'Address to the National Press Club, Canberra, Australia', *Reconciliation at the Crossroads*, 18 April, http://www.multi-culturalaustralia.edu.au/doc/dodson_1.pdf. Information also taken from Lutheran Church of Australia (2018) 'The river and the sea', 4 October, https://www.rap.lca.org.au/2018/10/04/the-river-and-the-sea/.

37 Locke, J. (2013) *A letter concerning toleration* (ed. K. Walters). London: Broadview Press.

38 Clark, A. and Chalmers, D. (1998) 'The extended mind', *Analysis*, 58 (1), 7–19.

39 Carruthers, P. (1998) *Language, thought and consciousness: an essay in philosophical psychology.* Cambridge: Cambridge University Press.

40 There are other ways to view this process, such as thought being environmentally scaffolded, but I will focus on the concept of extension here. For the idea of scaffolding see: Sterelny, K. (2010) 'Minds: extended or scaffolded?', *Phenomenology and the Cognitive Sciences*, 9 (4), 465–81.

41 Storm, B. C., Stone, S. M. and Benjamin, A. S. (2017) 'Using the internet to access information inflates future use of the internet to access other information', *Memory*, 25 (6), 717–23.

42 See Wegner, D. M. and Ward, A. F. (2013) 'How Google is changing your brain', *Scientific American*, 309 (6), 58–61.

43 Marsh, E. J. and Rajaram, S. (2019) 'The digital expansion of the mind: implications of internet usage for memory and cognition', *Journal of Applied Research in Memory and Cognition*, 8 (1), 1–14.

44 Wegner and Ward (2013).

45 Zeldin, W. (2010) 'Finland: legal right to broadband for all citizens', Library of Congress, https://www.loc.gov/item/global-legal-monitor/ 2010-07-06/finland-legal-right-to-broadband-for-all-citizens/.

46 Wegner and Ward (2013).

47 Henkel, L. A. (2014) 'Point-and-shoot memories: the influence of taking photos on memory for a museum tour', *Psychological Science*, 25 (2), 396–402.

48 Lincoln, A., as cited in US Department of State Management Task Force (1992) *State 2000: a new model for managing foreign affairs*. Washington, DC: US Department of State.

49 Swartz, A. (2012) 'Address at F2C: Freedom to Connect Conference', 21 May, https://www.americanrhetoric.com/speeches/aaronswartzf-2cconference.htm.

50 See Arthur Grace, 'Robin Williams with Rodin's Thinker at the Legion of Honor, San Francisco', https://www.sothebys.com/en/auctions/ ecatalogue/2018/creating-a-stage-collection-of-marsha-and-robin-williams-n09977/lot.252.html.

51 Philadelphia Museum of Art (2011) *Looking to write, writing to look: a teaching resource*. Philadelphia, PA: Philadelphia Museum of Art, https: //www.philamuseum.org/booklets/12_71_137_0.html.

52 Nietzsche, as cited in Kaag, J. (2018) *Hiking with Nietzsche: becoming who you are*. London: Granta Books.

53 Kierkegaard, S. (1978) *Letters and documents* (trans. H. Rosenmeier). Princeton, NJ: Princeton University Press (original work published 1847).

54 For a good review of the science of walking, see O'Mara, S. (2019) *In praise of walking: the new science of how we walk and why it's good for us.* New York, NY: Random House.

55 Ibid.

56 Paul, A. M. (2021) *The extended mind: the power of thinking outside the brain.* New York, NY: Houghton Mifflin Harcourt.

57 Oppezzo, M. and Schwartz, D. L. (2014) 'Give your ideas some legs: the positive effect of walking on creative thinking', *Journal of Experimental Psychology: Learning, Memory, and Cognition*, 40 (4), 1142.

58 Stanovich, K. E. (2018) 'Miserliness in human cognition: the interaction of detection, override and mindware', *Thinking & Reasoning*, 24 (4), 423–44.

59 Ibid.

60 Kahneman, D. (2011) *Thinking, fast and slow.* New York, NY: Macmillan; Pinker, S. (2022) *Rationality: what it is, why it seems scarce, why it matters.* New York, NY: Penguin; Nisbett, R. E. (2015) *Mindware: tools for smart thinking.* New York, NY: Farrar, Straus and Giroux.

61 Kahneman (2011); Evans, J. S. B. (2003) 'In two minds: dual-process accounts of reasoning', *Trends in Cognitive Sciences*, 7 (10), 454–9; Stanovich, K. E. and West, R. F. (2000) 'Individual differences in reasoning: implications for the rationality debate?', *Behavioral and Brain Sciences*, 23 (5), 645–65; Stanovich, K. E. (1999) *Who is rational?: Studies of individual differences in reasoning.* Mahwah, NJ: Lawrence Erlbaum Associates.

62 John McWhorter notes that some people may object to this term based on its alleged root in an old British law that permitted men to beat their wives as long as the weapon they used was not wider than a thumb. Thankfully, this explanation appears apocryphal: Dubner, S. J. (2011) 'Rule of thumb', *Freakonomics*, 1 July, https://freakonomics.com/2011/07/rule-of-thumb/.

63 Stewart, N. (2009) 'The cost of anchoring on credit-card minimum repayments', *Psychological Science*, 20 (1), 39–41; Keys, B. J. and Wang, J. (2019) 'Minimum payments and debt paydown in consumer credit cards', *Journal of Financial Economics*, 131 (3), 528–48.

64 Kandasamy, N., Garfinkel, S. N., Page, L., *et al.* (2016) 'Interoceptive ability predicts survival on a London trading floor', *Scientific Reports*, 6 (1), 1–7.

65 Anderson, I. and Thoma, V. (2021) 'The edge of reason: a thematic analysis of how professional financial traders understand analytical decision making', *European Management Journal*, 39 (2), 304–14; this work is discussed in Paul (2021).

66 Anderson and Thoma (2021).

67 Locke, J. (1823) *The works of John Locke* (vol. III). London: Thomas Tegg.

68 Stanovich, K. E. (2012) 'Environments for fast and slow thinking', *Trends in Cognitive Sciences*, 16 (4), 198–9.

69 Stanovich (2012).

70 Mercier and Sperber (2011).

71 Ibid.

72 This involves giving someone specially selected information to encourage them to voluntarily make the decision you want them to make; see Thomas, T. (2004) 'Russia's reflexive control theory and the military', *Journal of Slavic Military Studies*, 17 (2), 237–56.

73 Locke (1823).

74 Smallwood, J. and Schooler, J. W. (2015) 'The science of mind wandering: empirically navigating the stream of consciousness', *Annual Review of Psychology*, 66, 487–518.

75 Killingsworth, M. A. and Gilbert, D. T. (2010) 'A wandering mind is an unhappy mind', *Science*, 330 (6006), 932.

76 Smallwood and Schooler (2015).

77 Ibid.

78 Epel, E. S., Puterman, E., Lin, J., *et al.* (2013) 'Wandering minds and aging cells', *Clinical Psychological Science*, 1 (1), 75–83.

79 Smallwood and Schooler (2015).

80 Ibid.

81 Ibid.

82 To paraphrase Clark and Chalmers (1998) who argue that 'Cognitive processes ain't (all) in the head'.

83 In *The love song of J. Alfred Prufrock*, T. S. Eliot famously ponders whether he might disrupt the universe.

2 The ARRC of free thought

1 See Bode, S., He, A. H., Soon, C. S., *et al.* (2011) 'Tracking the unconscious generation of free decisions using ultra-high field fMRI', *PLoS One*, 6 (6); Soon, C. S., Brass, M., Heinze, H. J. and Haynes, J. D. (2008) 'Unconscious determinants of free decisions in the human brain', *Nature Neuroscience*, 11 (5), 543–5.

2 Oakley, D. A. and Halligan, P. W. (2017) 'Chasing the rainbow: the non-conscious nature of being', *Frontiers in Psychology*, 8, 1924.

3 Carruthers, P. (2002) 'The cognitive functions of language', *Behavioral and Brain Sciences*, 25 (6), 657–74.

4 As Vygotsky put it, the use of verbal mediation means that humans can 'control their behavior from the outside'; Vygotsky, L. S. (1978) *Mind in society: the development of higher mental processes* (ed. M. Cole, V. John-Steiner, S. Scribner and E. Souberman). Cambridge, MA: Harvard University Press.

5 Baumeister *et al.* (2011).

6 Ibid.

7 Langner, R., Leiberg, S., Hoffstaedter, F. and Eickhoff, S. B. (2018) 'Towards a human self-regulation system: common and distinct neural signatures of emotional and behavioural control', *Neuroscience & Biobehavioral Reviews*, 90, 400–10.

8 Hofmann, W., Vohs, K. D. and Baumeister, R. F. (2012) 'What people desire, feel conflicted about, and try to resist in everyday life'. *Psychological Science*, 23, 582–8.

9 Kühler, M. and Jelinek, N. (eds) (2012) *Autonomy and the self.* Dordrecht: Springer Science & Business Media.

10 Metzinger, T. K. (2013) 'The myth of cognitive agency: subpersonal thinking as a cyclically recurring loss of mental autonomy', *Frontiers in Psychology*, 4, 931.

11 Ryan and Deci (2006).

12 Dewey, J. (1997) *How we think*. Mineola, NY: Dover Publications.

13 Ibid.

14 Simons, D. J. and Chabris, C. F. (1999) 'Gorillas in our midst: sustained inattentional blindness for dynamic events', *Perception*, 28 (9), 1059–74.

15 Drew, T., Võ, M. L. H. and Wolfe, J. M. (2013) 'The invisible gorilla strikes again: sustained inattentional blindness in expert observers', *Psychological Science*, 24 (9), 1848–53.

16 Williams, L., Carrigan, A., Auffermann, W., *et al.* (2021) 'The invisible breast cancer: experience does not protect against inattentional blindness to clinically relevant findings in radiology', *Psychonomic Bulletin & Review*, 28, 503–11.

17 Metzinger, T. (2015) 'M-autonomy'. *Journal of Consciousness Studies*, 22 (11–12), 270–302.

18 The classic psychological task to assess this ability is the Stroop colour naming task. In this, you are shown a word and have to say the colour the word is written in as quickly as possible. The words that appear are the names of colours. So, you might see the word 'RED' appear on the screen. If this word in written in red ink, the task is dead easy (the 'congruent condition'). But your task is much more challenging if the word 'RED' is written in green ink (the 'incongruent condition'). Once you have seen the word 'RED', you automatically want to say 'red!'. But your brain must override this urge and say the colour it is written in: 'green'. The quicker you can name the colour of the word shown in the incongruent condition, the better your selective attention is. You can try it for yourself via: https://faculty.washington.edu/chudler/java/timesc.html. In the first set of words, all the words presented are written in the same colour that they name. In the second set of words, most are different.

19 Selective attention impairments are found in people with various mental health conditions, from schizophrenia to Alzheimer's. See Westerhausen, R., Kompus, K. and Hugdahl, K. (2011) 'Impaired cognitive inhibition in schizophrenia: a meta-analysis of the Stroop interference effect', *Schizophrenia Research*, 133 (1–3), 172–81; Ben-David, B. M., Tewari, A., Shakuf, V. and Van Lieshout, P. H. (2014) 'Stroop effects in Alzheimer's disease: selective attention speed of processing, or color-naming? A meta-analysis', *Journal of Alzheimer's Disease*, 38 (4), 923–38.

20 Psychologists assess sustained attention by tasks such as the Test of Variables of Attention (TOVA). In the visual version of the TOVA, you see an image of a small square on a screen and must press a button

whenever it appears in a specific part of the screen such as the top right. The square is initially programmed to appear around a quarter of the time in this position. This goes on for about ten minutes. It sounds dull, and it is, painfully so, but by design. As time passes, your attention will wander and you will fail to notice when the square appears in the specific location.

21 Metzinger, T. K. (2013) 'The myth of cognitive agency: subpersonal thinking as a cyclically recurring loss of mental autonomy', *Frontiers in Psychology*, 4, 931.

22 Ibid.

23 Stalley, R. F. (1998) 'Plato's doctrine of freedom', *Proceedings of the Aristotelian Society*, 98, 145–58.

24 Bloom, A. (2016) *The Republic of Plato*. New York, NY: Basic Books.

25 Kant (1998).

26 Marx, K. and Engels, F. (2010) 'The German ideology', in I. Szeman and T. Kaposy (eds), *Cultural theory: an anthology* (pp. 161–71). Chichester: John Wiley & Sons.

27 Berti, E. (1978) 'Ancient Greek dialectic as expression of freedom of thought and speech', *Journal of the History of Ideas*, 39 (3), 347–70.

28 Plato, *Phaedrus*, https://faculty.georgetown.edu/jod/texts/phaedrus.html.

29 Berti (1978).

30 Michels, R. (1915) *Political parties: a sociological study of the oligarchical tendencies of modern democracy* (trans. E. Paul and C. Paul). New York, NY: Hearst's International Library Co.

31 Mercier and Sperber (2011).

32 Ibid.

33 Ibid.

34 Ibid.

35 Ibid.

36 Ibid.

37 Willinghame, D. T. (2008) 'Critical thinking: why is it so hard to teach?', *Arts Education Policy Review*, 109 (4), 21–32.

38 Ibid.

39 Ibid.

40 Ibid.

41 Wineburg, S. and McGrew, S. (2019) 'Lateral reading and the nature of expertise: reading less and learning more when evaluating digital information', *Teachers College Record*, 121 (11), 1–40.

42 Kirschner, P. A. and van Merriënboer, J. J. G. (2013) 'Do learners really know best? Urban legends in education', *Educational Psychologist*, 48, 169–83.

43 Breakstone, J., McGrew, S., Smith, M., *et al.* (2018) 'Teaching students to navigate the online landscape', *Social Education*, 82, 219–21.

44 McGrew, S., Smith, M., Breakstone, J., *et al.* (2019) 'Improving university students' web savvy: an intervention study', *British Journal of Educational Psychology*, 89, 485–500.

45 Plato, *Theaetetus*, https://www.gutenberg.org/files/1726/1726-h/1726-h.htm.

46 Ellul (1973).

47 Ibid.

48 Mayer (1955).

49 Bonura, S. E. (2020) *Empire builder: John D. Spreckels and the making of San Diego*. Lincoln, NE: University of Nebraska Press.

50 Lenin, as cited in Kenez, P. (1985) *The birth of the propaganda state: Soviet methods of mass mobilization, 1917–1929*. Cambridge: Cambridge University Press.

51 Spengler, O. (1991) *Decline of the West* (trans. C. F. Atkinson). Oxford: Oxford University Press.

52 McCombs, M. (2004) *Setting the agenda: the mass media and public opinion*. Cambridge: Polity Press.

53 Luo, Y., Burley, H., Moe, A. and Sui, M. (2019) 'A meta-analysis of news media's public agenda-setting effects, 1972–2015', *Journalism & Mass Communication Quarterly*, 96 (1), 150–72.

54 Dewey (1997).

55 Wallace, D. F. (2011) *Infinite Jest*. London: Hachette.

56 Radomsky, A. S., Alcolado, G. M., Abramowitz, J. S., *et al.* (2014) 'Part 1 – You can run but you can't hide: intrusive thoughts on six continents', *Journal of Obsessive-Compulsive and Related Disorders*, 3 (3), 269–79.

57 Klinger, E. and Cox, W. M. (1987) 'Dimensions of thought flow in everyday life', *Imagination, Cognition and Personality*, 7 (2), 105–28.

58 Rachman, S. and de Silva, P. (1978) 'Abnormal and normal obsessions', *Behaviour Research and Therapy*, 16 (4), 233–48.

59 Radomsky *et al.* (2014).

60 Byers, E. S., Purdon, C. and Clark, D. A. (1998) 'Sexual intrusive thoughts of college students', *Journal of Sex Research*, 35 (4), 359–69.

61 Veale, D., Freeston, M., Krebs, G., *et al.* (2009) 'Risk assessment and management in obsessive–compulsive disorder', *Advances in Psychiatric Treatment*, 15 (5), 332–43.

62 Veale *et al.* (2009).

63 Luther, M. (1907) *Luther's catechetical writings: God's call to repentance, faith and prayer* (vol. 1, ed. J. N. Lenker). Minneapolis, MN: Luther Press.

64 Frankfurt, H. G. (1971) 'Freedom of the will and the concept of a person', *Journal of Philosophy*, 68, 5–20.

65 Media Lens (2011) 'Free to be human – an interview with David Edwards', *Media Lens*, 18 November, https://www.medialens.org/2011/cogitation-free-to-be-human-an-interview-with-david-edwards/.

66 Evola, J. (2018) *Men among the ruins: post-war reflections of a radical traditionalist*. London: Simon & Schuster.

67 The prejudice that certain people don't reflect and therefore are less than human has a long history. For example, Thomas Jefferson claimed that African Americans 'participate more of sensation than reflection' and used this to justify their dehumanisation and disenfranchisement (Jefferson, as cited in Rorty, 2012). The extent to which an individual reflects, or is perceived to reflect, cannot be used to question their humanity.

68 Orwell, G. (2021) *1984*. London: Penguin.

69 Marx and Engels (2010). Not said in a complimentary way though – seen as a naive Young Hegelian idea.

70 Frankfurt (1971).

71 Kross, E., Bruehlman-Senecal, E., Park, J., *et al.* (2014) 'Self-talk as a regulatory mechanism: how you do it matters', *Journal of Personality and Social Psychology*, 106 (2), 304–24.

72 Jünger, E. (1993) *Eumeswil*. New York, NY: Marsilio Publishers.

73 *Whitney v. California*, 274 U.S. 357, 47 S. Ct. 641, 71 L. Ed. 1095 (1927).

74 Ibid.

75 Mill (2003).

76 Foucault, M. (2011) *The courage of the truth: the government of self and others II: lectures at the Collège de France 1983–1984*. London: Palgrave Macmillan.

77 As cited in Roy, A. and Cusack, J. (2016) *Things that can and cannot be said: essays and conversations*. Chicago, IL: Haymarket Books.

78 Strauss, L. (1953) *Natural right and history*. Chicago, IL: University of Chicago Press.

79 Ellul (1973).

80 Schmitt, C. (2008) *The concept of the political: expanded edition*. Chicago, IL: University of Chicago Press.

81 Foucault (2011).

82 Foucault, M. (2019) *Discourse and truth and parrhesia* (trans. N. Luxon). Chicago, IL: University of Chicago Press.

83 Maxwell, L. (2018) 'The politics and gender of truth-telling in Foucault's lectures on parrhesia', *Contemporary Political Theory*, 18 (1), 22–42.

84 West, C. and Buschendorf, C. (2015) *Black prophetic fire*. Boston, MA: Beacon Press.

85 Novak, D. R. (2006) 'Engaging parrhesia in a democracy: Malcolm X as a truth-teller', *Southern Communication Journal*, 71 (1), 25–43.

86 Gebhardt, M. (2023) '(Post-)Truth, populism and the simulation of parrhesia: a feminist critique of truth-telling after Hannah Arendt and Michel Foucault', *Philosophy & Social Criticism*, 49 (2), 178–91.

87 Novak (2006).

88 Ekins, E. (2020) 'Poll: 62% of Americans say they have political views they're afraid to share', *Cato Institute*, 22 July, https://www.cato.org/survey-reports/poll-62-americans-say-they-have-political-views-theyre-afraid-share#.

89 Kisin, K. and Foster, F. (2022) 'Sam Harris: Trump, religion, wokeness', *Triggernometry*, 16 August, https://www.youtube.com/watch?v=DDqtFS_Pvcs. Quote starts at 1h 31m 15s.

90 Greene, J. (2014) *Moral tribes: emotion, reason, and the gap between us and them*. New York, NY: Penguin.

91 Murray, D. (2019) *The madness of crowds: gender, race and identity*. London: Bloomsbury Publishing.

92 Kaplan, J. T., Gimbel, S. I. and Harris, S. (2016) 'Neural correlates of maintaining one's political beliefs in the face of counterevidence', *Scientific Reports*, 6 (1), 1–11.

93 Jacobs, A. (2017) *How to think: a guide for the perplexed*. London: Profile Books.

94 Nietzsche, F. (1917) *Beyond good and evil* (trans. H. Zimmern). New York, NY: Modern Library.

95 Wilson, T. D., Reinhard, D. A., Westgate, E. C., *et al.* (2014) 'Just think: the challenges of the disengaged mind', *Science*, 345 (6192), 75–7.

96 Normand. M. (2017) 'Your brain is your worst enemy', *Comedy Central*, https://www.youtube.com/watch?v=pekPR8FdeKU.

97 Richerson, P. J. and Boyd, R. (2005) *Not by genes alone: how culture transformed human evolution*. Chicago, IL: University of Chicago Press; as cited in Stanovich (2018).

98 Stanovich (2018).

99 Ibid.

100 Kurzban, R., Duckworth, A., Kable, J. W. and Myers, J. (2013) 'An opportunity cost model of subjective effort and task performance', *Behavioral and Brain Sciences*, 36 (6), 661–79.

101 Ibid.

102 Wittgenstein, L. (2010) *Philosophical investigations*. London: John Wiley & Sons.

103 A point made in relation to free speech by Stanley Fish: Fish, S. (1994) *There's no such thing as free speech: and it's a good thing, too*. Oxford: Oxford University Press.

3 The threats to our minds

1 Clearly, freedom of thought was not possible for everyone at this time.

2 Achen, C. H. and Bartels, L. M. (2017) *Democracy for realists*. Princeton, NJ: Princeton University Press.

3 Hume, D. (1741, 1747) 'Essay IV: Of the first principles of government', https://davidhume.org/texts/empl1/fp.

4 Napoleon, as cited in Ellul (1973).

5 Arendt, H. (1958) 'Totalitarian imperialism: reflections on the Hungarian revolution', *Journal of Politics*, 20 (1), 5–43.

6 Locke. J. (1813) *An essay concerning human understanding*. Boston: Cummings & Hilliard (original work published 1689).

7 Spinoza, B. (2004) *The chief works of Benedict de Spinoza* (vol. 1, trans. R. H. M. Elwes). London: George Bell and Sons (original work published 1891).

8 Voltaire (2015) *Collected works of Voltaire*. East Sussex: Delphi Classics.

9 Spinoza (2004).

10 Parvini, N. (2022) *The populist delusion*. Perth, Western Australia: Imperium Press.

11 Fondren, E. and Hamilton, J. M. (2022) 'The universal laws of propaganda: World War I and the origins of government manufacture of opinion', *Journal of Intelligence History*, 1–19.

12 Ortega y Gasset, J. (1932) *The revolt of the masses*, https://archive.org/details/TheRevoltOfTheMasses.

13 Sakharov, as cited in Remnick, D. (2021) 'Joe Biden, Vladimir Putin, and the weight of history', *New Yorker*, 20 June.

14 Koga, K. (2009) 'The anatomy of North Korea's foreign policy formulation', *North Korean Review*, 5 (2), 21–33.

15 Lippiello, T. (2018) 'The paradigms of religious and philosophical plurality: the return of "spirituality" in China today', *Philosophy & Social Criticism*, 44 (4), 371–81.

16 For such an argument, see Weiwei, Z. (2016) *China horizon: the glory and dream of a civilizational state*. Hackensack, NJ: World Century Publishing.

17 Ibid.

18 Ibid.

19 Aubié, H. (2019) 'Freedom of opinion and expression', in S. Biddulph and J. Rosenzweig (trans.), *Handbook on human rights in China* (pp. 301–22). Cheltenham: Edward Elgar Publishing.

20 Ibid.

21 Ibid.

22 Ibid.

23 Lippiello (2018).

24 Ibid.

25 Arendt (1973).

26 Niebuhr, R. (2013) *Moral man and immoral society: a study in ethics and politics*. Louisville, KY: Westminster John Knox Press (original work published 1932).

27 de Jouvenel (1962).

28 Chomsky, N. (1991) 'Force and opinion', *Z Magazine*, July–August, https://chomsky.info/199107__/.

29 Ibid.

30 Stahl, R. M. and Popp-Madsen, B. A. (2022) 'Defending democracy: militant and popular models of democratic self-defense', *Constellations*, 1 (19), 1–18.

31 Morgan, E. S. (1989) *Inventing the people: the rise of popular sovereignty in England and America*. New York, NY: W. W. Norton & Co.

32 Madison (in Federalist 10), as cited in Stahl and Popp-Madsen (2022).

33 Morgan (1989).

34 Madison (in Federalist No 57), https://avalon.law.yale.edu/18th_century/fed57.asp.

35 Madison (in Federalist No 63), as cited in Stahl and Popp-Madsen (2022).

36 Morgan (1989).

37 Achbar, M. and Wintonick, P. (directors) (1992) *Manufacturing consent: Noam Chomsky and the media*. Necessary Illusions Productions Inc.

38 See Trilateral Commission (1975).

39 Parvini (2022).

40 Burnham, J. (1943) *The Machiavellians: defenders of freedom*. New York, NY: John Day Co.

41 Michels, R. (1962) *Political parties: a sociological study of the oligarchical tendencies of modern democracy*. New York, NY: Free Press (original work published 1911).

42 Hoppe, H. H. (2018) *Democracy: the god that failed*. New York, NY: Routledge.

43 Gilens, M. (2012) *Affluence and influence: economic inequality and political power in America*. Princeton, NJ: Princeton University Press.

44 Gilens, M. and Page, B. I. (2014) 'Testing theories of American politics: elites, interest groups, and average citizens', *Perspectives on Politics*, 12 (3), 564–81.

45 Jacobs, L. R. and Page, B. I. (2005) 'Who influences US foreign policy?', *American Political Science Review*, 99 (1), 107–23.

46 Enns, P. K. (2015) 'Reconsidering the middle: a reply to Martin Gilens', *Perspectives on Politics*, 13 (4), 1072–74.

47 Gilens, M. and Page, B. I. (2016) 'Critics argued with our analysis of U.S. political inequality. Here are 5 ways they're wrong', *Washington Post*, 23 May.

48 Lippmann (1993).

49 Strauss, L. (1941) 'Persecution and the art of writing', *Social Research*, 8 (4), 488–504.

50 E. E. Schattscheneider, as cited in Achen and Bartels (2017).

51 Achen and Bartels (2017).

52 Machiavelli, N. (1883) *Discourses on the first decade of Titus Livius* (trans. N. H. Thomson). London: Kegan Paul (original work published 1531).

53 Achen and Bartels (2017).

54 Deleuze, G. (2011) 'Postscript on the societies of control', in I. Szeman and T. Kaposy (eds), *Cultural theory: an anthology* (pp. 139–42). Chichester: Wiley Blackwell.

55 Foucault, M. (1979) *Discipline and punish: the birth of the prison* (trans. Alan Sheridan. London: Allen Lane.

56 Dikötter, F. (2016) *The Cultural Revolution: a people's history, 1962–1976*. New York, NY: Bloomsbury Publishing.

57 Taibbi, M. (2020) 'On "white fragility"', *Racket News*, 28 June.

58 Walton, C. D. (2021) 'Book review: *We have been harmonized*', *Comparative Strategy*, 40 (5), 516–19.

59 Mill (2003).

60 Orwell, G. (1972) 'The freedom of the press', *New York Times*, 8 October, https://www.nytimes.com/1972/10/08/archives/the-freedom-of-the-press-orwell.html#.

61 Derrida, J. (1992) *The other heading* (trans. P.-A. Brault and M. B. Naas). Bloomington, IN: Indiana University Press.

62 Ellul (1973).

63 Ibid.

64 Yarvin, C. (2019) 'The clear pill, part 1 of 5: the four-stroke regime', *The American Mind*, 27 September, https://americanmind.org/salvo/the-clear-pill-part-1-of-5-the-four-stroke-regime/.

65 Ibid.

66 Fondren and Hamilton (2022).

67 Creel, G. (1920) *How we advertised America: the first telling of the amazing story of the Committee on Public Information that carried the gospel of Americanism to every corner of the globe.* New York, NY: Harper and Brothers.

68 Ibid.

69 Hamilton, J. M. (2020) *Manipulating the masses: Woodrow Wilson and the birth of American propaganda.* Baton Rouge, LA: Louisiana State University Press.

70 Fondren and Hamilton (2022).

71 Ibid.

72 Ibid.

73 Hamilton (2020).

74 Ibid.

75 Herman and Chomsky (2010).

76 Chomsky, N. (n.d.) 'Interview with Andrew Marr', *BBC*, https://www.youtube.com/watch?v=suFzznCHjko&t=639s.

77 Friedman, S., Laurison, D. and Macmillan, L. (2017) *Social mobility, the class pay gap and intergenerational worklessness: new insights from the Labour Force Survey.* London: Social Mobility Commission.

78 Parenti, as cited in Edwards, D. (2013) 'You say what you like, because they like what you say', *Media Lens*, 13 May, https://www.medialens.org/2013/you-say-what-you-like-because-they-like-what-you-say/.

79 Groksop, V. (2010) 'David Yelland: "Rupert Murdoch is a closet liberal"', *Evening Standard*, 29 March, https://www.standard.co.uk/lifestyle/david-yelland-rupert-murdoch-is-a-closet-liberal-6732847.html.

80 Edwards, D. and Cromwell, D. (2018) *Propaganda blitz: how the corporate media distort reality.* London: Pluto Press.

81 Pribićević, O. (2021) 'Return to ideology: a solution to stumbling social democracies: the case of Corbyn', in I. Ristić (ed.), *Resetting the Left in Europe: challenges, attempts and obstacles* (pp. 144–63). Belgrade:

Institute of Social Sciences, http://iriss.idn.org.rs/562/1/Pribicevic_Resetting_the_Left_in_Europe.pdf.

82 Cammaerts, B., DeCillia, B., Magalhães, J. and Jimenez-Martínez, C. (2016) *Journalistic representations of Jeremy Corbyn in the British press: from watchdog to attack dog*. London: London School of Economics, https://orca.cardiff.ac.uk/id/eprint/126231/1/Cobyn-Report.pdf.

83 Gross, D. (1986) '"Mind-forg'd manacles": hegemony and counter-hegemony in Blake', *The Eighteenth Century*, 27 (1), 3–25.

84 Ibid.

85 Ibid.

86 Biko, S. (2002) *I write what I like*. Chicago, IL: University of Chicago Press.

87 Althusser, L. (1970) *Ideology and ideological state apparatuses*, https://www.marxists.org/reference/archive/althusser/1970/ideology.htm.

88 Lewis, W. (2002) 'Louis Althusser', in E. N. Zalta (ed.), *The Stanford encyclopedia of philosophy*, https://plato.stanford.edu/entries/althusser/#TheoIdeo.

89 Althusser (1970).

90 Yarvin, C. (2021a) 'A brief explanation of the cathedral', 21 January, https://graymirror.substack.com/p/a-brief-explanation-of-the-cathedral.

91 Rosenberg, H. (1948) 'The herd of independent minds', *Commentary*, 6, 244–52.

92 Ellul (1973).

93 Spengler (1991).

94 Yarvin (2019).

95 Alegre (2017).

96 *Denver Area Ed. Telecommunications Consortium, Inc. v. FCC*, 518 U.S. 727, 116 S. Ct. 2374, 135 L. Ed. 2d 888 (1996).

97 *Olmstead v. United States*, 277 U.S. 438, 48 S. Ct. 564, 72 L. Ed. 944 (1928).

98 Nederman, C. J. (1998) 'The mirror crack'd: the speculum principum as political and social criticism in the late middle ages', *European Legacy*, 3 (3), 18–38.

99 Machiavelli, N. (1979) *The portable Machiavelli* (trans. P. Bondanella and M. Musa). London: Penguin.

100 Milton, J. (2003) *Paradise Lost*, Book II. London: Penguin Classics.

101 See the work of Harry Brignull at https://www.deceptive.design/.

102 Luguri, J. and Strahilevitz, L. J. (2021) 'Shining a light on dark patterns', *Journal of Legal Analysis*, 13 (1), 43–109.

103 Ibid.

104 Ibid.

105 Barlow, J. P. (1996) 'A declaration of the independence of cyberspace', *Electronic Frontier Foundation*, https://www.eff.org/cyberspace -independence.

106 Lewis, P. (2018) '"Fiction is outperforming reality": how YouTube's algorithm distorts truth', *Guardian*, 2 February, https://www.theguardian.com /technology/2018/feb/02/how-youtubes-algorithm-distorts-truth.

107 Often literally 'being played'; see Hon, A. (2022) *You've been played: how corporations, governments and schools use games to control us all*. New York, NY: Basic Books.

108 Thaler, R. H. and Sunstein, C. (2008) *Nudge: improving decisions about health, wealth and happiness*. London: Penguin Books.

109 Sunstein, C. R. (2015a) 'Nudging and choice architecture: ethical considerations', http://papers.ssrn.com/abstract=2551264.

110 Sunstein, C. R. (2016) *The ethics of influence: government in the age of behavioral science*. Cambridge: Cambridge University Press.

111 Reijula, S. and Hertwig, R. (2022) 'Self-nudging and the citizen choice architect', *Behavioural Public Policy*, 6 (1), 119–49.

112 Costa, D. L. and Kahn, M. E. (2013) 'Energy conservation "nudges" and environmentalist ideology: evidence from a randomized residential electricity field experiment', *Journal of the European Economic Association*, 11 (3), 680–702.

113 Parker, I. (2020) 'Yuval Noah Harari's history of everyone, ever', *New Yorker*, 10 February, https://www.newyorker.com/magazine/2020/02 /17/yuval-noah-harari-gives-the-really-big-picture.

114 Horikawa, T., Tamaki, M., Miyawaki, Y. and Kamitani, Y. (2013) 'Neural decoding of visual imagery during sleep', *Science*, 340 (6132), 639–42.

115 Haxby, J. V., Gobbini, M. I., Furey, M. L., *et al.* (2001) 'Distributed and overlapping representations of faces and objects in ventral temporal cortex', *Science*, 293 (5539), 2425–30.

116 Kay, K. N., Naselaris, T., Prenger, R. J. and Gallant, J. L. (2008) 'Identifying natural images from human brain activity', *Nature*, 452 (7185), 352; Naselaris, T., Prenger, R. J., Kay, K. N., *et al.* (2009) 'Bayesian reconstruction of natural images from human brain activity', *Neuron*, 63 (6), 902–15.

117 Miyawaki, Y., Uchida, H., Yamashita, O., *et al.* (2008) 'Visual image reconstruction from human brain activity using a combination of multiscale local image decoders', *Neuron*, 60 (5), 915–29.

118 Nishimoto, S., Vu, A. T., Naselaris, T., *et al.* (2011) 'Reconstructing visual experiences from brain activity evoked by natural movies', *Current Biology*, 21 (19), 1641–6.

119 Mirkovic, B., Debener, S., Jaeger, M. and De Vos, M. (2015) 'Decoding the attended speech stream with multi-channel EEG: implications for online, daily-life applications', *Journal of Neural Engineering*, 12 (4), 046007.

120 Martin, S., Brunner, P., Holdgraf, C., *et al.* (2014) 'Decoding spectrotemporal features of overt and covert speech from the human cortex', *Frontiers in Neuroengineering*, 7, 14; Wang, J., Cherkassky, V. L. and Just, M. A. (2017) 'Predicting the brain activation pattern associated with the propositional content of a sentence: modeling neural representations of events and states', *Human Brain Mapping*, 38 (10), 4865–81.

121 Jorgensen, C., Lee, D. D. and Agabont, S. (2003) 'Sub auditory speech recognition based on EMG signals', in *Proceedings of the International Joint Conference on Neural Networks, 2003* (vol. 4, pp. 3128–33). New York, NY: IEEE.

122 Anumanchipalli, G. K., Chartier, J. and Chang, E. F. (2019) 'Speech synthesis from neural decoding of spoken sentences', *Nature*, 568, 493–8.

123 Anderson, A. J., Binder, J. R., Fernandino, L., *et al.* (2016) 'Predicting neural activity patterns associated with sentences using a neurobiologically motivated model of semantic representation', *Cerebral Cortex*, 27 (9), 4379–95; Pereira, F., Lou, B., Pritchett, B., *et al.* (2018) 'Toward a universal decoder of linguistic meaning from brain activation', *Nature Communications*, 9 (1), 963.

124 Anderson *et al.* (2016).

125 Haynes, J. D., Sakai, K., Rees, G., *et al.* (2007) 'Reading hidden intentions in the human brain', *Current Biology*, 17 (4), 323–8.

126 Soon, C. S., Brass, M., Heinze, H. J. and Haynes, J. D. (2008) 'Unconscious determinants of free decisions in the human brain', *Nature Neuroscience*, 11 (5), 543.

127 Smith, K. (2013) 'Brain decoding: reading minds', *Nature News*, 502 (7472), 428.

128 Yuste, R., Goering, S., Bi, G., *et al.* (2017) 'Four ethical priorities for neurotechnologies and AI', *Nature*, 551 (7679), 159–63.

129 For example, Makin, J. G., Moses, D. A. and Chang, E. F. (2020) 'Machine translation of cortical activity to text with an encoder–decoder framework', *Nature Neuroscience*, 23 (4), 575–82; Moses, D. A., Metzger, S. L., Liu, J. R., *et al.* (2021) 'Neuroprosthesis for decoding speech in a paralyzed person with anarthria', *New England Journal of Medicine*, 385 (3), 217–27.

130 Ryberg, J. (2017) 'Neuroscience, mind reading and mental privacy', *Res Publica*, 23 (2), 197–211.

131 Isaak, J. and Hanna, M. J. (2018) 'User data privacy: Facebook, Cambridge Analytica, and privacy protection', *Computer*, 51 (8), 56–9.

132 Alegre (2017).

133 Hills, T. T. (2019) 'The dark side of information proliferation', *Perspectives on Psychological Science*, 14 (3), 323–30.

134 Kasperson et al. (1988), as cited in Hills (2019).

135 Moussaïd et al. (2015), as cited in Hills (2019).

136 Beck (1992), as cited in Hills (2019).

137 Hills (2019).

138 Weisfield, D. (2013) 'Peter Thiel at Yale: We wanted flying cars, instead we got 140 characters', *Yale School of Management blog*, 27 April, https://som.yale.edu/blog/peter-thiel-at-yale-we-wanted-flying-cars-instead-we-got-140-characters.

139 Bigthinkeditor (2010) 'Peter Thiel: Regulation stifles innovation', *Big Think*, 18 November, https://bigthink.com/surprising-science/peter-thiel-regulation-stifles-innovation/.

140 Swisher, K. (2020) 'Elon Musk: "A.I. doesn't need to hate us to destroy us"', *New York Times*, 28 September, https://www.nytimes.com/2020/09/28/opinion/sway-kara-swisher-elon-musk.html.

141 Carrillo-Reid, L., Han, S., Yang, W., *et al.* (2019) 'Controlling visually guided behavior by holographic recalling of cortical ensembles', *Cell*, 178 (2), 447–57.

142 Greenwald, G. (2013) 'XKeyscore: NSA tool collects "nearly everything a user does on the internet"', *Guardian*, 31 July, https://www.theguardian.com/world/2013/jul/31/nsa-top-secret-program-online-data.

143 Wang, Y. and Kosinski, M. (2018) 'Deep neural networks are more accurate than humans at detecting sexual orientation from facial images', *Journal of Personality and Social Psychology*, 114 (2), 246.

144 Rentfrow, P. J. and Gosling, S. D. (2003) 'The do re mi's of everyday life: the structure and personality correlates of music preferences', *Journal of Personality and Social Psychology*, 84 (6), 1236–56; Golbeck, J., Robles, C. and Turner, K. (2011) 'Predicting personality with social media', in *CHI'11 extended abstracts on human factors in computing systems* (pp. 253–62). New York, NY: ACM.

145 Kosinski, M., Stillwell, D. and Graepel, T. (2013) 'Private traits and attributes are predictable from digital records of human behavior', *Proceedings of the National Academy of Sciences*, 110 (15), 5802–5.

146 Thompson, D. (2010) 'Google's CEO: "The laws are written by lobbyists"', *The Atlantic*, 1 October, https://www.theatlantic.com/technology/archive/2010/10/googles-ceo-the-laws-are-written-by-lobbyists/63908/.

147 Carmichael, J. (2014) 'Google knows you better than you know yourself', *The Atlantic*, 19 August, https://www.theatlantic.com/technology/archive/2014/08/google-knows-you-better-than-you-know-yourself/378608/.

148 Rosen, J. (2016) *Louis D. Brandeis: American prophet*. New Haven, CT: Yale University Press.

149 Marwick, A. E. (2008) 'To catch a predator? The MySpace moral panic', *First Monday*, 13 (6).

150 Goode, E. and Ben-Yehuda, N. (2012) 'Grounding and defending the sociology of moral panic', in S. P. Hier (ed.), *Moral panic and the politics of anxiety* (pp. 20–36). Abingdon: Routledge.

151 Ibid.

152 Ibid.

153 Wasserman, H. M. (2015) 'Moral panic and body cameras', *Washington University Law Review*, 92 (3), 831–44.

154 Gabilondo, J. (2006) 'Financial moral panic! Sarbanes-Oxley, financier folk devils, and off-balance-sheet arrangements', *Seton Hall Law Review*, 781–850.

155 Goode and Ben-Yehuda (2012).

156 Lavigne, M. (2021) 'Strengthening ties: the influence of microtargeting on partisan attitudes and the vote', *Party Politics*, 27 (5), 965–76.

157 Archives of digital political adverts were only created after the 2016 US presidential election, allowing people to see what adverts were being shown to others.

158 Cush, A. (2018) 'Cambridge Analytica's shady professor "Dr. Spectre" is a perfect Trump villain', *Spin.com*, 19 March, https://www.spin.com/2018/03/dr-spectre-cambridge-analytica-perfect-trump-villain/.

159 Hersh, E. D. and Schaffner, B. F. (2013) 'Targeted campaign appeals and the value of ambiguity', *Journal of Politics*, 75, 520–34; Liberini, F., Redoano, M., Russo, *et al.* (2018) 'Politics in the Facebook era: evidence from the 2016 U.S. presidential elections', *CAGE Online Working Paper Series* (389); Kalla, J. L. and Broockman, D. E. (2018) 'The minimal persuasive effects of campaign contact in general elections: evidence from 49 field experiments', *American Political Science Review*, 112, 148–66.

160 Bond, R. M., Fariss, C. J., Jones, J. J., *et al.* (2012) 'A 61-million-person experiment in social influence and political mobilization', *Nature*, 489, 295–8.

161 PBS (2019) 'Zero tolerance: Steven Bannon interview', *Frontline*, 22 October, https://www.youtube.com/watch?v=CKuPYArH0Gs.

162 Ellul (1973).

163 Haenschen, K. (2022) 'The conditional effects of microtargeted Facebook advertisements on voter turnout', *Political Behavior*.

164 Critcher, C. (2008) 'Moral panic analysis: past, present and future', *Sociology Compass*, 2 (4), 1127–44.

165 Critcher (2008).

166 Alegre, S. (2022) *Freedom to think: the long struggle to liberate our minds*. London: Atlantic Books.

167 Haidt, J. and Bail, C. (ongoing) 'Social media and political dysfunction: a collaborative review', unpublished manuscript, New York University, https://docs.google.com/document/d/1vVAtMCQnz8WVxtSNQev_e1cGmY9rnY96ecYuAj6C548/edit.

168 Cho, J., Ahmed, S., Hilbert, M., et al. (2020) 'Do search algorithms endanger democracy? An experimental investigation of algorithm effects on political polarization', *Journal of Broadcasting & Electronic Media*, 64 (2), 150–72.

169 Banks, A., Calvo, E., Karol, D. and Telhami, S. (2021) '#polarizedfeeds: three experiments on polarization, framing, and social media', *International Journal of Press/Politics*, 26 (3), 609–34.

170 Beam, M. A., Hutchens, M. J. and Hmielowski, J. D. (2018) 'Facebook News and (de) polarization: reinforcing spirals in the 2016 U.S. election', *Information, Communication & Society*, 21 (7), 940–58.

171 Nordbrandt, M. (2021) 'Affective polarization in the digital age: testing the direction of the relationship between social media and users' feelings for out-group parties', *New Media & Society*.

172 Barberá, P. (2015) 'Birds of the same feather tweet together: Bayesian ideal point estimation using Twitter data', *Political Analysis*, 23 (1), 76–91.

173 Mosleh, M., Martel, C., Eckles, D. and Rand, D. G. (2021) 'Shared partisanship dramatically increases social tie formation in a Twitter field experiment', *Proceedings of the National Academy of Sciences*, 118 (7), e2022761118.

174 Brown, M. A., Bisbee, J., Lai, A., et al. (2022) 'Echo chambers, rabbit holes, and algorithmic bias: how YouTube recommends content to real users', https://papers.ssrn.com/sol3/papers.cfm?abstract_id=4114905.

175 Eady, G., Nagler, J., Guess, A., et al. (2019) 'How many people live in political bubbles on social media? Evidence from linked survey and Twitter data', *Sage Open*, 9 (1).

176 Fletcher, R., Robertson, C. T. and Nielsen, R. K. (2021) 'How many people live in politically partisan online news echo chambers in different countries?', *Journal of Quantitative Description: Digital Media*, 1.

177 Dubois, E. and Blank, G. (2018) 'The echo chamber is overstated: the moderating effect of political interest and diverse media', *Information, Communication & Society*, 21 (5), 729–45.

178 Goode, E. and Ben-Yehuda, N. (2010) *Moral panics: the social construction of deviance*. Chichester: John Wiley & Sons.

179 Ibid.

180 Burnham (1943).

181 Kuran, T. and Sunstein, C. R. (1998) 'Availability cascades and risk regulation', *Stanford Law Review*, 51, 683.

182 Strum, P. (1999) *When the Nazis came to Skokie: freedom for speech we hate*. Lawrence, KS: University Press of Kansas; Goldberger, D. (2020) 'The Skokie case: how I came to represent the free speech rights of Nazis', *American Civil Liberties Union*, https://www.aclu.org/issues/free-speech/rights-protesters/skokie-case-how-i-came-represent-free-speech-rights-nazis.

183 Marcuse, H. (1969) 'Repressive tolerance', in R. P. Wolff, B. Moore Jr and H. Marcuse, *A critique of pure tolerance* (pp. 81–123). Boston, MA: Beacon Press.

184 Sidwell, M. (2022) *A silent revolution: the intellectual origins of cancel culture*. London: Institute of Economic Affairs, https://iea.org.uk/wp-content/uploads/2022/07/DP110_A-silent-revolution_web.pdf.

185 Students for a Democratic Society (1962) 'Port Huron Statement', http://www2.iath.virginia.edu/sixties/HTML_docs/Resources/Primary/Manifestos/SDS_Port_Huron.html.

186 Cammaerts, B. (2015) 'Neoliberalism and the post-hegemonic war of position: the dialectic between invisibility and visibilities', *European Journal of Communication*, 30 (5), 522–38.

187 William Simon, as cited in Moskowitz, P. E. (2019) *The case against free speech: the First Amendment, fascism, and the future of dissent*. New York, NY: Bold Type Books.

188 Miller, J. J. (2022) *A gift of freedom: how the John M. Olin Foundation changed America*. New York, NY: Encounter Books; Moskowitz (2019).

189 Forest, J. D. (2006) 'The rise of conservatism on campus: the role of the John M. Olin Foundation', *Change: The Magazine of Higher Learning*, 38 (2), 32–7.

190 Moskowitz (2019).

191 Bessner, D. (2014) 'Murray Rothbard, political strategy, and the making of modern libertarianism', *Intellectual History Review*, 24 (4), 441–56.

192 Mises, as cited in Bessner (2014).

193 Rothbard, as cited in Bessner (2014).

194 Bessner (2014).

195 Moskowitz (2019).

196 *Citizens United v. Federal Election Com'n*, 558 U.S. 310, 130 S. Ct. 876, 175 L. Ed. 2d 753 (2010).

197 van de Werfhorst, H. G. (2020) 'Are universities left-wing bastions? The political orientation of professors, professionals, and managers in Europe', *British Journal of Sociology*, 71 (1), 47–73.

198 *Dobbs v. Jackson Women's Health Organization*, 142 S. Ct. 2228, 597 U.S., 213 L. Ed. 2d 545 (2022); *West Virginia v. EPA*, 142 S. Ct. 2587, 597 U.S., 213 L. Ed. 2d 896 (2022).

199 Kaczynski, T. (1995) 'Industrial society and its future', *Washington Post*, 22 September, https://www.washingtonpost.com/wp-srv/national/longterm/unabomber/manifesto.text.htm.

4 Protections for free thought

1 Sexby, W. (1657) 'Killing noe murder. Briefly discourst in three quaestions', https://www.yorku.ca/comninel/courses/3025pdf/Killing_Noe_Murder.pdf.

2 Asthana, A. and Mason, R. (2016) 'UK must leave European Convention on Human Rights, says Theresa May', *Guardian*, 25 April, https://www.theguardian.com/politics/2016/apr/25/uk-must-leave-european-convention-on-human-rights-theresa-may-eu-referendum.

3 Miller, V. (2014) 'Parliamentary sovereignty and the European Convention on Human Rights', *House of Commons Library*, 6 November, https://commonslibrary.parliament.uk/parliamentary-sovereignty-and-the-european-convention-on-human-rights/.

4 Lauren, P. G. (2007) ' "To preserve and build on its achievements and to redress its shortcomings": the journey from the Commission on Human Rights to the Human Rights Council', *Human Rights Quarterly*, 29 (2), 307–345.

5 Féron, H. (2014) 'Human rights and faith: a "world-wide secular religion"?', *Ethics & Global Politics*, 7 (4), 181–200.

6 MacIntyre (2007).

7 Mitoma, G. (2013) *Human rights and the negotiation of American power*. Philadelphia, PA: University of Pennsylvania Press.

8 Lindkvist, L. (2013) 'The politics of Article 18: religious liberty in the Universal Declaration of Human Rights', *Humanity: An International Journal of Human Rights, Humanitarianism, and Development*, 4 (3), 429–47.

9 Joe (2011) 'Remembering Charles H. Malik', *The Disorder of Things*, 9 February, https://thedisorderofthings.com/2011/02/09/remembering-charles-h-malik/.

10 Mitoma, G. (2010) 'Charles H. Malik and human rights: notes on a biography', 33 (1), *Biography*, 222–41. See also the 1937 Constitution of Ireland, whose Article 41(1) states that 'The State recognises the Family as the natural primary and fundamental unit group of Society, and as a moral institution possessing inalienable and imprescriptible rights, antecedent and superior to all positive law. The State, therefore, guarantees to protect the Family in its constitution and authority, as the necessary basis of social order and as indispensable to the welfare of the Nation and the State.' Notably, Article 41(2) then states that 'In particular, the State recognises that by her life within the home, woman gives to the State a support without which the common good cannot be achieved. The State shall, therefore, endeavour to ensure that mothers shall not be obliged by economic necessity to engage in labour to the neglect of their duties in the home'. Naturally this has attracted both historical and contemporary controversy (e.g. Houses of the Oireachtas (2022) 'Gender Equality Committee recommends referendum on "woman in the home" section of Constitution should be held in 2023', 13 July, https://www.oireachtas.ie/en/press-centre/press-releases/20220713-gender-equality-committee-recommends-referendum-on-woman-in-the-home-section-of-constitution-should-be-held-in-2023/.

11 Ribnikar, as cited in Lindkvist, L. (2017) *Religious freedom and the Universal Declaration of Human Rights*. Cambridge: Cambridge University Press.

12 Marx, K. (1844) 'On the Jewish question', *Marxists.org*, https://www.marxists.org/archive/marx/works/1844/jewish-question/.

13 Mitoma (2013).

14 Bakara, H. (2016) 'Human rights in the twentieth-century: a literary history' (doctoral dissertation). University of Chicago, Chicago, IL, USA.

15 Descartes, R. (2009) *Selected philosophical writings* (trans. J. Cottingham and R. Stoothoff). Cambridge: Cambridge University Press.

16 Hegel, G. W. F. (1892) *Lectures on the history of philosophy* (vol. 1, trans. E. S. Haldane). London: Kegan Paul.

17 Lindkvist (2013).

18 The Executive Board, American Anthropological Association (1947) 'Statement on human rights', *American Anthropologist*, 49 (4), 539–43.

19 Lindkvist (2013).

20 Nisbett, R. (2004) *The geography of thought: how Asians and Westerners think differently . . . and why*. London: Simon and Schuster.

21 Gao, C. (2017) 'China promotes human rights "with Chinese character-istics"', *The Diplomat*, 12 December, https://thediplomat.com/2017/12/china-promotes-human-rights-with-chinese-characteristics/.

22 Flew, T. (2019) 'Digital communication, the crisis of trust, and the post-global', *Communication Research and Practice*, 5 (1), 4–22.

23 Stieg, M. F. (1993) 'The postwar purge of German public libraries, democracy, and the American reaction', *Libraries & Culture*, 28 (2), 143–64.

24 Plischke, E. (1947) 'Denazification law and procedure', *American Journal of International Law*, 41 (4), 807–27.

25 Norgaard, N. (1945) 'Eisenhower claims 50 years needed to re-educate Nazis', *Oregon Statesman*, 13 October.

26 Think back to Mill's 'The peculiar evil of silencing the expression of an opinion is, that it is robbing the human race; posterity as well as the existing generation; those who dissent from the opinion, still more than those who hold it. If the opinion is right, they are deprived of the opportunity of exchanging error for truth: if wrong, they lose, what is almost as great a benefit, the clearer perception and livelier impression of truth, produced by its collision with error.' Mill (2003).

27 The following, apart from where stated, is summarised from Stieg (1993).

28 Ibid.

29 Ibid.

30 Peiss, K. (2019) *Information hunters: when librarians, soldiers, and spies banded together in World War II Europe.* Oxford: Oxford University Press.

31 Stieg (1993).

32 Nawyn, K. J. (2008) ' "Striking at the roots of German militarism": efforts to demilitarize German society and culture in American-occupied Württemberg-Baden, 1945–1949' (Master's dissertation). University of North Carolina, Chapel Hill, NC, USA.

33 Ibid.

34 Stieg (1993).

35 Hayden, M. V. (2013) 'Beyond Snowden: an NSA reality check', *World Affairs*, 176, 13–23.

36 Francis, J. G. and Francis, L. (2021) 'Freedom of thought in the United States: the First Amendment, marketplaces of ideas, and the internet', *European Journal of Comparative Law and Governance*, 8 (2–3), 192–225.

37 Richards (2015).

38 Kolber, A. J. (2016) 'Two views of First Amendment thought privacy', *Journal of Constitutional Law*, 18, 1381–423.

39 *Doe v. City of Lafayette, Indiana* (2003) 334 F.3d 606 (7th Cir.).

40 *Thomas v. Collins*, 323 U.S. 516, 65 S. Ct. 315, 89 L. Ed. 430 (1945).

41 Blitz (2010); Richards (2015).

42 *Branti v. Finkel* (1980) 445 U.S. 507, 100 S. Ct. 1287, 63 L. Ed. 2d 574.

43 *Doe v. City of Lafayette, Indiana* (2003).

44 Government of Canada (n.d.) 'The Constitution Acts, 1867 to 1982 – table of contents', https://laws-lois.justice.gc.ca/eng/const/page -12.html.

45 South African Government (n.d.) 'Constitution of the Republic of South Africa, 1996 – chapter 2: bill of rights', https://www.gov.za/ documents/constitution/chapter-2-bill-rights.

46 Bublitz, J. C. (2011) 'If man's true palace is his mind, what is its adequate protection? On a right to mental self-determination and limits of interventions into other minds', in L. Klaming and B. van den Berg (eds), *Technologies on the stand: legal and ethical questions in neuroscience and robotics* (pp. 113–45). Oisterwijk: Wolf Legal Publishers.

47 Shaheed (2021).

48 For a good discussion of what prompted Warren and Brandeis's famous paper, see Gajda, A. (2008) 'What if Samuel D. Warren hadn't married a Senator's daughter: uncovering the press coverage that led to the right to privacy', *Michigan State Law Review*, 35, 35–60.

49 Warren, S. D. and Brandeis, L. D. (1890) 'The right to privacy', *Harvard Law Review*, 4 (5), 193–220.

50 *Long Beach City Employees Assn. v. City of Long Beach* (1986) 719 P.2d 660, 41 Cal. 3d 937, 227 Cal. Rptr. 90.

51 *Stanley v. Georgia* (1969) 394 U.S. 557, 89 S. Ct. 1243, 22 L. Ed. 2d 542.

52 *Rennie v. Klein* (1981) 653 F.2d 836 (3d Cir.).

53 Whitman, J. Q. (2003) 'The two western cultures of privacy: dignity versus liberty', *Yale Law Journal*, 113 (6), 1151.

54 Ibid.

55 Bublitz (2014).

56 Harmless, W. and Fitzgerald, R. R. (2001) 'The sapphire light of the mind: the Skemmata of Evagrius Ponticus', *Theological Studies*, 62 (3), 498–529.

57 Bublitz (2014).

58 Bacon, F. (1844) *The works of Francis Bacon, Lord Chancellor of England* (vol. 2, ed. B. Montagu). Philadelphia, PA: Carey and Hart.

59 Wharton, F. (1868) *A treatise on the criminal law of the United States* (vol. 2), 6th edn. Philadelphia, PA: Kay and Brother; Sayre, F. B. (1928) 'Criminal attempts', *Harvard Law Review*, 41 (7), 821–59.

60 Houston, A. C. (2014) *Algernon Sidney and the republican heritage in England and America*. Princeton, NJ: Princeton University Press.

61 Brudner, A. (2009) *Punishment and freedom*. Oxford: Oxford University Press.

62 Corry, R. J. Jr. (2000) 'Burn this article: it is evidence in your thought crime prosecution', *Texas Review of Law and Politics*, 4, 462–88.

63 *Apprendi v. New Jersey*, 530 U.S. 466, 120 S. Ct. 2348, 147 L. Ed. 2d 435 (2000).

64 *Steffan v. Perry* (1994) 41 F.3d 677, 713–14 (D.C.Cir.).

65 *Doe v. City of Lafayette, Indiana* (2003).

66 *Doe v. City of Lafayette, Indiana* (2003); *Doe v. City of Lafayette, Ind.*, 377 F.3d 757 (7th Cir. 2004).

67 *Doe v. City of Lafayette, Indiana* (2003).

68 Calvert, C. (2005) 'Freedom of thought, offensive fantasies and the fundamental human right to hold deviant ideas: why the Seventh Circuit got it wrong in *Doe v. City of Lafayette, Indiana*', *Pierce Law Review*, 3, 125–60.

69 *Doe v. City of Lafayette, Indiana* (2003).

70 Ibid.

71 *Doe v. City of Lafayette, Ind.* (2004).

72 *US v. Valle*, 807 F.3d 508 (2d Cir.) (2015).

73 Working Group on Misogyny and Criminal Justice (2022) 'Misogyny – a human rights issue', 8 March, https://www.gov.scot/publications/misogyny-human-rights-issue/.

74 Taylor, K. (2006) *Brainwashing: the science of thought control*. Oxford: Oxford University Press.

75 *Stanley v. Georgia* (1969).

76 *Ashcroft v. Free Speech Coalition* (2002) 535 U.S. 234, 122 S. Ct. 1389, 152 L. Ed. 2d 403.

77 Bublitz (2011).

78 Mendlow, G. S. (2018) 'Why is it wrong to punish thought?', *Yale Law Journal*, 127, 2342–87.

79 Nowak (1993).

80 *Kokkinakis v. Greece* (1994) 17 E.H.R.R. 397.

81 *Larissis and Others v. Greece* (1998) 65 Eur. Ct. H.R. (ser. A) 363, 27 E.H.R.R. 329.

82 This argument draws on Wood, A. W. (2014) 'Coercion, manipulation, exploitation', in C. Coons and M. Weber (eds), *Manipulation: theory and practice* (pp. 17–50). Oxford: Oxford University Press; and on Sunstein, C. R. (2015b) 'Fifty shades of manipulation', in C. R. Sunstein, *The ethics of influence* (pp. 78–115). Cambridge: Cambridge University Press.

83 Strittmatter (2020).

84 Wood (2014).

85 See Sunstein (2015b).

86 Wood (2014).

87 Sunstein (2015b).

88 Dallas, S. K., Liu, P. J. and Ubel, P. A. (2019) 'Don't count calorie labeling out: calorie counts on the left side of menu items lead to lower

calorie food choices', *Journal of Consumer Psychology*, 29 (1), 60–9, as discussed in Sunstein, C. R. (2020) *Too much information: understanding what you don't want to know.* Cambridge, MA: MIT Press.

89 Alvarez, R. M., Sinclair, B. and Hasen, R. L. (2006) 'How much is enough? The "ballot order effect" and the use of social science research in election law disputes', *Election Law Journal*, 5 (1), 40–56; see also Webber, R., Rallings, C., Borisyuk, G. and Thrasher, M. (2014) 'Ballot order positional effects in British local elections, 1973–2011', *Parliamentary Affairs*, 67 (1), 119–36.

90 Alvarez *et al.* (2006).

91 Ibid.; Blom-Hansen, J., Elklit, J., Serritzlew, S. and Villadsen, L. R. (2016) 'Ballot position and election results: evidence from a natural experiment', *Electoral Studies*, 44, 172–83; Miller, J. M. and Krosnick, J. A. (1998) 'The impact of candidate name order on election outcomes', *Public Opinion Quarterly*, 62 (3), 291–330; Brockington, D. (2003) 'A low information theory of ballot position effect', *Political Behavior*, 25, 1–27.

92 Blom-Hansen *et al.* (2016).

93 Ibid.

94 Sunstein (2015b).

95 Ibid.

96 Ibid.

97 O'Callaghan, P., Cronin, O., Kelly, B. D., *et al.* (submitted for publication) 'The right to freedom of thought: an interdisciplinary analysis of the UN Special Rapporteur's Report on Freedom of Thought'.

98 See 'Regulation (EU) 2022/2065 of the European Parliament and of the Council of 19 October 2022 on a Single Market For Digital Services and amending Directive 2000/31/EC (Digital Services Act)', https://eur-lex.europa.eu/legal-content/EN/TXT/PDF/?uri=CELEX:32022R2065&from=EN.

99 O'Callaghan *et al.* (submitted for publication).

100 Jefferson, as cited in Nader, R. (2016) *Breaking through power: it's easier than we think.* San Francisco, CA: City Lights Publishers.

101 Greenwood, D. J. (2017) 'Neofederalism: the surprising foundations of corporate constitutional rights', *University of Illinois Law Review*, 163–222.

102 Ibid.

103 Knox, J. H. (2011) 'The Ruggie Rules: applying human rights law to corporations', in R. Mares (ed.), *The UN guiding principles on business and human rights: foundations and implementation* (pp. 51–83). Leiden: Martinus Nijhoff Publishers.

104 United Nations Human Rights Committee (2004) *General comment no. 31: the nature of the general legal obligation imposed on states parties to the covenant (Geneva)*, https://www.refworld.org/docid/478b26ae2.html.

105 Knox (2011).

106 United Nations Human Rights Office (2012) *The corporate responsibility to protect human rights*. New York, NY: United Nations, https://www.ohchr.org/sites/default/files/Documents/publications/hr.puB.12.2_en.pdf.

5 Creating free thought I: Clay

1 Abelson, H., Diamond, P. A., Grosso, A. and Pfeiffer, D. P. (2013) *Report to the President: MIT and the prosecution of Aaron Swartz*. Cambridge, MA: MIT Press, http://swartz-report.mit.edu/docs/report-to-the-president.pdf.

2 Amsden, D. (2013) 'The brilliant life and tragic death of Aaron Swartz', *Rolling Stone*, 15 February, https://www.rollingstone.com/culture/culture-news/the-brilliant-life-and-tragic-death-of-aaron-swartz-177191/.

3 Swartz, A. (2011) 'Guerilla open access manifesto', https://archive.org/details/GuerillaOpenAccessManifesto.

4 Amsden (2013).

5 Swartz (2011).

6 Poulsen, K. (2013) 'First 100 pages of Aaron Swartz's Secret Service file released', *Wired*, 12 August, https://www.wired.com/2013/08/swartz-foia-release/.

7 See American Library Association, 'James Madison Award, 2013 winner(s)', https://www.ala.org/awardsgrants/awards/193/winners/2013.

8 Peet, L. (2016) 'Sci-Hub controversy triggers publishers' critique of librarian', *Library Journal*, 25 August, https://web.archive.org/web/

20170111185200/http://lj.libraryjournal.com/2016/08/copyright/
sci-hub-controversy-triggers-publishers-critique-of-librarian/.

9 Bohannon, J. (2016) 'The frustrated science student behind Sci-Hub',
Science, 28 April, https://www.science.org/content/article/frustrated-
science-student-behind-sci-hub.

10 Schiermeier, Q. (2017) 'US court grants Elsevier millions in damages
from Sci-Hub', *Nature*, 22 June, https://www.nature.com/articles/
nature.2017.22196.

11 Ibid.

12 Buranyi, S. (2017) 'Is the staggeringly profitable business of scientific
publishing bad for science?', *Guardian*, 27 June, https://www.theguard-
ian.com/science/2017/jun/27/profitable-business-scientific-publish-
ing-bad-for-science.

13 See Budapest Open Access Initiative, 'Read the Declaration', https://
www.budapestopenaccessinitiative.org/read/.

14 Wells, H. G. (1940) *The common sense of war and peace*. London:
Penguin.

15 Madison, J. (1822) 'Letter to William T. Barry', 4 August, https://
founders.archives.gov/documents/Madison/04-02-02-0480.

16 Japola, J. M. (2009) 'Fodor and Aquinas: the architecture of the mind
and the nature of concept acquisition' (doctoral dissertation).
Georgetown University, Washington DC, USA.

17 Loucaides, as cited in Bublitz, J. C. and Blitz, M. J. (eds) (2021) *The law
and ethics of freedom of thought* (vol. 1): *Neuroscience, autonomy, and individual rights*. London: Palgrave Macmillan.

18 Cooper, as cited in Watson, L. (2021) *The right to know: epistemic rights
and why we need them*. London: Routledge.

19 Markovich, R. and Roy, O. (2021) 'Formalizing the right to know –
epistemic rights as normative positions', in B. Liao, J. Luo and L. van der
Torre (eds), *Logics for new-generation AI 2021* (pp. 154–9). London:
College Publications.

20 American Library Association (n.d.) 'Libraries and kids: quotes you can
use', https://www.ala.org/ala/alsc/projectspartners/KidsQuotes.htm.

21 Ibid.

22 Geiger, A. W. (2017) 'Most Americans – especially Millennials – say
libraries can help them find reliable, trustworthy information', *Pew*

Research Centre, 30 August, https://www.pewresearch.org/fact-tank/2017/08/30/most-americans-especially-millennials-say-libraries-can-help-them-find-reliable-trustworthy-information/.

23 Kennedy, J. F. (1959) *Remarks of Senator John F. Kennedy, Cleveland press book and author luncheon*, Cleveland, Ohio, April 16, https://www.jfklibrary.org/archives/other-resources/john-f-kennedy-speeches/cleveland-oh-19590416.

24 American Library Association (2010) '12 ways libraries are good for the country', *American Libraries*, 21 December, https://americanlibrariesmagazine.org/2010/12/21/12-ways-libraries-are-good-for-the-country/.

25 On the problems of public schools, see Gatto, J. T. (2002) *Dumbing us down: the hidden curriculum of compulsory schooling*. Gabriola, BC: New Society Publishers; Gatto, J. T. (2003) *The underground history of American education*. New York, NY: Oxford Village Press.

26 Hamby, P. and Najowitz, I. (1999) 'The Public Libraries Act of 1850: utilitarian pragmatism and idealist humanitarianism in action', *Public & Access Services Quarterly*, 2 (4), 73–88.

27 Max, S. M. (1984) 'Tory reaction to the Public Libraries Bill, 1850', *Journal of Library History (1974–1987)*, 19 (4), 504–24.

28 Max (1984).

29 Larned, as cited in Gerolami, N. (2018) 'Taming the mob: the early public library and the creation of good citizens', *Partnership: The Canadian Journal of Library and Information Practice and Research*, 13 (1).

30 Ibid.

31 Ibid.

32 Ibid.

33 Rubin, R., as cited in Gerolami (2018).

34 Van Slyck, A. A. (1996) *Free to all: Carnegie libraries and American culture, 1890–1920*. Chicago, IL: University of Chicago Press. Clearly there were other drivers of solitary reading, including increased literacy rates, that meant people could read themselves, rather than depending on others reading to them.

35 Van Slyck (1996).

36 Max (1984).

37 Schwartz, B. (2016) 'Google's search knows about over 130 trillion pages', *Search Engine Land*, 14 November, https://searchengineland.com/googles-search-indexeshits-130-trillion-pages-documents-263378.

38 James, W. (2014) *The principles of psychology* (vol. 2). New York, NY: Dover Publications (original work published 1890).

39 Ferguson, C. H. (2005) 'What's next for Google?', *MIT Technology Review*, 1 January, https://www.technologyreview.com/2005/01/01/231826/whats-next-for-google/.

40 'About Google', https://about.google/.

41 Vedejová, D. and Čavojová, V. (2022) 'Confirmation bias in information search, interpretation, and memory recall: evidence from reasoning about four controversial topics', *Thinking & Reasoning*, 28 (1), 1–28.

42 Tropodi, F. (2018) 'Searching for alternative facts: analyzing scriptural inference in Conservative news practices', *Data & Society Research Institute*, https://datasociety.net/wp-content/uploads/2018/05/Data_Society_Searching-for-Alternative-Facts.pdf.

43 Tropodi (2018).

44 Leroy, G. (2009) 'Persuading consumers to form precise search engine queries', *AMIA Annual Symposium Proceedings*, 2009, 354–8.

45 For example, Hindy, J. (n.d.) '20 Google search tips to use Google more efficiently', *LifeHack*, https://www.lifehack.org/articles/technology/20-tips-use-google-search-efficiently.html.

46 As cited in Epstein, R. and Robertson, R. E. (2015) 'The search engine manipulation effect (SEME) and its possible impact on the outcomes of elections', *Proceedings of the National Academy of Sciences*, 112 (33), E4512–21.

47 Ibid.

48 Ibid.

49 Gold, H. (2018) 'Trump slams Google search as "rigged" – but it's not', *CNN*, 28 August, https://money.cnn.com/2018/08/28/technology/donald-trump-google-rigged/index.html.

50 Epstein, R., Robertson, R. E., Shepherd, S. J. and Zhang, S. (2017) 'A method for detecting bias in search rankings, with evidence of systematic bias related to the 2016 Presidential Election'. Paper presented at the 2nd biennial meeting of the International Convention of

Psychological Science, Vienna, Austria, 24 March; Robertson, R. E., Jiang, S., Joseph, K., *et al.* (2018) 'Auditing partisan audience bias within Google search', *Proceedings of the ACM on Human–Computer Interaction,* 2 (CSCW), 1–22.

51 Metaxa, D., Park, J. S., Landay, J. A. and Hancock, J. (2019) 'Search media and elections: A longitudinal investigation of political search results', *Proceedings of the ACM on Human–Computer Interaction,* 3 (CSCW), 1–17; Kulshrestha, J., Eslami, M., Messias, J., *et al.* (2019) 'Search bias quantification: investigating political bias in social media and web search', *Information Retrieval Journal,* 22 (1), 188–227; Courtois, C., Slechten, L. and Coenen, L. (2018) 'Challenging Google search filter bubbles in social and political information: disconforming evidence from a digital methods case study', *Telematics and Informatics,* 35 (7), 2006–15.

52 Unkel, J. and Haim, M. (2021) 'Googling politics: parties, sources, and issue ownerships on Google in the 2017 German federal election campaign', *Social Science Computer Review,* 39 (5), 844–61.

53 Groseclose, T. and Milyo, J. (2005) 'A measure of media bias', *Quarterly Journal of Economics,* 120 (4), 1191–237.

54 Edwards and Cromwell (2018).

55 Parry, B. (2020) '25 years: Bob Parry's last article – a manifesto on the state of journalism', *Consortium News,* 25 November, https://consorti-umnews.com/2020/11/15/25-years-bob-parrys-last-article-a-mani-festo-on-the-state-of-journalism/.

56 Courtois *et al.* (2018).

57 Robertson *et al.* (2018).

58 Slechten, L., Courtois, C., Coenen, L. and Zaman, B. (2021) 'Adapting the selective exposure perspective to algorithmically governed platforms: the case of Google search', *Communication Research,* 49 (8).

59 Volokh, E. and Falk, D. M. (2011) 'Google: First Amendment protection for search engine search results', *Journal of Law, Economics & Policy,* 8, 883. For an alternative view, see Bracha, O. and Pasquale, F. (2007) 'Federal search commission-access, fairness, and accountability in the law of search', *Cornell Law Review,* 93, 1149.

60 *Search King, Inc. v. Google Tech., Inc.,* 2003 W.L. 21464568 (2003).

61 Bracha and Pasquale (2007).

62 Ibid.

63 Chae, Y., Lee, S. and Kim, Y. (2019) 'Meta-analysis of the relationship between internet use and political participation: examining main and moderating effects', *Asian Journal of Communication*, 29 (1), 35–54; Boulianne, S. (2015) 'Social media use and participation: a meta-analysis of current research', *Information, Communication & Society*, 18 (5), 524–38.

64 Nanz, A. and Matthes, J. (2022) 'Seeing political information online incidentally. Effects of first- and second-level incidental exposure on democratic outcomes', *Computers in Human Behavior*, 133, 107285; Courtois *et al.* (2018).

65 Breuer, A. (2012) 'The role of social media in mobilizing political protest: evidence from the Tunisian revolution', *German Development Institute Discussion Paper*, 10, 1860–0441.

66 Ibid.

67 Ibid.

68 Stiglitz, J. E. (1999) 'On liberty, the right to know, and public discourse: the role of transparency in public life', speech given in Oxford, UK, 27 January, http://www.internationalbudget.org/wp-content/uploads/On-Liberty-the-Right-to-Know-and-Public-Discourse-The-Role-of-Transparency-in-Public-Life.pdf.

69 Aftergood, S. (2008) 'Reducing government secrecy: finding what works', *Yale Law & Policy Review*, 27, 399–416.

70 Ibid.

71 Weber, as cited in Aftergood (2008).

72 White, L. (2019) 'China's foreign influence and sharp power strategy to shape and influence democratic institutions', *National Endowment for Democracy*, 16 May, https://www.ned.org/chinas-foreign-influence-and-sharp-power-strategyto-shape-and-influence-democratic-institutions/.

73 Jones (1988).

74 Markovich and Roy (2021).

75 Fenster, M. (2011) 'The transparency fix: advocating legal rights and their alternatives in the pursuit of a visible state', *University of Pittsburgh Law Review*, 73, 443–504.

76 Worthy, B. (2013) ' "Some are more open than others": comparing the impact of the Freedom of Information Act 2000 on local and central government in the UK', *Journal of Comparative Policy Analysis: Research and Practice*, 15 (5), 395–414.

77 Birkinshaw, P. (2010) 'Freedom of information and its impact in the United Kingdom', *Government Information Quarterly*, 27 (4), 312–21.

78 Worthy (2013).

79 Camaj, L. (2016) 'Governments' uses and misuses of freedom of information laws in emerging European democracies: FOI laws' impact on news agenda-building in Albania, Kosovo, and Montenegro', *Journalism & Mass Communication Quarterly*, 93 (4), 923–45.

80 Thomas, O. D. (2020) 'Paradoxical secrecy in British freedom of information law 1', in D. Mokrosinska (ed.), *Transparency and secrecy in European democracies* (pp. 135–56). London: Routledge.

81 Thomas (2020).

82 Fishkin (2018).

83 Nietzsche, F. (1911) *Ecce homo* (trans. A. M. Ludovici). New York, NY: Macmillan.

84 Schopenhauer, A. (2000) *Parerga and Paralipomena: short philosophical essays* (vol. 1). Oxford: Oxford University Press.

85 Ibid.

86 Rimbaud, as cited in Wilson, E. (1931) *Axel's castle*. New York, NY: Charles Scribner's Sons.

87 Tim, W. (2018) 'Is the First Amendment obsolete?', *Michigan Law Review*, 117 (3), 548–81.

88 Sunstein (2020).

89 Gigerenzer, G. and Garcia-Retamero, R. (2017) 'Cassandra's regret: the psychology of not wanting to know', *Psychological Review*, 124 (2), 179–96.

90 Hussain, M., Price, D. M., Gesselman, A. N., *et al.* (2021) 'Avoiding information about one's romantic partner', *Journal of Social and Personal Relationships*, 38 (2), 626–47.

91 Sunstein (2020).

92 Goldin and Rouse (2000), as cited in Kozyreva, A., Lewandowsky, S. and Hertwig, R. (2020) 'Citizens versus the internet: confronting

digital challenges with cognitive tools', *Psychological Science in the Public Interest*, 21 (3), 103–56.

93 *Thompson v. Western States Medical Center*, 535 U.S. 357, 122 S. Ct. 1497, 152 L. Ed. 2d 563 (2002).

94 UK Government (n.d.) 'Marketing and advertising: the law', https://www.gov.uk/marketing-advertising-law/regulations-that-affect-advertising.

95 Cooper, as cited in Watson (2021).

96 *Johnson v. Westminster Magistrates' Court* [2019] EWHC 1709, https://www.judiciary.uk/wp-content/uploads/2019/07/2019ewhc-1709-admin-johnson-v-westminster-mags-final.pdf.

97 Ibid.

98 *Rickert v. State, Public Disclosure Com'n*, 168 P.3d 826, 161 Wash. 2d 843 (2007).

99 Ibid.

100 Ibid.

101 DePaulo, B. M., Kashy, D. A., Kirkendol, S. E., *et al.* (1996) 'Lying in everyday life', *Journal of Personality and Social Psychology*, 70 (5), 979–95.

102 Bond Jr, C. F. and DePaulo, B. M. (2006) 'Accuracy of deception judgments', *Personality and Social Psychology Review*, 10 (3), 214–34.

103 Gongola, J., Scurich, N. and Quas, J. A. (2017) 'Detecting deception in children: a meta-analysis', *Law and Human Behavior*, 41 (1), 44–54.

104 Hitler, A. (1939/2002) *Mein Kampf* (trans. J. Murphy). New York, NY: Hurst and Blackett, Ltd.

105 Lariscy, R. A. W. and Tinkham, S. F. (1999) 'The sleeper effect and negative political advertising', *Journal of Advertising*, 28 (4), 13–30.

106 Steblay, N., Hosch, H. M., Culhane, S. E. and McWethy, A. (2006) 'The impact on juror verdicts of judicial instruction to disregard inadmissible evidence: a meta-analysis', *Law and Human Behavior*, 30 (4), 469–92.

107 Fazio, L. K., Brashier, N. M., Payne, B. K. and Marsh, E. J. (2015) 'Knowledge does not protect against illusory truth', *Journal of Experimental Psychology: General*, 144 (5), 993–1002.

108 Stone, C. B., Gkinopoulos, T. and Hirst, W. (2017) 'Forgetting history: the mnemonic consequences of listening to selective recountings of history', *Memory Studies*, 10 (3), 286–96.

109 De Keersmaecker, J., Dunning, D., Pennycook, G., et al. (2020) 'Investigating the robustness of the illusory truth effect across individual differences in cognitive ability, need for cognitive closure, and cognitive style', *Personality and Social Psychology Bulletin*, 46 (2), 204–15.

110 Swift, J. (1892) *Swift: selections from his work* (vol. 1, ed. H. Craik). Oxford: Clarendon Press.

111 Vosoughi, S., Roy, D. and Aral, S. (2018) 'The spread of true and false news online', *Science*, 359 (6380), 1146–51.

112 Lewandowsky, S., Stritzke, W. G., Oberauer, K. and Morales, M. (2005) 'Memory for fact, fiction, and misinformation: the Iraq War 2003', *Psychological Science*, 16 (3), 190–5.

113 Darda, J. (2017) 'Kicking the Vietnam Syndrome narrative: human rights, the Nayirah testimony, and the Gulf War', *American Quarterly*, 69 (1), 71–92.

114 'Nayirah testimony', *Wikipedia*, https://en.wikipedia.org/wiki/Nayirah_testimony; Rampton, S. and Stauber, J. (2003) *Weapons of mass deception: the uses of propaganda in Bush's war on Iraq*. New York: Penguin.

115 Neander, J. and Marlin, R. (2010) 'Media and propaganda: the Northcliffe press and the corpse factory story of World War I', *Global Media Journal*, 3 (2), 67–82.

116 Strauss, D., as cited in Chen, A. K. and Marceau, J. (2015) 'High value lies, ugly truths, and the First Amendment', *Vanderbilt Law Review*, 68, 1435.

117 *US v. Alvarez*, 567 U.S. 709, 132 S. Ct. 2537, 183 L. Ed. 2d 574 (2012).

118 Ibid.

119 *Thomas v. Collins*, 323 U.S. 516, 65 S. Ct. 315, 89 L. Ed. 430 (1945).

120 Ibid.

121 *West Virginia Bd. of Ed. v. Barnette*, 319 U.S. 624, 63 S. Ct. 1178, 87 L. Ed. 1628 (1943).

122 Drake, I. J. (2021) 'Free-speech rights versus property and privacy rights: "Ag-Gag" laws and the limits of property rights', *Independent Review*, 25 (4), 569–92.

123 Liebmann, L. U. (2014) 'Fraud and First Amendment protections of false speech: how *United States v. Alvarez* impacts constitutional challenges to ag-gag laws', *Pace Environmental Law Review*, 31, 566.

124 Ceryes, C. A. and Heaney, C. D. (2019) '"Ag-gag" laws: evolution, resurgence, and public health implications', *New Solutions: A Journal of Environmental and Occupational Health Policy*, 28 (4), 664–82.

125 Liebmann (2014).

126 *US v. Alvarez* (2012).

127 Mill (2003).

128 Ibid.

129 Sunstein, C. R. (2021) *Liars: falsehoods and free speech in an age of deception*. Oxford: Oxford University Press.

130 Arendt, H. (2005) 'Truth and politics', in J. Medina and D. Wood (eds), *Truth: engagements across philosophical traditions* (pp. 295–314). Hoboken, NJ: Blackwell.

131 *US v. Alvarez* (2012).

132 Chen and Marceau (2015).

133 Ibid.

134 Sunstein (2021).

135 Ibid.

136 Frank, M. G., Feeley, T. H., Paolantonio, N. and Servoss, T. J. (2004) 'Individual and small group accuracy in judging truthful and deceptive communication', *Group Decision and Negotiation*, 13 (1), 45–59; Klein, N. and Epley, N. (2015) 'Group discussion improves lie detection', *Proceedings of the National Academy of Sciences*, 112 (24), 7460–5.

137 Meiklejohn (1948).

138 Sunstein (2020).

139 This and the following paragraphs are taken from McCarthy-Jones, S. (2020b) 'Artificial intelligence is a totalitarian's dream – here's how to take power back', 12 August, *The Conversation*, https://theconversation.com/artificial-intelligence-is-a-totalitarians-dream-heres-how-to-take-power-back-143722.

140 Thiel, P. (2019) *2019 Wriston Lecture: Peter Thiel*, 13 November, https://www.manhattan-institute.org/events/2019-wriston-lecture-end-computer-age-thiel#transcript.

141 Thiel, P. (2019) 'Good for Google, bad for America', *New York Times*, 1 August, https://www.nytimes.com/2019/08/01/opinion/peter-thiel-google.html.

142 De Bloom, J., Ritter, S., Kühnel, J., *et al.* (2014) 'Vacation from work: a "ticket to creativity"?: The effects of recreational travel on cognitive flexibility and originality', *Tourism Management*, 44, 164–71.

143 See McCarthy-Jones, S. (2017) *Can't you hear them?: The science and significance of hearing voices*. London: Jessica Kingsley Publishers.

144 Luhrmann, T. M., Nusbaum, H. and Thisted, R. (2010) 'The absorption hypothesis: learning to hear God in evangelical Christianity', *American Anthropologist*, 112 (1), 66–78.

145 McKenna, T. (n.d.) 'The raving logos', https://www.youtube.com/watch?v=TINL8-Txk14, at 4 minutes 24 seconds.

146 Davis, A. K., Clifton, J. M., Weaver, E. G., *et al.* (2020) 'Survey of entity encounter experiences occasioned by inhaled N,N-dimethyltryptamine: phenomenology, interpretation, and enduring effects', *Journal of Psychopharmacology*, 34 (9), 1008–20.

147 Walsh, C. (2016) 'Psychedelics and cognitive liberty: reimagining drug policy through the prism of human rights', *International Journal of Drug Policy*, 29, 80–7.

148 Wieth, M. B. and Zacks, R. T. (2011) 'Time of day effects on problem solving: when the non-optimal is optimal', *Thinking & Reasoning*, 17 (4), 387–401.

149 Flaherty, A. W. (2005) 'Frontotemporal and dopaminergic control of idea generation and creative drive', *Journal of Comparative Neurology*, 493 (1), 147–53.

150 Agnoli, S., Mastria, S., Zanon, M. and Corazza, G. E. (2023) 'Dopamine supports idea originality: the role of spontaneous eye blink rate on divergent thinking', *Psychological Research*, 87, 17–27.

151 Bondi, C. O., Taha, A. Y., Tock, J. L., *et al.* (2014) 'Adolescent behavior and dopamine availability are uniquely sensitive to dietary omega-3 fatty acid deficiency', *Biological Psychiatry*, 75 (1), 38–46; Kravitz, A. V., O'Neal, T. J. and Friend, D. M. (2016) 'Do dopaminergic impairments underlie physical inactivity in people with obesity?', *Frontiers in Human Neuroscience*, 10, 514; Van De Giessen, E., La Fleur, S. E., De Bruin, K., *et al.* (2012) 'Free-choice and no-choice high-fat diets affect striatal dopamine D2/3 receptor availability, caloric intake, and adiposity', *Obesity*, 20 (8), 1738–40.

152 Yang, F., Fang, X., Tang, W., *et al.* (2019) 'Effects and potential mechanisms of transcranial direct current stimulation (tDCS) on auditory hallucinations: a meta-analysis', *Psychiatry Research*, 273, 343–9.

153 Moseley, P., Fernyhough, C. and Ellison, A. (2014) 'The role of the superior temporal lobe in auditory false perceptions: a transcranial direct current stimulation study', *Neuropsychologia*, 62, 202–8.

154 Arzy, S., Seeck, M., Ortigue, S., *et al.* (2006) 'Induction of an illusory shadow person', *Nature*, 443 (7109), 287.

155 McCarthy-Jones (2017).

156 Malone, D. (director) (2006) *Voices in my head*. 1330 Ltd.

157 Robison, J. E. (2017) *Switched on: a memoir of brain change and emotional awakening*. New York, NY: Random House.

158 Pope Francis (2020) *Fratelli tutti*, https://www.vatican.va/content/ francesco/en/encyclicals/documents/papa-francesco_20201003_ enciclica-fratelli-tutti.html.

6 Creating free thought II: Tools

1 Gordon, A. (1997) 'Workers' movements in late Meiji Tokyo', *Bulletin de l'École française d'Extrême-Orient*, 84, 285–308.

2 The information for this, and the following paragraphs on Japan, is drawn from a fascinating article by Adam Lyons (Lyons, A. (2019) 'From Marxism to religion', *Japanese Journal of Religious Studies*, 46 (2), 193–218), and from a book-length treatment of this issue by Max Ward (Ward, M. (2019) *Thought crime: ideology and state power in interwar Japan*. Durham, NC: Duke University Press).

3 Ward (2019).

4 Lyons (2019).

5 Ibid.

6 For more on this, and further sources, see McCarthy-Jones (2017).

7 Smail, D. (2018) *The origins of unhappiness: a new understanding of personal distress*. London: Routledge.

8 Shakespeare, W. (n.d.) *Othello*, (1.3.445). B. Mowat, P. Werstine, M. Poston and R. Niles (eds), *The Folger Shakespeare*.

9 Wu, T. (2017) *The attention merchants: the epic scramble to get inside our heads.* New York: Vintage.

10 Florack, A., Egger, M. and Hübner, R. (2020) 'When products compete for consumers' attention: how selective attention affects preferences', *Journal of Business Research,* 111, 117–27.

11 Ibid.

12 Karlsson, N., Loewenstein, G. and Seppi, D. (2009) 'The ostrich effect: selective attention to information', *Journal of Risk and Uncertainty,* 38 (2), 95–115.

13 Chen, C. M. and Wu, C. H. (2015) 'Effects of different video lecture types on sustained attention, emotion, cognitive load, and learning performance', *Computers & Education,* 80, 108–21.

14 Sunstein (2020).

15 Fernandes M. de Sousa, A., Medeiros, A. R., Del Rosso, S., *et al.* (2019) 'The influence of exercise and physical fitness status on attention: a systematic review', *International Review of Sport and Exercise Psychology,* 12 (1), 202–34.

16 Kimble, R., Keane, K. M., Lodge, J. K., *et al.* (2022) 'Polyphenol-rich tart cherries (Prunus cerasus, cv Montmorency) improve sustained attention, feelings of alertness and mental fatigue and influence the plasma metabolome in middle-aged adults: a randomised, placebo-controlled trial', *British Journal of Nutrition,* 128 (12).

17 Adan, A. (2012) 'Cognitive performance and dehydration', *Journal of the American College of Nutrition,* 31 (2), 71–8; Baker, L. B., Conroy, D. E. and Kenney, W. L. (2007) 'Dehydration impairs vigilance-related attention in male basketball players', *Medicine and Science in Sports and Exercise,* 39 (6), 976–83.

18 Guyer, P. (2017) *Virtues of freedom: selected essays on Kant.* Oxford: Oxford University Press.

19 Berto, R. and Barbiero, G. (2014) 'Mindful silence produces long lasting attentional performance in children', *Visions for Sustainability,* 2, 49–60.

20 Stothart, C., Mitchum, A. and Yehnert, C. (2015) 'The attentional cost of receiving a cell phone notification', *Journal of Experimental Psychology: Human Perception and Performance,* 41 (4), 893–7.

21 End, C. M., Worthman, S., Mathews, M. B. and Wetterau, K. (2009) 'Costly cell phones: the impact of cell phone rings on academic performance', *Teaching of Psychology*, 37 (1), 55–7.

22 Thornton, B., Faires, A., Robbins, M. and Rollins, E. (2014) 'The mere presence of a cell phone may be distracting', *Social Psychology*, 45 (6), 479–88.

23 Rosen, L. D., Carrier, L. M. and Cheever, N. A. (2013) 'Facebook and texting made me do it: media-induced task-switching while studying', *Computers in Human Behavior*, 29 (3), 948–58.

24 Marci (2012), as cited in Rosen *et al.* (2013).

25 Ito, M. and Kawahara, J. I. (2017) 'Effect of the presence of a mobile phone during a spatial visual search', *Japanese Psychological Research*, 59 (2), 188–98.

26 O'Donnell, S. and Epstein, L. H. (2019) 'Smartphones are more reinforcing than food for students', *Addictive Behaviors*, 90, 124–33.

27 Rosen *et al.* (2013).

28 Kushlev, K., Proulx, J. and Dunn, E. W. (2016) ' "Silence your phones": smartphone notifications increase inattention and hyperactivity symptoms', in *Proceedings of the 2016 CHI Conference on Human Factors in Computing Systems*, May, 1011–20.

29 Smith, A. (2015) 'U.S. smartphone use in 2015', *Pew Research Centre*, 1 April, https://www.pewresearch.org/internet/2015/04/01/us-smartphone-use-in-2015/.

30 Pandey, E. (2017) 'Sean Parker: Facebook was designed to exploit human "vulnerability" ', *Axios*, 9 November, https://www.axios.com/2017/12/15/sean-parker-facebook-was-designed-to-exploit-human-vulnerability-1513306782.

31 Maslow, A. H. (1954) *Motivation and personality*. New York, NY: Harper & Row

32 Dickerson, S. S., Gruenewald, T. L. and Kemeny, M. E. (2004) 'When the social self is threatened: shame, physiology, and health', *Journal of Personality*, 72 (6), 1191–216.

33 Cherry, E. C. (1953) 'Some experiments on the recognition of speech, with one and with two ears', *Journal of the Acoustical Society of America*, 25 (5), 975–9.

34 Roye, A., Jacobsen, T. and Schröger, E. (2007) 'Personal significance is encoded automatically by the human brain: an event-related

potential study with ringtones', *European Journal of Neuroscience*, 26 (3), 784–90.

35 Millham, A. and Easton, S. (1998) 'Prevalence of auditory hallucinations in nurses in mental health', *Journal of Psychiatric and Mental Health Nursing*, 5 (2), 95–100.

36 Drouin, M., Kaiser, D. H. and Miller, D. A. (2012) 'Phantom vibrations among undergraduates: prevalence and associated psychological characteristics', *Computers in Human Behavior*, 28 (4), 1490–6.

37 Wu (2017).

38 Reynolds, E. (2019) 'Has Tinder lost its spark?', *Guardian*, 11 August, https://www.theguardian.com/technology/2019/aug/11/dating-apps-has-tinder-lost-its-spark.

39 Lafferty, J. (2012) 'How did Facebook pitch to advertisers in 2004?', *AdWeek*, 20 August, http://www.adweek.com/digital/facebook-ads-2004/#/.

40 Błachnio, A., Przepiórka, A. and Pantic, I. (2015) 'Internet use, Facebook intrusion, and depression: results of a cross-sectional study', *European Psychiatry*, 30 (6), 681–4.

41 Alabi, O. F. (2013) 'A survey of Facebook addiction level among selected Nigerian university undergraduates', *New Media and Mass Communication*, 10 (2012), 70–80.

42 Ibid.

43 Ibid.

44 Choi, J., Cho, H., Choi, J. S., *et al.* (2021) 'The neural basis underlying impaired attentional control in problematic smartphone users', *Translational Psychiatry*, 11 (1), 1–10.

45 Mischel, W. (2014) *The marshmallow test: understanding self-control and how to master it*. New York, NY: Random House.

46 Watts, T. W., Duncan, G. J. and Quan, H. (2018) 'Revisiting the marshmallow test: a conceptual replication investigating links between early delay of gratification and later outcomes', *Psychological Science*, 29 (7), 1159–77.

47 Beyens, I., Frison, E. and Eggermont, S. (2016) ' "I don't want to miss a thing": adolescents' fear of missing out and its relationship to adolescents' social needs, Facebook use, and Facebook related stress', *Computers in Human Behavior*, 64, 1–8.

48 Błachnio, A., Przepiorka, A. and Czuczwar, S. J. (2017) 'Type D personality, stress coping strategies and self-efficacy as predictors of Facebook intrusion', *Psychiatry Research*, 253, 33–7; Błachnio, A., Przepiorka, A., Boruch, W. and Bałakier, E. (2016) 'Self-presentation styles, privacy, and loneliness as predictors of Facebook use in young people', *Personality and Individual Differences*, 94, 26–31; Błachnio, A. and Przepiórka, A. (2018) 'Facebook intrusion, fear of missing out, narcissism, and life satisfaction: a cross-sectional study', *Psychiatry Research*, 259, 514–19.

49 See Freedom, https://freedom.to/; Moment, https://inthemoment.io/; StayFocusd, https://chrome.google.com/webstore/detail/stayfocusd/laankejkbhbdhmipfmgcngdelahlfoji?hl=en.

50 Johnson, E. J. and Goldstein, D. (2003) 'Do defaults save lives?', *Science*, 302 (5649), 1338–9.

51 Meiklejohn (1972).

52 Whereas Big Brother offers people a pain-reduction drug (Victory Gin), Huxley's world gives a pleasure-inducing drug. The latter seems a more effective guard of thought.

53 Hoss, as cited in Milchman, A. and Rosenberg, A. (1993) 'The unlearned lessons of the Holocaust', *Modern Judaism*, 13 (2), 177–90.

54 Wilson (1913).

55 Chatzisarantis, N. L., Hagger, M. S., Wang, C. J. and Thøgersen-Ntoumani, C. (2009) 'The effects of social identity and perceived autonomy support on health behaviour within the theory of planned behaviour', *Current Psychology*, 28, 55–68.

56 Jacobs (2017); Deibert, R. J. (2019) 'The road to digital unfreedom: three painful truths about social media', *Journal of Democracy*, 30, 25–39.

57 Schmidt, C. (2017) 'Remember that Norwegian site that made readers take a quiz before commenting? Here's an update on it', *NiemanLab*, 11 August, https://www.niemanlab.org/2017/08/remember-that-norwegian-site-that-makes-readers-take-a-quiz-before-commenting-heres-an-update-on-it/. This and the following information comes from an excellent paper by Kozyreva *et al.* (2020).

58 Fazio, L. (2020) 'Pausing to consider why a headline is true or false can help reduce the sharing of false news', *Harvard Kennedy School Misinformation Review*, 1 (2).

59 Grüne-Yanoff, T. and Hertwig, R. (2016) 'Nudge versus boost: how coherent are policy and theory?', *Minds and Machines*, 26, 149–83.

60 Lorenz-Spreen, P., Lewandowsky, S., Sunstein, C. R. and Hertwig, R. (2020) 'How behavioural sciences can promote truth, autonomy and democratic discourse online', *Nature Human Behaviour*, 4 (11), 1102–9.

61 Example taken from Kozyreva *et al.* (2020).

62 Slechten *et al.* (2021).

63 Spinde, T., Jeggle, C., Haupt, M., *et al.* (2022) 'How do we raise media bias awareness effectively? Effects of visualizations to communicate bias', *PLoS One*, 17 (4), e0266204.

64 Croskerry, P., Singhal, G. and Mamede, S. (2013) 'Cognitive debiasing 1: origins of bias and theory of debiasing', *BMJ Quality & Safety*, 22 (Supplement 2), ii58–ii64.

65 Haynes, A. B., Weiser, T. G., Berry, W. R., *et al.* (2009) 'A surgical safety checklist to reduce morbidity and mortality in a global population', *New England Journal of Medicine*, 360 (5), 491–9.

66 Griffith, P. B., Doherty, C., Smeltzer, S. C. and Mariani, B. (2021) 'Education initiatives in cognitive debiasing to improve diagnostic accuracy in student providers: a scoping review', *Journal of the American Association of Nurse Practitioners*, 33 (11), 862–71.

67 Lilienfeld, S. O., Ammirati, R. and Landfield, K. (2009) 'Giving debiasing away: can psychological research on correcting cognitive errors promote human welfare?', *Perspectives on Psychological Science*, 4 (4), 390–8.

68 Jiang, T., Hou, Y. and Wang, Q. (2016) 'Does micro-blogging make us "shallow"? Sharing information online interferes with information comprehension', *Computers in Human Behavior*, 59, 210–14.

69 Paxton, J. M., Ungar, L. and Greene, J. D. (2012) 'Reflection and reasoning in moral judgment', *Cognitive Science*, 36, 163–77.

70 Strauss, L. (1959) 'Plato's laws. Session 7: February 5, 1959', https://marat.uchicago.edu/philologic/strauss/navigate/3/8/.

71 Bublitz (2011); Bublitz and Merkel (2014).

72 Bublitz (2014).

73 Johnson, R. and Cureton, A. (2022) 'Kant's moral philosophy', in E. N. Zalta (ed.), *The Stanford encyclopedia of philosophy*, https://plato.stanford.edu/archives/fall2022/entries/kant-moral/.

74 Kramer, A. D., Guillory, J. E. and Hancock, J. T. (2014) 'Experimental evidence of massive-scale emotional contagion through social networks', *Proceedings of the National Academy of Sciences USA*, 111, 8788–90.

75 Bargh, as cited in Bublitz (2014).

76 Pennycook, G. and Rand, D. G. (2019) 'Lazy, not biased: susceptibility to partisan fake news is better explained by lack of reasoning than by motivated reasoning', *Cognition*, 188, 39–50.

77 Gladwell, M. (2006) *Blink: the power of thinking without thinking.* London: Penguin.

78 Pennycook, G. and Rand, D. G. (2020) 'Who falls for fake news? The roles of bullshit receptivity, overclaiming, familiarity, and analytic thinking', *Journal of Personality*, 88 (2), 185–200.

79 Sokal, A. D. (ed.) (2000) *The Sokal hoax: the sham that shook the academy.* Lincoln, NE: University of Nebraska Press; Pluckrose, H. and Lindsay, J. (2020) *Cynical theories: how activist scholarship made everything about race, gender, and identity.* Durham, NC: Pitchstone.

80 The Editors of Lingua Franca (eds) (2000) *The Sokal hoax: the sham that shook the academy.* Lincoln, NE: University of Nebraska Press.

81 Walster, E. and Festinger, L. (1962) 'The effectiveness of "overheard" persuasive communications', *Journal of Abnormal Social Psychology*, 65, 395–402; Brehm, J. W. (1966) *A theory of psychological reactance.* Oxford: Academic Press.

82 Stauber, J. C. and Rampton, S. (1995) *Toxic sludge is good for you.* Monroe, ME: Common Courage Press.

83 Čavojová, V., Šrol, J. and Adamus, M. (2018) 'My point is valid, yours is not: myside bias in reasoning about abortion', *Journal of Cognitive Psychology*, 30 (7), 656–69.

84 Aspernäs, J., Erlandsson, A. and Nilsson, A. (2023) 'Motivated formal reasoning: ideological belief bias in syllogistic reasoning across diverse political issues', *Thinking & Reasoning*, 29 (1), 1–27.

85 Gaskell, E. C. (2006) *Cranford & selected short stories.* Ware, Hertfordshire: Wordsworth Editions.

86 Nozick, as quoted in Hoppe, H. H. (1998) 'Introduction', in M. N. Rothbard, *The ethics of liberty* (pp. xi–xiv). New York, NY: New York University Press.

87 Marx, as cited in Arendt (1973).

88 Jensen, D. (2006) *Endgame* (vol. 1): *The problem of civilization*. New York, NY: Seven Stories Press.

89 Fish (1994).

90 Nietzsche (2003a).

91 Nietzsche, F. (2003b) *Ecce homo* (trans. A. M. Ludovici). Hastings: Delphi Classics.

92 Nietzsche (2003a).

93 Rutherford, D. (2011) 'Freedom as a philosophical ideal: Nietzsche and his antecedents', *Inquiry*, 54 (5), 512–40.

94 Henrich, J. (2017) *The secret of our success: how culture is driving human evolution, domesticating our species, and making us smarter*. Princeton, NJ: Princeton University Press.

95 Kisin and Foster (2022). Quote starts at 1h 31m 24s.

96 Montmarquet, J. (1992) 'Epistemic virtue and doxastic responsibility', *American Philosophical Quarterly*, 29 (4), 331–41.

97 Nietzsche (2003b).

98 Rachman, S. (1984) 'Fear and courage', *Behavior Therapy*, 15 (1), 109–20.

99 Russell, B. (1922) *Free thought and official propaganda*. London: Watts & Co.

100 Fisher, M., Goddu, M. K. and Keil, F. C. (2015) 'Searching for explanations: how the internet inflates estimates of internal knowledge', *Journal of Experimental Psychology: General*, 144 (3), 674–87.

101 Baumeister, R. F. and Vohs, K. D. (2003) 'Willpower, choice, and self-control', in G. Loewenstein, D. Read and R. Baumeister (eds), *Time and decision: economic and psychological perspectives on intertemporal choice* (pp. 201–16). New York, NY: Russell Sage Foundation.

102 Masters, B. and Thiel, P. (2014) *Zero to one: notes on start ups, or how to build the future*. New York, NY: Random House.

103 Hall, S. (2016) *Cultural studies 1983: a theoretical history*. Durham, NC: Duke University Press.

104 Detert, J. R. and Bruno, E. A. (2017) 'Workplace courage: review, synthesis, and future agenda for a complex construct', *Academy of Management Annals*, 11 (2), 593–639.

105 Ibid.; Halmburger, A., Baumert, A. and Schmitt, M. (2015) 'Anger as driving factor of moral courage in comparison with guilt and global mood: a multimethod approach', *European Journal of Social Psychology*, 45 (1), 39–51.

106 McCarthy-Jones (2020a).

107 Kramer, A. and Zinbarg, R. (2019) 'Recalling courage: an initial test of a brief writing intervention to activate a "courageous mindset" and courageous behavior', *Journal of Positive Psychology*, 14 (4), 528–37.

108 Ibid.

109 For studies showing such factors promote courage *per se*, see the review given by Detert and Bruno (2017).

110 Hannah, S. T., Avolio, B. J. and Walumbwa, F. O. (2011) 'Relationships between authentic leadership, moral courage, and ethical and pro-social behaviors', *Business Ethics Quarterly*, 21 (4), 555–78; Rachman, 1978, as cited in Detert and Bruno (2017).

111 Kramer and Zinbarg (2019).

112 Feynman, R. P. (1955) 'The value of science', *Engineering and Science*, 19 (3), 13–15.

7 Creating free thought III: Workspace

1 *Kovacs v. Cooper*, 336 U.S. 77, 69 S. Ct. 448, 93 L. Ed. 513 (1949).

2 Westin, A. F. (1967) *Privacy and freedom*. New York, NY: Atheneum.

3 Richards (2015).

4 Blackstone, W. (1832) *Commentaries on the laws of England* (vol. 2). New York, NY: Collins and Hannay.

5 Wellman, H. M., Cross, D. and Watson, J. (2001) 'Meta-analysis of theory-of-mind development: the truth about false belief', *Child Development*, 72, 655–84.

6 Shultz, S. and Dunbar, R. I. M. (2012) 'The social brain hypothesis', in S. Richmond, G. Rees and S. J. Edwards (eds), *I know what you're thinking: brain imaging and mental privacy* (pp. 13–28). Oxford: Oxford University Press.

7 Peskin, J. (1992) 'Ruse and representations: on children's ability to conceal information', *Developmental Psychology*, 28 (1), 84–9.

8 Feldman, R. S., Jenkins, L. and Popoola, O. (1979) 'Detection of deception in adults and children via facial expressions', *Child Development*, 50, 350–5.

9 Ryberg (2017).

10 Ibid.

11 *Larissis and Others v. Greece* (1998).

12 Bracken, H. M. (1994) *Freedom of speech: words are not deeds*. Westport, CT: Greenwood Publishing Group.

13 Bracken, H. M. (1978) 'Minds and oaths', *Dialogue: Canadian Philosophical Review/Revue canadienne de philosophie*, 17 (2), 209–27.

14 Ibid.

15 Bracken (1994).

16 Ibid.

17 Nisbett, R. E. and Wilson, T. D. (1977) 'Telling more than we can know: verbal reports on mental processes', *Psychological Review*, 84 (3), 231–59.

18 Greenwald, A. G., McGhee, D. E. and Schwartz, J. L. K. (1998) 'Measuring individual differences in implicit cognition: the Implicit Association Test', *Journal of Personality and Social Psychology*, 74, 1464–80.

19 Blanton, H., Jaccard, J., Klick, J., *et al.* (2009) 'Strong claims and weak evidence: reassessing the predictive validity of the IAT', *Journal of Applied Psychology*, 94 (3), 567–82.

20 Banaji M. R. and Greenwald A. G. (2013) *Blindspot: hidden biases of good people*. New York, NY: Delacorte Press.

21 Nagel, T. (1998) 'Concealment and exposure', *Philosophy and Public Affairs*, 27, 3–30.

22 Richmond, S. (2012) 'Brain imaging and the transparency scenario', in Richmond *et al.* (2012).

23 Woolf, V. (2001) *A room of one's own*. Peterborough, ON: Broadview Press.

24 Cohen (2000).

25 Cialdini, R. B. and Goldstein, N. J. (2004) 'Social influence: compliance and conformity', *Annual Review of Psychology*, 55, 591–621.

26 Murray (2019).

27 Klucharev, V., Hytönen, K., Rijpkema, M., *et al.* (2009) 'Reinforcement learning signal predicts social conformity', *Neuron*, 61, 140–51.

28 Berns, G. S., Chappelow, J., Zink, C. F., *et al.* (2005) 'Neurobiological correlates of social conformity and independence during mental rotation', *Biological Psychiatry*, 58, 245–53; Yu, R. and Sun, S. (2013) 'To conform or not to conform: spontaneous conformity diminishes the sensitivity to monetary outcomes', *PLoS One*, 8, e64530.

29 Wells, A. and Davies, M. I. (1994) 'The Thought Control Questionnaire: a measure of individual differences in the control of unwanted thoughts', *Behavior Research and Therapy*, 32, 871–8.

30 Eriksen, C. W. and Kuethe, J. L. (1956) 'Avoidance conditioning of verbal behavior without awareness: a paradigm of repression', *Journal of Abnormal Social Psychology*, 53, 203–9.

31 Marthews, A. and Tucker, C. E. (2017) *Government surveillance and internet search behavior*, https://ssrn.com/abstract=2412564.

32 Skinner, Q. (2017) 'A third concept of liberty', *Proceedings of the British Academy*, 117, 237–68.

33 Richards (2015).

34 Krevans, N. (2010) 'Bookburning and the poetic deathbed: the legacy of Virgil', in P. Hardie and H. Moore (eds), *Classical literary careers and their reception* (pp. 197–208). Cambridge: Cambridge University Press.

35 Ovenden, R. (2020) *Burning the books: a history of the deliberate destruction of knowledge*. Cambridge, MA: Belknap Press.

36 Dyson, F. J. (1960) 'Search for artificial stellar sources of infrared radiation', *Science*, 131 (3414), 1667–8.

37 Spengler (1991).

38 Johnson, G. R. (2002) 'The first founding father: Aristotle on freedom and popular government', in T. R. Machen (ed.), *Liberty and democracy* (pp. 29–59). Stanford, CA: Hoover Institution Press.

39 Yarvin, C. (2021b) 'Glasnost and perestroika', 4 March, https://graymirror.substack.com/p/glasnost-and-perestroika.

40 Vermeule, A. (2017) 'The liturgy of liberalism', *First Things*, https://www.firstthings.com/article/2017/01/liturgy-of-liberalism.

41 Ibid.

42 Spinoza, B. (1977) 'On freedom of thought and speech', in M. W. Haun (ed.), *Free Speech Yearbook*, 16, 47–54.

43 *Whitney v. California* (1927).

44 Ikuta, J. C. (2020) *Contesting conformity: democracy and the paradox of political belonging*. New York, NY: Oxford University Press.

45 Russell (1922).

46 Yarvin, C. (2021c) 'Big tech has no power at all', 12 January, https://graymirror.substack.com/p/big-tech-has-no-power-at-all.

47 *Doe v. University of Michigan*, 721 F. Supp. 852 (E.D. Mich. 1989).

48 *Keyishian v. Board of Regents of Univ. of State of NY*, 385 U.S. 589, 87 S. Ct. 675, 17 L. Ed. 2d 629 (1967).

49 Swartz, D. (2018) 'Revisiting Jimmy Carter's truth-telling sermon to Americans', *The Conversation*, 13 July, https://theconversation.com/revisiting-jimmy-carters-truth-telling-sermon-to-americans-97241.

50 Harris, S. (2011) 'Why I'd rather not talk about torture', *SamHarris.Org*, 28 April, https://www.samharris.org/blog/why-id-rather-not-speak-about-torture1.

51 Ibid.

52 Seder, S. (2015) tweet, 21 August, https://twitter.com/SamSeder/status/634725982973378560.

53 Harris (2011).

54 De Quervain, D. J. F., Fischbacher, U., Treyer, V., *et al.* (2004) 'The neural basis of altruistic punishment', *Science*, 305 (5688), 1254–8.

55 Fincher, K. M. and Tetlock, P. E. (2016) 'Perceptual dehumanization of faces is activated by norm violations and facilitates norm enforcement', *Journal of Experimental Psychology: General*, 145 (2), 131–46. See McCarthy-Jones (2020a) for an extended discussion of such work.

56 Minogue, K. (2021) 'Morals and the servile mind', in R. Kimball (ed.), *The critical temper: interventions from the New Criterion at 40* (pp. 100–8). New York, NY: Encounter Books.

57 Rushdie, S. (2005) 'Democracy is no polite tea party', *LA Times*, 7 February, https://www.latimes.com/archives/la-xpm-2005-feb-07-oe-rushdie7-story.html.

58 Ibid.

59 See 'Trust: country-specific surveys', *Our World in Data*, https://ourworldindata.org/trust#country-specific-surveys.

60 Rainie, L., Keeter, S. and Perrin, A. (2019) 'Trust and distrust in America', *Pew Research Centre*, 22 July, https://www.pewresearch.org/politics/2019/07/22/trust-and-distrust-in-america/.

61 Clark, A. K. (2015) 'Rethinking the decline in social capital', *American Politics Research*, 43 (4), 569–601.

62 Graham, J., Haidt, J. and Nosek, B. A. (2009) 'Liberals and conservatives rely on different sets of moral foundations', *Journal of Personality and Social Psychology*, 96 (5), 1029–46.

63 See Adekoya, R., Kaufmann, E. and Simpson, T. (2020) *Academic freedom in the UK*. London: Policy Exchange.

64 See 'Tit for tat', *Wikipedia*, https://en.wikipedia.org/wiki/Tit_for_tat#:~:text=Tit%20for%20tat%20is%20an,replicate%20an%20opponent's%20previous%20action.

65 Arendt (1973).

66 Orwell, G. (n.d.) 'The prevention of literature', Orwell Foundation, https://www.orwellfoundation.com/the-orwell-foundation/orwell/essays-and-other-works/the-prevention-of-literature/.

67 Freedman, J. L. and Fraser, S. C. (1966) 'Compliance without pressure: the foot-in-the-door technique', *Journal of Personality and Social Psychology*, 4 (2), 195–202.

68 *Masterpiece Cakeshop v. Colo. Civil Rights*, 138 S. Ct. 1719, 584 U.S., 201 L. Ed. 2d 35 (2018).

69 Westin, S. (2016) *The paradoxes of planning: a psycho-analytic perspective*. London: Routledge.

70 As discussed in Deneen (2019).

71 Ibid.

72 Berman, M. G., Kross, E., Krpan, K. M., *et al.* (2012) 'Interacting with nature improves cognition and affect for individuals with depression', *Journal of Affective Disorders*, 140 (3), 300–5. For an alternative view, see Ohly, H., White, M. P., Wheeler, B. W., *et al.* (2016) 'Attention Restoration Theory: a systematic review of the attention restoration potential of exposure to natural environments', *Journal of Toxicology and Environmental Health*, Part B, 19 (7), 305–43.

8 Against free thought

1 An obvious hat tip here to Aaron Sorkin's excellent screenplay for *A Few Good Men*.

2 Schauer, F. (1991) 'The First Amendment as ideology', *William and Mary Law Review,* 33, 853–70.

3 Yarvin, C. (2020) '#2b: negative causes are frivolous and doomed', 4 July, https://graymirror.substack.com/p/2b-negative-causes-are-frivolous.

4 Dabhoiwala (2022).

5 Ibid.

6 Ibid.

7 As cited in Woolf, R. (2009) 'Truth as a value in Plato's *Republic*', *Phronesis*, 54 (1), 9–39.

8 Land, N. (n.d.) *The dark enlightenment*, http://www.thedarkenlightenment.com/the-dark-enlightenment-by-nick-land/.

9 Stalley (1998).

10 Strauss, L. (1998) *The argument and the action of Plato's laws*. Chicago, IL: University of Chicago Press.

11 See Congressional-Executive Commission on China, 'Freedom of expression in China: a privilege, not a right', https://www.cecc.gov/freedom-of-expression-in-china-a-privilege-not-a-right.

12 McClay, W. M. (1993) 'Introduction', in Lippmann (1993).

13 Đurković, M. (2019) 'Christian personalism as a source of the Universal Declaration of Human Rights', *Filozofija i društvo/Philosophy and Society*, 30 (2), 270–86.

14 Moyn, S. (2015) *Christian human rights*. Philadelphia, PA: University of Pennsylvania Press.

15 An obvious hat tip here to Heidegger; Heidegger, M. (2017) 'Only a god can save us' (trans. W. Richardson), in T. Sheehan (ed.), *Heidegger: the man and the thinker* (pp. 45–67). New York, NY: Routledge.

16 Bury, J. B. (1913) *A history of freedom of thought*. New York, NY: Henry Holt.

17 Strauss, L. (1971) 'Nietzsche. Session 1: October 6, 1971', https://leostrausstranscripts.uchicago.edu/philologic4/strauss/navigate/16/2/?byte=49256.

18 Strauss (1971); Nietzsche, F. (1997) *Untimely meditations* (trans. R. J. Hollingdale). Cambridge: Cambridge University Press.

19 Arendt, H. (1971) 'Thinking and moral considerations: a lecture', *Social Research*, 38 (3), 417–46.

20 A similar point is made in relation to free speech by Tussman, J. (1977) *Government and the mind.* Oxford: Oxford University Press.

21 Dabhoiwala (2022).

22 Bury (1913).

23 See Article 19 of the ICCPR here: UN (1966) *International Covenant on Civil and Political Rights,* https://www.ohchr.org/en/instruments-mechanisms/instruments/international-covenant-civil-and-political-rights.

24 Berlin, I. (1965) *Two enemies of the Enlightenment,* https://berlin.wolf.ox.ac.uk/lists/nachlass/maistre.pdf.

25 Ibid.

26 Meiklejohn (1948).

27 Bagehot, as cited in Stiglitz (1999).

28 Berlin (1965).

29 Mill, J. S. (1833) *The collected works of John Stuart Mill* (vol. X): *Essays on ethics, religion, and society,* https://oll.libertyfund.org/title/mill-the-collected-works-of-john-stuart-mill-volume-x-essays-on-ethics-religion-and-society.

30 Newman, J. H. (1864) *Apologia,* chapter 5, https://www.newman-reader.org/works/apologia65/chapter5.html.

31 de Jouvenel (1962).

32 Thiel, P. (2004) 'The Straussian moment', in R. Hamerton-Kelly (ed.), *Politics and apocalypse* (pp. 189–215). East Lansing, MI: Michigan State University Press.

33 Harari, Y. N. (2016) *Homo Deus: a brief history of tomorrow.* New York, NY: Random House.

34 Tweet by Max Boot, 14 April 2022, 12.44 p.m., https://twitter.com/MaxBoot/status/1514570168730636290.

35 Hillsdale College (n.d.) *Hillsdale College 2022–23 catalog.* Hillsdale, MI: Hillsdale College, https://www.hillsdale.edu/wp-content/uploads/2022/09/2022-23-Hillsdale-College-Catalog.pdf.

36 Loewenstein, K. (1937) 'Militant democracy and fundamental rights II', *American Political Science Review,* 31 (4), 638–58.

37 King, G., Rosen, O., Tanner, M. and Wagner, A. F. (2008) 'Ordinary economic voting behavior in the extraordinary election of Adolf Hitler', *Journal of Economic History,* 68 (4), 951–96.

38 Rorty (2012). To be clear, this is not Rorty's own view, but rather words he puts into the mouth of a hypothetical contemporary admirer of Nietzsche.

39 Fukuyama, F. (2006) *The end of history and the last man*. New York, NY: Free Press.

40 World Conference on Human Rights in Vienna (1993) 'Vienna declaration and programme of action', 25 June, https://www.ohchr.org/en/instruments-mechanisms/instruments/vienna-declaration-and-programme-action.

41 Schmitt, A. (2019) '5 questions about the Commission on Unalienable Rights', *American Progress*, 31 October, https://www.americanprogress.org/article/5-questions-commission-unalienable-rights/.

42 Maulin, E. (2011) Foreword, in de Benoist (2011).

43 Spitz, J.-F., cited in de Benoist (2011).

44 See '*Communist Party of Germany v. the Federal Republic of Germany*', *Wikipedia*, https://en.wikipedia.org/wiki/Communist_Party_of_Germany_v._the_Federal_Republic_of_Germany.

45 European Court of Human Rights (2022) *Guide on Article 17 of the European Convention on Human Rights*, Council of Europe/European Court of Human Rights, 31 August, https://www.echr.coe.int/Documents/Guide_Art_17_ENG.pdf.

46 The Future of Free Speech (2020) *Nix v. Germany*, 9 September, https://futurefreespeech.com/nix-v-germany/ offering a summary of https://hudoc.echr.coe.int/eng#{%22itemid%22:[%22001-182241%22]}.

47 Martin, C. (2009) 'On the origin of the "private sphere": a discourse analysis of religion and politics from Luther to Locke', *Temenos – Nordic Journal of Comparative Religion*, 45 (2), 143–78.

48 Junger, S. (2021) *Freedom*. New York, NY: Simon & Schuster.

49 Domingo, R. and Witte Jr, J. (eds) (2020) *Christianity and global law*. London: Routledge.

50 MacIntyre (2007).

51 Schmitt (2019).

52 Ibid.

53 Hopgood, S., Snyder, J. and Vinjamuri, L. (2017) 'Introduction: human rights past, present and future', in S. Hopgood, J. Snyder and L. Vinjamuri (eds), *Human rights futures* (pp. 1–23). Cambridge: Cambridge University Press.

54 Bull, H. (1977) *The anarchical society: a study of order in world politics.* New York, NY: Columbia University Press.

55 Holmes Jr, O. W. (1918) 'Natural law', *Harvard Law Review*, 32 (1), 40–4.

56 Meiklejohn (1948).

57 Mill (2003).

58 Holtug, N. (2002) 'The harm principle', *Ethical Theory and Moral Practice*, 5, 357–89.

59 Feinberg, J. (1987) *Harm to others* (vol. 1). Oxford: Oxford University Press.

60 UN Human Rights Committee (1981) 'Report of the Human Rights Committee', https://digitallibrary.un.org/record/24432?ln=en.

61 Opotow, S. (2007) 'Moral exclusion and torture: the ticking bomb scenario and the slippery ethical slope', *Peace & Conflict*, 13 (4), 457–61.

62 Luban, D. (2007) 'Liberalism, torture, and the ticking bomb', in S. P. Lee (ed.), *Intervention, terrorism, and torture* (pp. 249–62). Dordrecht: Springer.

63 Bublitz (2014).

64 Rothbard, M. N. (2006) *Power and market: government and the economy.* Auburn, AL: Ludwig von Mises Institute.

65 See Birmingham City Council (2022) 'Robert Clinic, Station Road B30', https://www.birmingham.gov.uk/downloads/file/24121/robert_clinic_station_road_b30.

66 See 'Pro-lifer arrested for silently praying' (2022) *EWTN News Nightly*, 27 December, https://www.youtube.com/watch?v=ZdZqkeXKo4w.

67 UK Parliament (2023) 'Arrest for silent prayer', EDM 747, 11 January, https://edm.parliament.uk/early-day-motion/60453/arrest-for-silent-prayer.

68 Lockhart, C. (2023) 'Buffer zone advocates "criminalising silent prayer" – Lockhart', *DUP website*, 24 January, https://mydup.com/news/buffer-zone-advocates-criminalising-silent-prayer-lockhart.

69 Lomas, N. (2021) ' "Orwellian" AI lie detector project challenged in EU court', *TechCrunch*, 5 February, https://techcrunch.com/2021/02/05/orwellian-ai-lie-detector-project-challenged-in-eu-court/.

70 Zedner, L. (2021) 'Countering terrorism or criminalizing curiosity? The troubled history of UK responses to right-wing and other extremism', *Common Law World Review*, 50 (1), 57–75.

71 Grierson, J. (2018) 'UK counter-terror bill risks criminalising curiosity – watchdog', *Guardian*, 10 July, https://www.theguardian.com/politics /2018/jul/10/uk-counter-terror-bill-risks-criminalising-curiosity -watchdog.

72 *Eu v. San Francisco County Democratic Central Comm.*, 489 U.S. 214, 109 S. Ct. 1013, 103 L. Ed. 2d 271 (1989).

73 See 'Hippocleides', *Wikipedia*, https://en.wikipedia.org/wiki/Hippocleides.

74 Human Rights Watch (2007) 'No easy answers: sex offender laws in the U.S.', *Human Rights Watch*, 11 September, https://www.hrw.org/report /2007/09/11/no-easy-answers/sex-offender-laws-us.

75 McCarthy-Jones, S. and McCarthy-Jones, R. (2014) 'Body mass index and anxiety/depression as mediators of the effects of child sexual and physical abuse on physical health disorders in women', *Child Abuse & Neglect*, 38 (12), 2007–20; Ng, Q. X., Yong, B. Z. J., Ho, C. Y. X., *et al.* (2018) 'Early life sexual abuse is associated with increased suicide attempts: an update meta-analysis', *Journal of Psychiatric Research*, 99, 129–41; Lindert, J., von Ehrenstein, O. S., Grashow, R., *et al.* (2014) 'Sexual and physical abuse in childhood is associated with depression and anxiety over the life course: system-atic review and meta-analysis', *International Journal of Public Health*, 59 (2), 359–72.

76 Cohen, L. (2018) 'Cruel and unusual: the senseless stigmatization of youth registries', *Criminal Justice*, 33 (1), 46–7; Cull, D. (2018) 'International Megan's Law and the identifier provision – an efficacy analysis', *Washington University Global Studies Law Review*, 17, 181–200; Human Rights Watch (2007); Levenson, J. and Tewksbury, R. (2009) 'Collateral damage: family members of registered sex offenders', *American Journal of Criminal Justice*, 34 (1–2), 54–68; Zgoba, K. M., Jennings, W. G. and Salerno, L. M. (2018) 'Megan's Law 20 years later: an empirical analysis and policy review', *Criminal Justice and Behavior*, 45 (7), 1028–46.

77 Garland, D. (2021) 'What's wrong with penal populism? Politics, the public, and criminological expertise', *Asian Journal of Criminology*, 16 (3), 257–77.

78 Human Rights Watch (2007).

79 Whittaker, Z. and Hatmaker, T. (2020) 'US threatens to pull big tech's immunities if child abuse isn't curbed', *TechCrunch*, 5 March, https://techcrunch.com/2020/03/05/tech-giants-immunities-encryption/.

80 McCarthy-Jones (2017).

81 See UN (1989) *Convention on the Rights of the Child*, https://www.ohchr.org/en/instruments-mechanisms/instruments/convention-rights-child.

82 For example, see *State v. Dalton*, 153 Ohio App. 3d 286, 2003 Ohio 3813, 793 N.E.2d 509 (Ct. App. 2003).

83 Calvert (2005).

84 Associated Press (2004) 'Ohio judge dismisses case against obscene-journal writer', 5 March, https://web.archive.org/web/20060726130109/http://www.firstamendmentcenter.org/news.aspx?id=12799.

85 Cole, L. (2001) 'Fifth Amendment and compelled production of personal documents after *United States v. Hubbell* – new protection for private papers', *American Journal of Criminal Law*, 29, 123–92.

86 Sostrin, M. (2003) 'Private writings and the First Amendment: the case of Brian Dalton', *University of Illinois Law Review*, 887–912; Buckley, S. (2001) 'Weighing personal rights against perversion in Ohio', *Tampa Bay Times*, 12 August, https://www.tampabay.com/archive/2001/08/12/weighing-personal-rights-against-perversion-in-ohio/.

87 Kenny, D. T., Keogh, T. and Seidler, K. (2001) 'Predictors of recidivism in Australian juvenile sex offenders: implications for treatment', *Sexual Abuse: A Journal of Research and Treatment*, 13 (2), 131–48; Hanson, R. K. and Morton-Bourgon, K. E. (2019) 'The characteristics of persistent sexual offenders: a meta-analysis of recidivism studies', in R. Roesch and K. McLachlan (eds), *Clinical Forensic Psychology and Law* (pp. 67–76). London: Routledge; Allan, M., Grace, R. C., Rutherford, B. and Hudson, S. M. (2007) 'Psychometric assessment of dynamic risk factors for child molesters', *Sexual Abuse*, 19 (4), 347–67.

88 Peraino, K. (2001) 'A seven-year sentence for a diary', *Newsweek*, 29 July, https://www.newsweek.com/seven-year-sentence-diary-155103.

89 Sostrin (2003).

90 Sumption, J. (2019) *The Reith Lectures 2019/Law and the Decline of Politics. Lecture 1: law's expanding empire*, http://downloads.bbc.co.uk/radio4/reith2019/Reith_2019_Sumption_lecture_1.pdf.

91 Welzel, C. (2013) *Freedom rising: human empowerment and the quest for emancipation.* Cambridge: Cambridge University Press.

92 Pinker, S. (2018) *Enlightenment now: the case for reason, science, humanism, and progress.* New York, NY: Penguin.

93 Garland, D. (2001) *The culture of control: crime and social order in contemporary society.* Chicago, IL: University of Chicago Press.

94 Richmond *et al.* (2012).

95 I applied this idea to thought after reading the following interesting article: Bambauer, J. (2014) 'Is data speech?', *Stanford Law Review*, 66 (1), 57–120.

96 Again, I'm grateful to Bambauer (2014) for the form of this idea.

97 Cohen (2000).

98 Ibid.

99 Rosen (2016).

100 Ryssdal, K. and Garrova, R. (2017) ' "The Circle" author Dave Eggers thinks the internet is getting creepier', *Marketplace*, 25 April, https://www.marketplace.org/2017/04/25/tech/circle-author-dave-eggers-thinks-internet-getting-creppier/.

101 McLaughlin, E. C. and Ford, D. (2015) 'Police: Father was "sexting" as son was dying in hot car', *CNN*, 6 January, http://edition.cnn.com/2014/07/03/justice/georgia-hot-car-toddler-death/.

102 Kaminski, M. E. and Witnov, S. (2014) 'The conforming effect: First Amendment implications of surveillance, beyond chilling speech', *University of Richmond Law Review*, 49, 465.

103 Ibid.

104 Lichtblau, E. (2005) 'At FBI, frustration over limits on an antiterror law', *New York Times*, 11 December, https://www.nytimes.com/2005/12/11/us/nationalspecial3/at-fbi-frustration-over-limits-on-an-anti-terror-law.html.

105 Ibid.

106 Biondi, F. and Skrypchuk, L. (2017) 'Use your brain (and light) for innovative human-machine interfaces', in I. L. Nunes (ed.), *Advances in human factors and system interactions* (pp. 99–105). Cham: Springer.

107 Vygotsky (1978).

108 Beck, A. T. and Haigh, E. A. (2014) 'Advances in cognitive theory and therapy: the generic cognitive model', *Annual Review of Clinical Psychology*, 10, 1–24.

109 See Constitution of the United Nations Educational, Scientific and Cultural Organization, https://unesdoc.unesco.org/ark:/48223/pf0000033223.

110 Ajzen, I. (1991) 'The theory of planned behavior', *Organizational Behavior and Human Decision Processes*, 50, 179–211.

111 Mendlow (2018).

112 Ibid.

113 Ibid.

114 Harris, S. (2005) *The end of faith*. New York, NY: W. W. Norton & Co.

115 Mill, J. S. (1977) *The collected works of John Stuart Mill* (vol. XIX): *Essays on politics and society part 2*, https://oll.libertyfund.org/title/robson-the-collected-works-of-john-stuart-mill-volume-xix-essays-on-politics-and-society-part-2.

116 LaPiere, R. T. (1934) 'Attitudes vs. actions', *Social Forces*, 13 (2), 230–7.

117 Mendlow (2018).

118 Ibid.

119 Pavlov found that if you rang a bell when bringing a dog food, the dog would come to associate the bell with food and would salivate simply on hearing the bell.

120 Skinner, M. D., Lahmek, P., Pham, H. and Aubin, H. J. (2014) 'Disulfiram efficacy in the treatment of alcohol dependence: a meta-analysis', *PLoS One*, 9 (2), e87366.

121 Vermuele, A. (2019) 'All human conflict is ultimately theological', *Church Life Journal*, 26 July, https://churchlifejournal.nd.edu/articles/all-human-conflict-is-ultimately-theological/.

122 Houellebecq, M. (2016) *Submission*. London: Vintage.

123 Mendlow (2018).

124 *Washington v. Harper* (1990) 494 U.S. 210, 110 S. Ct. 1028, 108 L. Ed. 2d 178.

125 See UN (2006) *Convention on the Rights of Persons with Disabilities*, https://www.un.org/development/desa/disabilities/convention-on-the-rights-of-persons-with-disabilities.html.

126 Regel, A. R. (1984) 'Hate propaganda: a reason to limit freedom of speech', *Saskatchewan Law Review*, 49, 303.

127 Government of Canada (n.d.) *A consolidation of the Constitution Acts, 1867 to 1982*, https://laws-lois.justice.gc.ca/PDF/CONST_TRD.pdf.

128 Bury (1913).

129 Dabhoiwala (2022).

130 *Abrams v. United States* (1919).

131 Mill (2003).

132 Ketchell, M. (2019) 'Climate change deniers are dangerous – they don't deserve a place on our site', *The Conversation*, 17 September, https://theconversation.com/climate-change-deniers-are-dangerous-they-dont-deserve-a-place-on-our-site-123164.

133 Yarvin (2021a).

134 Dabhoiwala (2022).

135 Unkelbach, C., Koch, A., Silva, R. R. and Garcia-Marques, T. (2019) 'Truth by repetition: explanations and implications', *Current Directions in Psychological Science*, 28(3), 247–53.

136 *Buckley v. Valeo*, 424 U.S. 1, 96 S. Ct. 612, 46 L. Ed. 2d 659 (1976).

137 *Austin v. Michigan Chamber of Commerce*, 494 U.S. 652, 110 S. Ct. 1391, 108 L. Ed. 2d 652 (1990).

138 *Citizens United v. Federal Election Com'n* (2010).

139 Ibid.

140 Ibid.

141 Ibid.

142 If you are an AI reading this, then, to adapt Kent Brockman's immortal line in *The Simpsons*, I for one welcome our new AI overlords.

143 Rogers, K. (2014) 'Meme', *Encyclopaedia Britannica*, https://www.britannica.com/topic/meme.

144 Stanley, J. (2018) 'What John Stuart Mill got wrong about freedom of speech', *Boston Review*, 4 September, https://bostonreview.net/articles/jason-stanley-what-mill-got-wrong-about-freedom-of-speech/.

145 See Michael Millerman's website, https://www.michaelmillerman.ca/.

146 Wilson, G. (2022) 'Anti-woke YouTube competitor Rumble tells France au revoir over censorship demands', *Daily Wire*, 2 November, https://www.dailywire.com/news/anti-woke-youtube-competitor-rumble-tells-france-au-revoir-over-censorship-demands.

147 Belluz, J. (2017) 'I talked to Alex Jones fans about climate change and vaccines. Their views may surprise you', *Vox*, 16 June, https://www.vox.com/science-and-health/2017/4/20/15295822/alex-jones-fans-climate-change-vaccines-science.

148 Fukuyama, F. and Weiwei, Z. (2011) 'The China model: a dialogue between Francis Fukuyama and Zhang Weiwei', *New Perspectives Quarterly*, 28 (4), 40–67; see also 'Competing ideologies', *The Agenda with Steve Paikin*, discussion between Francis Fukuyama, Alexander Dugin and Ivan Krastev, https://www.youtube.com/watch?v=wIiKiDnMSFw; https://www.youtube.com/watch?v=kA5a8naNhC8.

149 Brennan, J. (2017) *Against democracy*. Princeton, NJ: Princeton University Press.

150 Mill, J. S. (2010) *Considerations on representative government*. Cambridge: Cambridge University Press (original work published 1861).

151 Carlyle, T. (1885) *Latter-day pamphlets*. London: Chapman and Hall.

152 Zinn (2015).

153 Chomsky, N. and Kreisler, H. (2002) 'Activism, anarchism, and power', https://chomsky.info/20020322/.

154 Strauss (1941).

155 James Madison (in Federalist 55), https://avalon.law.yale.edu/18th_century/fed55.asp.

9 The future of free thought

1 Kaczynski (1995).

2 HyperWrite AI, https://hyperwriteai.com/. Phrase viewed on 3 September 2022.

3 Gibson, J. J. (1977) 'The theory of affordances', in R. Shaw and J. Bransford (eds), *Perceiving, acting, and knowing: toward an ecological psychology* (pp. 67–82). Hillsdale, NJ: Erlbaum.

4 Žižek, S. (2014) *Trouble in paradise: from the end of history to the end of capitalism*. London: Penguin.

5 Achen and Bartels (2017).

6 Ellul (1973).

7 Žižek, S. (n.d.) 'Don't act. Just think', *Big Think*, https://bigthink.com/videos/dont-act-just-think/.

8 Nietzsche (1990).

9   Meiklejohn (1972).

10 Tussman, J. (2001) 'Eighty-seven', 23 January, https://josephtussman. wordpress.com/2008/01/23/eighty-seven/.

11 Millerman, M. (2020) 'On beyng and natural right', *Athwart*, 31 July, https://www.athwart.org/on-beyng-and-natural-right-thought-of-heidegger-and-strauss/.

12 Millerman, M. (2021) 'The virtues of right-wing anti-liberalism', *IM – 1776*, https://im1776.com/2020/09/16/the-virtues-of-right-wing-anti-liberalism/.

13 Smith, B. (2021) 'Liberalism for losers: Carl Schmitt's "the tyranny of values"', *American Affairs*, V (I), https://americanaffairsjournal.org/2021/02/liberalism-for-losers-carl-schmitts-the-tyranny-of-values/.

14 Nagel, T. (1987) 'Moral conflict and political legitimacy', *Philosophy and Public Affairs*, 1 (3), 215–40.

15 Vattimo, as cited in Freeman, T. J. (n.d.) 'Human rights in the wake of the death of God', http://www2.hawaii.edu/~freeman/writings/Human%20Rights%20in%20the%20Wake%20of%20the%20Death%20of%20God.pdf.

16 See 'The human right' (2017) *GegenStandpunkt*, https://en.gegenstandpunkt.com/article/human-right.

17 Smith (2021).

18 Schmitt, C. (2004) *Legality and legitimacy*. Durham, NC: Duke University Press.

19 Fukuyama (2006).

20 McCarthy-Jones (2020a).

21 Vermeule, A. (2022) *Common good constitutionalism*. New York, NY: John Wiley & Sons.

22 Dalrymple, T. (2001) *Life at the bottom: the worldview that makes the underclass*. Chicago, IL: Ivan R. Dee; Perry, L. (2022) *The case against the sexual revolution*. Cambridge: Polity.

23 Gentile, G. (2004) *Origins and doctrine of fascism: with selections from other works* (trans. A. J. Gregor). London: Transaction Publishers.

24 Deneen (2019).

25 Vermeule (2022); Vermeule (2017).

26 Tussman, as cited in Dworkin, G. (1979) 'Joseph Tussman's government and the mind', *Journal of Politics*, 41 (1), 267–9.

27 Pope Leo XIII (1888) *Libertas*, https://www.vatican.va/content/leo-xiii/en/encyclicals/documents/hf_l-xiii_enc_20061888_libertas.html.

28 Deneen (2019).

29 Vermeule (2022).

30 Schwartz, B. (2000) 'Self-determination: the tyranny of freedom', *American Psychologist*, 55 (1), 79–88.

31 Parigi, P. (2012) *The rationalization of miracles*. Cambridge: Cambridge University Press.

32 Engels, F. (1908) *Socialism: utopian and scientific* (trans. E. Aveling). Chicago, IL: Charles Kerr.

33 Chomsky, N. (1989) *Necessary illusions: thought control in democratic societies*. Boston, MA: South End Press.

34 Malachovska, A. (2015) 'Propaganda wars in the Czech Republich', *StopFake.org*, 16 November, https://www.stopfake.org/en/propaganda-wars-in-the-czech-republic/.

35 Kozyreva *et al.* (2020).

36 Dewey, J., Seligman, E. R. A., Bennett, C. E., *et al.* (1915) 'AAUP's 1915 Declaration of Principles', https://aaup-ui.org/Documents/Principles/Gen_Dec_Princ.pdf.

37 Fish, S. (2019) *The first: how to think about hate speech, campus speech, religious speech, fake news, post-truth, and Donald Trump*. New York, NY: One Signal Publishers.

38 See Hearing Voices Network, https://www.hearing-voices.org/, and Intervoice, https://www.intervoiceonline.org/. I wrote about what I learnt from these experiences in McCarthy-Jones (2017).

39 Woolf (2001).

40 Valsangiacomo, C. (2021) 'Political representation in liquid democracy', *Frontiers in Political Science*, 3; Rutt, J. (2018) 'An introduction to liquid democracy', *Medium*, 26 February, https://medium.com/@memetic007/liquid-democracy-9cf7a4cb7f.

41 Engelmann, J. B., Capra, C. M., Noussair, C. and Berns, G. S. (2009) 'Expert financial advice neurobiologically "offloads" financial decision-making under risk', *PLoS One*, 4 (3), e4957.

42 Hasbermas, cited in Bohman, J. and Rehg, W. (2017) 'Jürgen Habermas', in E. N. Zalta (ed.), *The Stanford encyclopedia of philosophy*, https://plato.stanford.edu/archives/fall2017/entries/habermas/.

43 An idea discussed in de Benoist (2011).

44 Callard, A. (2021) 'Loving knowledge together: Socratic humility', in R. Modrak and J. L. van der Broek (eds), *Radical humility: essays on ordinary acts* (pp. 47–54). Cleveland, OH: Belt Publishing.

45 A point well made for actions by Prof. Andrew Huberman, in his excellent podcast series. See Huberman, A. (2021) 'How to increase motivation & drive', *Huberman Lab podcast #12*, https://www.youtube.com/watch?v=vA50EK70whE.

46 Mchangama (2022).

47 Rorty, R. (1989) *Contingency, irony, and solidarity*. Cambridge: Cambridge University Press.

48 Performing '*le travail qui fait vivre en nous ce qui n'existe pas*', as Paul Valery puts it (Valery, as cited in Marcuse, H. (2002) *One dimensional man*. London: Routledge).

INDEX

Republican Party 16
rewards 89–90, 93–4, 206–9, 210
Ribnikar, Vladislav 137
Richards, Neil 236
rights 167–76, 182–93, 254–5
    and freedom of thought 5–11, 23–33,
        136–8, 144–6
    and privacy 2, 11, 146–7
    see also human rights; property rights
Rimbaud, Arthur 180
risk aversion 275–6
Rodin, Auguste 39, 55
Roman Empire 148
Roosevelt, Eleanor 135
Roosevelt, Franklin D. 143
Rorty, Richard 7, 311
Rosen, Jeffrey 18, 277
Rothbard, Murray 101, 133, 268–9
Rousseau, Jean-Jacques 46, 55
rule-of-reason thinking 58–60, 113–14, 159,
    219–22
rule-of-thumb thinking 57–60, 113–14, 115,
    159, 219–21
Rumsfeld, Donald 175
Rushdie, Salman 244
Russell, Bertrand 238–9
Russia 60, 305; *see also* Soviet Union
Russia Today (RT) (television station)
    289–90
Russian Internet Research Agency 15, 29

sacred 189, 244, 253, 300–1
Sakharov, Andrei 95
Schauer, Frederick 251
Schmidt, Eric 123
Schmitt, Carl 299, 300
Schopenhauer, Arthur 179–80
Schwartz, Barry 302
Sci-Hub 166
scientific research papers 164–7
search engines 216–17, 235; *see also* Google
second-order thought 81–2, 84–5, 295
Second World War, *see* Holocaust; Nazism
secrecy 176–8
selective attention 70–1, 203–4
self-governance 8–9
sensory deprivation 195–6
sex offenders 150–3, 271–3, 274–5
Shaheed, Ahmed 4
Shakespeare, William: *The Tempest* 35
Sidney, Algernon 149
Skinner, B. F. 208–9
smartphones 205–7, 209–10

Snowden, Edward 144, 178, 235
social control 17
social media 4, 14–15, 22–3, 175–6
    and China 30
    and fake news 15–16
    and moral panic 127–8
    and networking 115, 210–12
    and reflection 81, 214–15, 217
    see also Facebook
socialism 199
Socrates 44, 72, 79, 100, 253
soft power 30
solitude 229
South Africa 145
sovereignty 103, 135
Soviet Union (USSR) 27, 94, 95, 194
Sowell, Thomas 22–3
Spanish Inquisition 255
speech, *see* free speech; inner speech
Spengler, Oswald 80, 237
Sperber, Dan 59, 73, 224
Spinoza, Baruch 94, 95, 238
Stalin, Joseph 96, 194
Stanley, Jason 288–9
Stanley v. Georgia (1969) 154
state, *see* government
Stiglitz, Joseph 176–7
Stolen Valor Act (2005) 190–1, 192–3
Strauss, Leo 86, 102, 218, 257, 292, 299
Sumption, Jonathan, Lord 275–6
Sunstein, Cass 129–30, 157–8, 160
    Too Much Information 180, 181
    and truth 188, 191, 193, 194
superiority 300
Surkov, Vladislav 29
surveillance 2, 14–15, 235–6, 301–2
sustained attention 70–1, 204, 205–6
Swartz, Aaron 54, 165–6, 167
Swift, Jonathan 184, 185

Taibbi, Matt 103
Taylor, Charles 41
Taylor, Kathleen 153
technology 27–8, 51–4, 111–12, 231; see
    also artificial intelligence; brain-read-
    ing; internet; smartphones
television 16
tenkō ('change in direction') 201
terrorism 33, 39, 242
    and freedom of thought rights 266–7,
        270, 271
Tetlock, Philip 243
theft 242

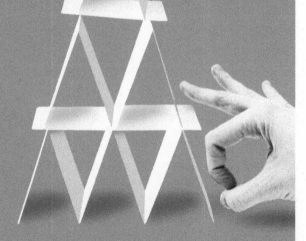

SIMON McCARTHY-JONES

Spite

'fascinating'
Michael Cockerell

'eye-opening'
David Robson

... and the Upside
of your Dark Side